Intelligent Systems Reference Library

Volume 108

Series editors

Janusz Kacprzyk, Polish Academy of Sciences, Warsaw, Poland
e-mail: kacprzyk@ibspan.waw.pl

Lakhmi C. Jain, Bournemouth University, Fern Barrow, Poole, UK, and
University of Canberra, Canberra, Australia
e-mail: jainlc2002@yahoo.co.uk

About this Series

The aim of this series is to publish a Reference Library, including novel advances and developments in all aspects of Intelligent Systems in an easily accessible and well structured form. The series includes reference works, handbooks, compendia, textbooks, well-structured monographs, dictionaries, and encyclopedias. It contains well integrated knowledge and current information in the field of Intelligent Systems. The series covers the theory, applications, and design methods of Intelligent Systems. Virtually all disciplines such as engineering, computer science, avionics, business, e-commerce, environment, healthcare, physics and life science are included.

More information about this series at http://www.springer.com/series/8578

Roumen Kountchev · Kazumi Nakamatsu
Editors

New Approaches
in Intelligent Image Analysis

Techniques, Methodologies and Applications

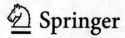 Springer

Editors
Roumen Kountchev
Department of Radio Communications and
  Video Technologies
Technical University of Sofia
Sofia
Bulgaria

Kazumi Nakamatsu
School of Human Science and Environment
University of Hyogo
Himeji
Japan

ISSN 1868-4394   ISSN 1868-4408   (electronic)
Intelligent Systems Reference Library
ISBN 978-3-319-81219-9   ISBN 978-3-319-32192-9   (eBook)
DOI 10.1007/978-3-319-32192-9

This Springer imprint is published by Springer Nature
The registered company is Springer International Publishing AG Switzerland

Preface

This book represents the advances in the development of new approaches, used for the intelligent image analysis. It introduces various aspects of the image analysis, related to the theory for their processing, and to some practical applications.

The book comprises 11 chapters, whose authors are researchers from different countries: USA, Russia, Bulgaria, Japan, Brazil, Romania, Ukraine, and Egypt. Each chapter is a small monograph, which represents the recent research work of the authors in the corresponding scientific area. The object of the investigation is new methods, algorithms, and models, aimed at the intelligent analysis of signals and images—single and sequences of various kinds: natural, medical, multispectral, multi-view, sound pictures, acoustic maps of sources, etc.

New Approaches for Hierarchical Image Decomposition, Based on IDP, SVD, PCA, and KPCA

In Chap. 1 the basic methods for hierarchical decomposition of grayscale and color images, and of sequences of correlated images are analyzed. New approaches are introduced for hierarchical image decomposition: the Branched Inverse Difference Pyramid (BIDP) and the Hierarchical Singular Value Decomposition (HSVD) with tree-like computational structure for single images; the Hierarchical Adaptive Principle Component Analysis (HAPCA) for groups of correlated images and the Hierarchical Adaptive Kernel Principal Component Analysis (HAKPCA) for color images. In the chapter the evaluation of the computational complexity of the algorithms used for the implementation of these decompositions is also given. The basic application areas are defined for efficient image hierarchical decomposition, such as visual information redundancy reduction; noise filtration; color segmentation; image retrieval; image fusion; dimensionality reduction, where the following is executed: the objects classification; search enhancement in large-scale image databases, etc.

Intelligent Digital Signal Processing and Feature Extraction Methods

The goal of Chap. 2 is to present well-known signal processing methods and the way they can be combined with intelligent systems in order to create powerful feature extraction techniques. In order to achieve this, several case studies are presented to illustrate the power of hybrid systems. The main emphasis is on the instantaneous time–frequency analysis, since it is proven to be a powerful method in several technical and scientific areas. The oldest and most utilized method is the Fourier transform, which has been applied in several domains of data processing, but it has very strong limitations due to the constraints it imposes on the analyzed data. Then the short-time Fourier transform and the wavelet transform are presented as they provide both temporal and frequency information as opposed to the Fourier transform. These methods form the basis of most applications, as they offer the possibility of time–frequency analysis of signals. The Hilbert–Huang transform is presented as a novel signal processing method, which introduces the concept of the instantaneous frequency that can be determined for every time point, making it possible to have a deeper look into different phenomena. Several applications are presented where fuzzy classifiers, support vector machines, and artificial neural networks are used for decision-making. Interconnecting these intelligent methods with signal processing will result in hybrid intelligent systems capable of solving computationally difficult problems.

Multi-dimensional Data Clustering and Visualization via Echo State Networks

Chapter 3 summarizes the proposed recently approach for multidimensional data clustering and visualization. It uses a special kind of recurrent networks called Echo State Networks (ESN) to generate multiple 2D projections of the multidimensional original data. The 2D projections are subjected to selection based on different criteria depending on the aim of particular clustering task to be solved. The selected projections are used to cluster and/or to visualize the original data set. Several examples demonstrate the possible ways to apply the proposed approach to variety of multidimensional data sets: steel alloys discrimination by their composition; Earth cover classification from hyperspectral satellite images; working regimes classification of an industrial plant using data from multiple measurements; discrimination of patterns of random dot motion on the screen; and clustering and visualization of static and dynamic "sound pictures" by multiple randomly placed microphones.

Unsupervised Clustering of Natural Images in Automatic Image Annotation Systems

Chapter 4 is devoted to automatic annotation of natural images joining the strengths of the text-based and the content-based image retrieval. The automatic image annotation is based on the semantic concept models, which are built from large number of patches received from a set of images. In this case, image retrieval is implemented by keywords called Visual Words (VWs) that is similar to text document retrieval. The task involves two main stages: a low-level segmentation based on color, texture, and fractal descriptors and a high-level clustering of received descriptors into the separated clusters corresponding to the VWs set. The enhanced region descriptor including color, texture, and fractal features has been proposed. For the VWs generation, the unsupervised clustering is a suitable approach. The Enhanced Self-Organizing Incremental Neural Network (ESOINN) was chosen due to its main benefits as a self-organizing structure and online implementation. The preliminary image segmentation permitted to change a sequential order of descriptors entering the ESOINN as associated sets. Such approach simplified, accelerated, and decreased the stochastic variations of the ESOINN. The experiments demonstrate acceptable results of the VWs clustering for a non-large natural image sets. This approach shows better precision values and execution time as compared to the fuzzy c-means algorithm and the classic ESOINN. Also issues of parallel implementation of unsupervised segmentation in OpenMP and Intel Cilk Plus environments were considered for processing of HD-quality images.

An Evolutionary Optimization Control System for Remote Sensing Image Processing

Chapter 5 provides an evolutionary control system via two Darwinian Particle Swarm Optimizations (DPSO)—one novel application of DPSO—coupled with remote sensing image processing to help in the image data analysis. The remote sensing image analysis has been a topic of ongoing research for many years and has led to paradigm shifts in the areas of resource management and global biophysical monitoring. Due to distortions caused by variations in signal/image capture and environmental changes, there is not a definite model for image processing tasks in remote sensing and such tasks are traditionally approached on a case-by-case basis. Intelligent control, however, can streamline some of the case-by-case scenarios and allows faster, more accurate image processing to support the more accurate remote sensing image analysis.

Tissue Segmentation Methods Using 2D Histogram Matching in a Sequence of MR Brain Images

In Chap. 6 a new transductive learning method for tissue segmentation using a 2D histogram modification, applied to Magnetic Resonance (MR) image sequence, is introduced. The 2D histogram is produced from a normalized sum of co-occurrence matrices of each MR image. Two types of model 2D histograms are constructed for each subsequence: intra-tissue 2D histogram to separate tissue regions and an inter-tissue edge 2D histogram. First, the MR image sequence is divided into few subsequences, using wave hedges distance between the 2D histograms of the consecutive MR images. The test 2D histogram segments are modified in the confidence interval and the most representative entries for each tissue are extracted, which are used for the kNN classification after distance learning. The modification is applied by using LUT and two ways of distance metric learning: large margin nearest neighbor and neighborhood component analysis. Finally, segmentation of the test MR image is performed using back projection with majority vote between the probability maps of each tissue region, where the inter-tissue edge entries are added with equal weights to corresponding tissues. The proposed algorithm has been evaluated with free access data sets and has showed results that are comparable to the state-of-the-art segmentation algorithms, although it does not consider specific shape and ridges of brain tissues.

Multistage Approach for Simple Kidney Cysts Segmentation in CT Images

In Chap. 7 a multistage approach for segmentation of medical objects in Computed Tomography (CT) images is presented. Noise reduction with consecutive applied median filter and wavelet shrinkage packet decomposition, and contrast enhancement based on Contrast limited Adaptive Histogram Equalization (CLAHE) are applied in the preprocessing stage. As a next step a combination of two basic methods is used for image segmentation such as the split and merge algorithm, followed by the color-based K-mean clustering. For refining the boundaries of the detected objects, additional texture analysis is introduced based on the limited Haralick's feature set and morphological filters. Due to the diminished number of components for the feature vectors, the speed of the segmentation stage is higher than that for the full feature set. Some experimental results are presented, obtained by computer simulation. The experimental results give detailed information about the detected simple renal cysts and their boundaries in the axial plane of the CT images. The proposed approach can be used in real time for precise diagnosis or in disease progression monitoring.

Audio Visual Attention Models in Mobile Robots Navigation

In Chap. 8, it is proposed to use the exiting definitions and models for human audio and visual attention, adapting them to the models of mobile robots audio and visual attention, and combining with the results from mobile robots audio and visual perception in the mobile robots navigation tasks. The mobile robots are equipped with sensitive audio visual sensors (usually microphone arrays and video cameras). They are the main sources of audio and visual information to perform suitable mobile robots navigation tasks modeling human audio and visual perception. The audio and visual perception algorithms are widely used, separately or in audio visual perception, in mobile robot navigation, for example to control mobile robots motion in applications like people and objects tracking, surveillance systems, etc. The effectiveness and precision of the audio and visual perception methods in mobile robots navigation can be enhanced combining audio and visual perception with audio and visual attention. There exists relative sufficient knowledge describing the phenomena of human audio and visual attention.

Local Adaptive Image Processing

Three methods for 2D local adaptive image processing are presented in Chap. 9. In the first one, the adaptation is based on the local information from the four neighborhood pixels of the processed image and the interpolation type is changed to zero or bilinear. The analysis of the local characteristics of images in small areas is presented, from which the optimal selection of thresholds for dividing into homogeneous and contour blocks is made and the interpolation type is changed adaptively. In the second one, the adaptive image halftoning is based on the generalized 2D Last Mean Square (LMS) error-diffusion filter for image quantization. The thresholds for comparing the input image levels are calculated from the gray values dividing the normalized histogram of the input halftone image into equal parts. In the third one, the adaptive line prediction is based on the 2D LMS adaptation of coefficients of the linear prediction filter for image coding. An analysis of properties of 2D LMS filters in different directions was made. The principal block schemes of the developed algorithms are presented. An evaluation of the quality of the processed images was made on the base of the calculated objective criteria and the subjective observation. The given experimental results, from the simulation for each of the developed algorithms, suggest that the effective use of local information contributes to minimize the processing error. The methods are suitable for different types of images (fingerprints, contour images, cartoons, medical signals, etc.). The developed algorithms have low computational complexity and are suitable for real-time applications.

Machine Learning Techniques for Intelligent Access Control

In Chap. 10 several biometric techniques, their usage, advantages and disadvantages are introduced. The access control is the set of regulations used to access certain areas or information. By access we mean entering a specific area, or logging on a machine (PC, or another device). The access regulated by a set of rules that specifies who is allowed to get access, and what are the restrictions on such access. Over the years several basic kinds of access control systems have been developed. With advancement of technology, older systems are now easily bypassed with several methods, thus the need to have new methods of access control. Biometrics is referred to as an authentication technique that relies on a computer system to electronically validate a measurable biological characteristic that is physically unique and cannot be duplicated. Biometrics has been used for ages as an access control security system.

Experimental Evaluation of Opportunity to Improve the Resolution of the Acoustic Maps

Chapter 11 is devoted to generation of acoustic maps. The experimental work considers the possibility to increase the maps resolution. The work uses 2D microphone array with randomly spaced elements to generate acoustic maps of sources located in its near-field region. In this region the wave front is not flat and the phase of the input signals depends on the arrival direction, and on the range as well. The input signals are partially distorted by the indoor multipath propagation and the related interference of sources emissions. For acoustic mapping with improved resolution an algorithm in the frequency domain is proposed. The algorithm is based on the modified method of Capon. Acoustic maps of point-like noise sources are generated. The maps are compared with the maps generated using other famous methods including built-in equipment software. The obtained results are valuable in the estimation of direction of arrival for Noise Exposure Monitoring.

This book will be very useful for students and Ph.D. students, researchers, and software developers, working in the area of digital analysis and recognition of multidimensional signals and images.

Sofia, Bulgaria   Roumen Kountchev
Himeji, Japan   Kazumi Nakamatsu
2015

Contents

1 New Approaches for Hierarchical Image Decomposition, Based on IDP, SVD, PCA and KPCA 1

Roumen Kountchev and Roumiana Kountcheva

1.1   Introduction 2

1.2   Related Work 3

1.3   Image Representation Based on Branched Inverse
Difference Pyramid 4

      1.3.1   Principles for Building the Inverse Difference
Pyramid 4

      1.3.2   Mathematical Representation of n-Level IDP 5

      1.3.3   Reduced Inverse Difference Pyramid 8

      1.3.4   Main Principle for Branched IDP Building 8

      1.3.5   Mathematical Representation for One
BIDP Branch 9

      1.3.6   Transformation of the Retained Coefficients
into Sub-blocks of Size 2×2 11

      1.3.7   Experimental Results 13

1.4   Hierarchical Singular Value Image Decomposition 15

      1.4.1   SVD Algorithm for Matrix Decomposition 17

      1.4.2   Particular Case of the SVD for Image Block of Size
2×2 18

      1.4.3   Hierarchical SVD for a Matrix of Size $2^n \times 2^n$ 19

      1.4.4   Computational Complexity of the Hierarchical
SVD of Size $2^n \times 2^n$ 23

      1.4.5   Representation of the HSVD Algorithm Through
Tree-like Structure 25

1.5   Hierarchical Adaptive Principal Component Analysis
for Image Sequences 27

      1.5.1   Principle for Decorrelation of Image Sequences
by Hierarchical Adaptive PCA 30

| | 1.5.2 | Description of the Hierarchical Adaptive PCA Algorithm | 30 |

1.5.2   Description of the Hierarchical Adaptive PCA
        Algorithm   30
1.5.3   Setting the Number of the Levels and the Structure
        of the HAPCA Algorithm   35
1.5.4   Experimental Results   39
1.6   Hierarchical Adaptive Kernel Principal Component Analysis
      for Color Image Segmentation   42
        1.6.1   Mathematical Representation of the Color Adaptive
                Kernel PCA   42
        1.6.2   Algorithm for Color Image Segmentation
                by Using HAKPCA   47
        1.6.3   Experimental Results   50
1.7   Conclusions   53

2   **Intelligent Digital Signal Processing and Feature
    Extraction Methods**   59
    János Szalai and Ferenc Emil Mózes
    2.1   Introduction   59
    2.2   The Fourier Transform   60
          2.2.1   Application of the Fourier Transform   62
    2.3   The Short-Time Fourier Transform   64
          2.3.1   Application of the Short-Time Fourier Transform   65
    2.4   The Wavelet Transform   67
          2.4.1   Application of the Wavelet Transform   70
    2.5   The Hilbert-Huang Transform   71
          2.5.1   Introducing the Instantaneous Frequency ..   71
          2.5.2   Computing the Instantaneous Frequency   72
          2.5.3   Application of the Hilbert-Huang Transform   76
    2.6   Hybrid Signal Processing Systems   80
          2.6.1   The Discrete Wavelet Transform and Fuzzy
                  C-Means Clustering   80
          2.6.2   Automatic Sleep Stage Classification   82
          2.6.3   The Hilbert-Huang Transform and Support
                  Vector Machines   85
    2.7   Conclusions   88
    References   88

3   **Multi-dimensional Data Clustering and Visualization
    via Echo State Networks**   93
    Petia Koprinkova-Hristova
    3.1   Introduction   93
    3.2   Echo State Networks and Clustering Procedure   95
          3.2.1   Echo State Networks Basics   95
          3.2.2   Effects of IP Tuning Procedure   97
          3.2.3   Clustering Algorithms 101

3.3 Examples 105
   3.3.1 Clustering of Steel Alloys in Dependence
       on Their Composition 105
   3.3.2 Clustering and Visualization of Multi-spectral
       Satellite Images 106
   3.3.3 Clustering of Working Regimes of an Industrial
       Plant. 109
   3.3.4 Clustering of Time Series from Random Dots
       Motion Patterns 111
   3.3.5 Clustering and 2D Visualization
       of "Sound Pictures" 114
3.4 Summary of Results and Discussion. 117
3.5 Conclusions 119
References 120

4 Unsupervised Clustering of Natural Images in Automatic
  Image Annotation Systems 123
  Margarita Favorskaya, Lakhmi C. Jain and Alexander Proskurin
  4.1 Introduction. 124
  4.2 Related Work 125
     4.2.1 Unsupervised Segmentation of Natural Images 125
     4.2.2 Unsupervised Clustering of Images. 128
  4.3 Preliminary Unsupervised Image Segmentation 129
  4.4 Feature Extraction Using Parallel Computations. 131
     4.4.1 Color Features Representation 133
     4.4.2 Calculation of Texture Features 134
     4.4.3 Fractal Features Extraction 135
     4.4.4 Enhanced Region Descriptor 137
     4.4.5 Parallel Computations of Features. 138
  4.5 Clustering of Visual Words by Enhanced SOINN 139
     4.5.1 Basic Concepts of ESOINN. 140
     4.5.2 Algorithm of ESOINN Functioning 141
  4.6 Experimental Results 143
  4.7 Conclusion and Future Development 151
  References 152

5 An Evolutionary Optimization Control System for Remote
  Sensing Image Processing. 157
  Victoria Fox and Mariofanna Milanova
  5.1 Introduction. 157
  5.2 Background Techniques 159
     5.2.1 Darwinian Particle Swarm Optimization 159
     5.2.2 Total Variation for Texture-Structure Separation. 162
     5.2.3 Multi-phase Chan-Vese Active Contour Without
       Edges 166

5.3   Evolutionary Optimization of Segmentation. 167
    5.3.1   Darwinian PSO for Thresholding 167
    5.3.2   Novel Darwinian PSO for Relative
              Total Variation. 169
    5.3.3   Multi-phase Active Contour Without Edges
              with Optimized Initial Level Mask 170
    5.3.4   Workflow of Proposed System. 173
5.4   Experimental Results 174
    5.4.1   Results 174
    5.4.2   Discussion. 179
    5.4.3   Conclusion and Future Research. 179
References 180

**6   Tissue Segmentation Methods Using 2D Histogram
Matching in a Sequence of MR Brain Images** 183
Vladimir Kanchev and Roumen Kountchev
6.1   Introduction. 184
6.2   Related Works. 185
6.3   Overview of the Developed Segmentation Algorithm 188
6.4   Preprocessing and Construction of a Model
    and Test 2D Histograms 189
    6.4.1   Transductive Learning. 190
    6.4.2   MRI Data Preprocessing 190
    6.4.3   Construction of a 2D Histogram. 191
    6.4.4   Separation into MR Image Subsequences 192
    6.4.5   Types of 2D Histograms and Preprocessing. 194
6.5   Matching and Classification of a 2D Histogram 196
    6.5.1   Construct Train 2D Histogram Segments
              Using 2D Histogram Matching. 197
    6.5.2   2D Histogram Classification After Distance
              Metric Learning 199
6.6   Segmentation Through Back Projection. 204
6.7   Experimental Results 207
    6.7.1   Test Data Sets and Parameters of the Developed
              Algorithm 207
    6.7.2   Segmentation Results 209
6.8   Discussion 217
6.9   Conclusion 219
References 219

**7   Multistage Approach for Simple Kidney Cysts
Segmentation in CT Images** 223
Veska Georgieva and Ivo Draganov
7.1   Introduction. 224
    7.1.1   Medical Aspect of the Problem for Kidney
              Cyst Detection 224

	7.1.2	Review of Segmentation Methods	225
	7.1.3	Proposed Approach	229
7.2	Preprocessing Stage of CT Images		230
	7.2.1	Noise Reduction with Median Filter	230
	7.2.2	Noise Reduction Based on Wavelet Packet Decomposition and Adaptive Threshold	231
	7.2.3	Contrast Limited Adaptive Histogram Equalization (CLAHE)	232
7.3	Segmentation Stage		232
	7.3.1	Segmentation Based on Split and Merge Algorithm	232
	7.3.2	Clustering Classification of Segmented CT Image	234
	7.3.3	Segmentation Based on Texture Analysis	234
7.4	Experimental Results		239
7.5	Discussion		247
7.6	Conclusion		248
	References		249

8   Audio Visual Attention Models in the Mobile Robots Navigation 253
Snejana Pleshkova and Alexander Bekiarski

8.1	Introduction		254
8.2	Related Work		254
8.3	The Basic Definitions of the Human Audio Visual Attention		256
8.4	General Probabilistic Model of the Mobile Robot Audio Visual Attention		257
8.5	Audio Visual Attention Model Applied in the Audio Visual Mobile Robot System		263
	8.5.1	Room Environment Model for Description of Indoor Initial Audio Visual Attention	263
	8.5.2	Development of the Algorithm for Definition of the Mobile Robot Initial Audio Visual Attention Model	266
	8.5.3	Definition of the Initial Mobile Robot Video Attention Model with Additional Information from the Laser Range Finder Scan	272
	8.5.4	Development of the Initial Mobile Robot Video Attention Model Localization with Additional Information from a Speaker to the Mobile Robot Initial Position	274
8.6	Definition of the Probabilistic Audio Visual Attention Mobile Robot Model in the Steps of the Mobile Robot Navigation Algorithm		276

8.7   Experimental Results from the Simulations of the Mobile
       Robot Motion Navigation Algorithm Applying the
       Probabilistic Audio Visual Attention Model. 279
       8.7.1   Experimental Results from the Simulations
                of the Mobile Robot Motion Navigation
                Algorithm Applying Visual Perception Only 280
       8.7.2   Experimental Results from the Simulations
                of the Mobile Robot Motion Navigation
                Algorithm Using Visual Attention in Combination
                with the Visual Perception. 282
       8.7.3   Quantitative Comparison of the Simulations Results
                Applying Visual Perception Only, and Visual
                Attention with Visual Perception 283
       8.7.4   Experimental Results from Simulations Using Audio
                Visual Attention in Combination with Audio Visual
                Perception 285
       8.7.5   Quantitative Comparison of the Results Achieved
                in Simulations Applying Audio Visual Perception
                Only, and Visual Attention Combined with Visual
                Perception 287
8.8   Conclusion 289
References 291

9   **Local Adaptive Image Processing** 295
    Rumen Mironov
    9.1   Introduction. 296
    9.2   Method for Local Adaptive Image Interpolation. 297
           9.2.1   Mathematical Description of Adaptive
                    2D Interpolation. 297
           9.2.2   Analysis of the Characteristics of the Filter
                    for Two-Dimensional Adaptive Interpolation 298
           9.2.3   Evaluation of the Error of the Adaptive 2D
                    Interpolation 302
           9.2.4   Functional Scheme of the 2D Adaptive
                    Interpolator 305
    9.3   Method for Adaptive 2D Error Diffusion Halftoning. 306
           9.3.1   Mathematical Description of Adaptive
                    2D Error-Diffusion 306
           9.3.2   Determining the Weighting Coefficients
                    of the 2D Adaptive Halftoning Filter 308
           9.3.3   Functional Scheme of 2D Adaptive
                    Halftoning Filter. 309
           9.3.4   Analysis of the Characteristics of the 2D Adaptive
                    Halftoning Filter. 311

9.4   Method for Adaptive 2D Line Prediction
        of Halftone Images. 314
        9.4.1   Mathematical Description of Adaptive 2D Line
                Prediction 314
        9.4.2   Synthesis and Analysis of Adaptive 2D
                LMS Codec for Linear Prediction. 316
9.5   Experimental Results 320
        9.5.1   Experimental Results from the Work
                of the Developed Adaptive 2D Interpolator 320
        9.5.2   Experimental Results from the Work
                of the Developed Adaptive 2D Halftoning Filter 325
        9.5.3   Experimental Results from the Work
                of the Developed Codec for Adaptive 2D
                Linear Prediction 326
9.6   Conclusion 328
References 329

10   **Machine Learning Techniques for Intelligent Access Control** 331
     Wael H. Khalifa, Mohamed I. Roushdy and Abdel-Badeeh M. Salem
     10.1   Introduction. 331
     10.2   Machine Learning Methodology for Biometrics 333
             10.2.1 Signal Capturing 333
             10.2.2 Feature Extraction 334
             10.2.3 Classification 334
     10.3   User Authentication Techniques. 335
     10.4   Physiological Biometrics Taxonomy. 336
             10.4.1 Finger Print. 336
             10.4.2 Face. 337
             10.4.3 Iris 337
     10.5   Behavioral Biometrics Taxonomy. 339
             10.5.1 Keystroke Dynamics. 339
             10.5.2 Voice 340
             10.5.3 EEG. 340
     10.6   Multimodal Biometrics 341
     10.7   Applications 342
     10.8   Machine Learning Techniques for Biometrics 343
             10.8.1 Fisher's Discriminant Analysis. 343
             10.8.2 Linear Discriminant Classifier 345
             10.8.3 LVQ Neural Net 346
             10.8.4 Neural Networks 347
     10.9   Conclusion 349
     References 351

**11 Experimental Evaluation of Opportunity to Improve
     the Resolution of the Acoustic Maps** 353
Volodymyr Kudriashov
11.1 Introduction. 353
11.2 Theoretical Part 354
        11.2.1 Signal Model Limitations 354
        11.2.2 Signal Model 355
        11.2.3 Acoustic Mapping Methods 357
11.3 The Experimental Acoustic Camera Equipment 359
11.4 Experimental Results 361
        11.4.1 Microphone Array Patterns Generated
                with the Delay-and-Sum Beamforming Method 362
        11.4.2 Microphone Array Patterns Generated
                with the Christensen Beamforming Method 363
        11.4.3 Microphone Array Patterns Generated with the
                Modified Capon-Based Beamforming Method 365
        11.4.4 Microphone Array Responses for Two Point-like
                Emitters 368
        11.4.5 The Acoustic Camera Responses for Two Point-like
                Emitters 371
11.5 Conclusions 372
References 373

Contributors

Alexander Bekiarski Department of Telecommunications, Technical University of Sofia, Sofia, Bulgaria

Ivo Draganov Department of Radio Communications and Video Technologies, Technical University of Sofia, Sofia, Bulgaria

Margarita Favorskaya Institute of Informatics and Telecommunications, Siberian State Aerospace University, Krasnoyarsk, Russian Federation

Victoria Fox Department of Mathematics, University of Arkansas at Monticello, Monticello, Arkansas, USA

Veska Georgieva Department of Radio Communications and Video Technologies, Technical University of Sofia, Sofia, Bulgaria

Lakhmi C. Jain Bournemouth University, Fern Barrow, Poole, UK; University of Canberra, Canberra, Australia

Vladimir Kanchev Department of Radio Communications and Video Technologies, Technical University of Sofia, Sofia, Bulgaria

Wael H. Khalifa Artificial Intelligence and Knowledge Engineering Research Labs, Computer Science Department, Faculty of Computer and Information sciences, Ain Shams University, Cairo, Egypt

Petia Koprinkova-Hristova Bulgarian Academy of Sciences, Institute of Information and Communication Technologies, Sofia, Bulgaria

Roumen Kountchev Department of Radio Communications and Video Technologies, Technical University of Sofia, Sofia, Bulgaria

Roumiana Kountcheva T&K Engineering Co., Sofia, Bulgaria

Volodymyr Kudriashov Mathematical Methods for Sensor Information Processing Department, Institute of Information and Communication Technologies, Bulgarian Academy of Sciences, Sofia, Bulgaria

Mariofanna Milanova Computer Science Department, University of Arkansas at Little Rock, Little Rock, Arkansas, USA

Ferenc Emil Mózes Petru Maior University of Târgu Mures, Târgu Mures, Romania

Rumen Mironov Department of Radio Communications and Video Technologies, Technical University of Sofia, Sofia, Bulgaria

Snejana Pleshkova Department of Telecommunications, Technical University of Sofia, Sofia, Bulgaria

Alexander Proskurin Institute of Informatics and Telecommunications, Siberian State Aerospace University, Krasnoyarsk, Russian Federation

Mohamed I. Roushdy Artificial Intelligence and Knowledge Engineering Research Labs, Computer Science Department, Faculty of Computer and Information sciences, Ain Shams University, Cairo, Egypt

Abdel-Badeeh M. Salem Artificial Intelligence and Knowledge Engineering Research Labs, Computer Science Department, Faculty of Computer and Information sciences, Ain Shams University, Cairo, Egypt

János Szalai Technical University of Cluj Napoca, Cluj Napoca, Romania

Chapter 1
New Approaches for Hierarchical Image Decomposition, Based on IDP, SVD, PCA and KPCA

Roumen Kountchev and Roumiana Kountcheva

Abstract The contemporary forms of image representation vary depending on the application. There are well-known mathematical methods for image representation, which comprise: matrices, vectors, determined orthogonal transforms, multi-resolution pyramids, Principal Component Analysis (PCA) and Independent Component Analysis (ICA), Singular Value Decomposition (SVD), wavelet sub-band decompositions, hierarchical tensor transformations, nonlinear decompositions through hierarchical neural networks, polynomial and multiscale hierarchical decompositions, multidimensional tree-like structures, multi-layer perceptual and cognitive models, statistical models, etc. In this chapter are analyzed the basic methods for hierarchical decomposition of grayscale and color images, and of sequences of correlated images of the kind: medical, multispectral, multi-view, etc. Here is also added one expansion and generalization of the ideas of the authors from their previous publications, regarding the possibilities for the development of new, efficient algorithms for hierarchical image decompositions with various purposes. In this chapter are presented and analyzed the following four new approaches for hierarchical image decomposition: the Branched Inverse Difference Pyramid (BIDP), based on the Inverse Difference Pyramid (IDP); the Hierarchical Singular Value Decomposition (HSVD) with tree-like computational structure; the Hierarchical Adaptive Principle Component Analysis (HAPCA) for groups of correlated images; and the Hierarchical Adaptive Kernel Principal Component Analysis (HAKPCA) for color images. In the chapter are given the algorithms, used for the implementation of these decompositions, and their computational complexity is evaluated. Some experimental results, related to selected applications are also given, and various possibilities for the creation of new hybrid algorithms for hierarchical decomposition of multidimensional images are specified. On the basis

R. Kountchev (✉)
Department of Radio Communications and Video Technologies,
Technical University of Sofia, 8 Kl. Ohridski Blvd., 1000 Sofia, Bulgaria
e-mail: rkountch@tu-sofia.bg

R. Kountcheva
T&K Engineering Co., Drujba 2, Bl. 404/2, 1582 Sofia, Bulgaria
e-mail: kountcheva_r@yahoo.com

© Springer International Publishing Switzerland 2016
R. Kountchev and K. Nakamatsu (eds.), *New Approaches in Intelligent
Image Analysis*, Intelligent Systems Reference Library 108,
DOI 10.1007/978-3-319-32192-9_1

1

of the results obtained from the executed analysis, the basic application areas for efficient image processing are specified, such as: reduction of the information surplus; noise filtration; color segmentation; image retrieval; image fusion; dimensionality reduction for objects classification; search enhancement in large scale image databases, etc.

Keywords Hierarchical image decomposition · Branched inverse difference pyramid · Hierarchical singular value decomposition · Hierarchical principal component analysis for groups of images · Hierarchical adaptive kernel principal component analysis for color images

1.1 Introduction

The methods for image processing, transmission, registration, restoration, analysis and recognition, are defined at high degree by the corresponding mathematical forms and models for their representation. On the other hand, they all depend on the way the image was created, and on their practical use. The primary forms for image representation depend on the used sources, such as: photo and video cameras, scanners, ultrasound sensors, X-ray, computer tomography, etc. The matrix descriptions are related to the *primary discrete forms*. Each still halftone image is represented by one matrix; the color RGB image—by three matrices; the multi-spectral, hyper spectral and multi-view images, and also some kinds of medical images (for example, computer tomography, IMR, etc.)—by N matrices (for $N > 3$), while the moving images are represented through M temporal sequences, of N matrices each. There are already many *secondary forms* created for image representation, obtained from the primary forms, after reduction of the information surplus, and depending on the application. Various mathematical methods are used to transform the image matrices into reduced (secondary) forms by using: vectors, for each image block, through which are composed vector fields; deterministic and statistical orthogonal transforms; multi-resolution pyramids; wavelet sub-band decompositions; hierarchical tensor transforms; nonlinear decompositions through hierarchical neural networks, polynomial and multiscale hierarchical decompositions, multi-dimensional tree-like structures, multi-layer perceptual and cognitive models, statistical models, fuzzy hybrid methods for image decomposition, etc.

The decomposition methods permit each image matrix to be represented as the sum of the matrix components with different weights, defined by the image contents. Besides, the description of each matrix in the decomposition is much simpler than that of the original (primary) matrix. The number of the matrices in the decomposition could be significantly reduced through analyzing their weights, without significant influence on the approximation accuracy of the primary matrix. To this group could be related the methods for linear orthogonal transforms [1]: the Discrete Fourier Transform (DFT), the Discrete Cosine Transform (DCT), the Walsh-Hadamard Transform (WHT), the Hartley Transform (HrT), the Haar

Transform (HT), etc.; the pyramidal decompositions [2]: the Gaussian Pyramid (GP), the Laplacean Pyramid (LP), the Discrete Wavelet Transform (DWT), the Discrete Curvelet Transform (DCuT) [3], the Inverse Difference Pyramid (IDP) [4], etc.; the statistical decompositions [5]: the Principal Component Analysis (PCA), the Independent Component Analysis (ICA) and the Singular Value Decomposition (SVD); the polynomial and multiscale hierarchical decompositions [6, 7]; multi-dimensional tree-like structures [8]; hierarchical tensor transformations [9]; the decompositions based on hierarchical neural networks [10]; etc.

The aim of this chapter is to be analyzed the basic methods and algorithms for *hierarchical image decomposition*. Here are also generalized the following new approaches for hierarchical decomposition of multi-component matrix images: the Branched Inverse Difference Pyramid (BIDP), based on the Inverse Difference Pyramid (IDP), the Hierarchical Singular Value Decomposition (HSVD)—for the representation of single images; the Hierarchical Adaptive Principal Component Analysis (HAPCA)—for the decorrelation of sequences of images, and the Hierarchical Adaptive Kernel Principal Component Analysis (HAKPCA)—for the analysis of color images.

1.2 Related Work

One of the contemporary methods for hierarchical image decomposition is called multiscale decomposition [7]. It is used for noise filtration in the image f, represented by the sum of the clean part u, and the noisy part, v. In accordance to Rudin, Osher and Fatemi (ROF) [11], to define the components u and v it is necessary to calculate the total variation of the functional Q, defined by the relation:

$$Q(f,\lambda) = inf\left\{ \int_{\Omega} |\nabla u| + \lambda ||v||_{L^2}^2, f = u + v \right\},$$

where $\lambda > 0$ is a scale parameter; and $f \in L^2(\Omega)$—the image function, defined in the space $L^2(\Omega)$. The minimization of Q leads to decomposition, in result of which the visual information is divided into a part u that extracts the edges of f, and a part v that captures the texture. Denoising at different scales λ generates a multiscale image representation. In [6], Tadmor, Nezzar and Vese proposed a multiscale image decomposition which offers a hierarchical and adaptive representation for different features in the analyzed images. The image is hierarchically decomposed into the sum of simpler atoms u_k, where u_k extracts more refined information from the previous scale u_{k-1}. To this end, the atoms u_k are obtained as dyadically scaled minimizers of the ROF functionals at increasing λ_k scales. Thus, starting with $v_{-1} := f$ and letting v_k denote the residual at a given dyadic scale, $\lambda_k = 2^k$, the recursive step $[u_k, v_k] = arg\{inf[Q_T(v_{k-1}, k)]\}$ leads to the desired hierarchical decomposition, $f = \Sigma T(u_k)$ (here T is a blurring operator).

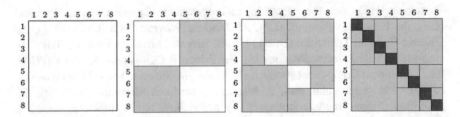

Fig. 1.1 Representation of the matrix of size 8×8 through three hierarchical matrices, or H-matrices

Another well-known approach for hierarchical decomposition is based on the hierarchical matrices [12]. The concept of hierarchical, or H-matrices, is based on the observation that submatrices of a full rank matrix may be of low rank, and respectively—to have low rank approximations. On Fig. 1.1 is given an example for the representation of a matrix of size 8×8 through H-matrices, which contain sub-matrices of three different sizes: 4×4, 2×2 and 1×1.

This observation is used for the matrix-skeleton approximation. The inverses of finite element matrices have, under certain assumptions, submatrices with exponentially decaying singular values. This means that these submatrices have also good low rank approximations. The hierarchical matrices permit decomposition by QR or Cholesky algorithms, which are iterative. Unlike them, the new approaches for hierarchical image decomposition, given in this chapter (BIDP and HSVD—for single images, HAPCA—for groups of correlated images, and HAKPCA—for color images), are not based on iterative algorithms.

1.3 Image Representation Based on Branched Inverse Difference Pyramid

1.3.1 Principles for Building the Inverse Difference Pyramid

In this section is given a short description of the inverse difference pyramid, IDP [4, 13], used as a basis for building its modifications. Unlike the famous Gaussian (GP) and Laplacian (LP) pyramids, the IDP represents the image in the spectral domain. After the decomposition, the image energy is concentrated in its first components, which permits to achieve very efficient compression, by cutting off the low-energy components. As a result, the main part of the energy of the original image is retained, despite the limited number of decomposition components used. For the decomposition implementation various kinds of orthogonal transforms could be used. In order to reduce the number of decomposition levels and the computational complexity, the image is initially divided into blocks and for each is then built the corresponding IDP.

In brief, the IDP is executed as follows: At the lowest (initial) level, on the matrix $[B]$ of size $2^n \times 2^n$ is applied the pre-selected "Truncated" Orthogonal Transform (TOT) and are calculated the values of a relatively small number of "retained" coefficients, located in the high-energy area of the so calculated transformed (spectrum) matrix $[S_0]$. These are usually the coefficients with spatial frequencies $(0, 0)$, $(0, 1)$, $(1, 0)$ and $(1, 1)$. After Inverse Orthogonal Transform (IOT) of the "truncated" spectrum matrix $[\hat{S}_0]$, which contains the retained coefficients only, is obtained the matrix $[\hat{B}_0]$ for the initial IDP level $(p = 0)$, which approximates the matrix $[B]$. The accuracy of the approximation depends on: the positions of the retained coefficients in the matrix $[S_0]$; the values, used to substitute the missing coefficients from the approximating matrix $[\hat{S}_0]$ for the zero level, and on the selected orthogonal transform. In the next decomposition level $(p = 1)$, is calculated the difference matrix $[E_0] = [B] - [\hat{B}_0]$. The resulting matrix is then split into 4 sub-matrices of size $2^{n-1} \times 2^{n-1}$ and on each is applied the corresponding TOT. The total number of retained coefficients for level $p = 1$ is 4 times larger than that in the zero level. In case, that Walsh-Hadamard Transform (WHT) is used for this level, the values of coefficients $(0, 0)$ in the IDP decomposition levels 1 and higher are always equal to zero, which permits to reduce the number of retained coefficients with ¼. On each of the four spectrum matrices $[\hat{S}_1]$ for the IDP level $p = 1$ is applied IOT and as a result, four sub-matrices are obtained, which build the approximating difference matrix $[\hat{E}_0]$. In the next IDP level $(p = 2)$ is calculated the difference matrix $[E_1] = [E_0] - [\hat{E}_0]$. After that, each difference sub-matrix is divided in similar way as in level 1, into four matrices of size $2^{n-2} \times 2^{n-2}$, and for each is performed TOT, etc. In the last (highest) IDP level is obtained the "residual" difference matrix. In case that the image should be losslessly coded, each block of the residual matrix is processed with full orthogonal transform and no coefficients are omitted.

1.3.2 Mathematical Representation of n-Level IDP

The digital image is represented by a matrix of size $(2^n m) \times (2^n m)$. For the processing, the matrix is first divided into blocks of size $2^n \times 2^n$ and on each is applied the IDP decomposition. The matrix $[B(2^n)]$ of each block is represented by the equation:

$$[B(2^n)] = [\hat{B}_0(2^n)] + \sum_{p=1}^{r} [\hat{E}_{p-1}(2^n)] + [E_r(2^n)] \quad \text{for } r = 1, 2, \ldots, n-1. \quad (1.1)$$

Here the number of decomposition components, which are matrices of size $2^n \times 2^n$, is equal to $(r + 2)$. The maximum possible number of decomposition levels for one block is $n + 1$ (for $r = n - 1$). The last component $[E_r(2^n)]$ defines the approximation error for the block $[B(2^n)]$ for the case, when the decomposition is limited up to level $p = r$. The first component $[\hat{B}_0(2^n)]$ for the level $p = 0$ is the coarse approximation of the block $[B(2^n)]$. It is obtained through 2D IOT on the block $[\hat{S}_0(2^n)]$ in correspondence with the relation:

$$[\hat{B}_0(2^n)] = [T_0(2^n)]^{-1}[\hat{S}_0(2^n)][T_0(2^n)]^{-1} \text{ for } p = 0, \qquad (1.2)$$

where $[T_0(2^n)]^{-1}$ is a matrix of size $2^n \times 2^n$, used for the inverse orthogonal transform of $[\hat{S}_0(2^n)]$.

The matrix $[\hat{S}_0(2^n)] = [m_0(u, v)s_0(u, v)]$ is the "truncated" orthogonal transform of the block $[B(2^n)]$. Here $m_0(u, v)$ are the elements of the binary matrix-mask $[M_0(2^n)]$, used to define the retained coefficients of $[\hat{S}_0(2^n)]$ in correspondence to the relation:

$$m_0(u, v) = \begin{cases} 1, & \text{if } s_0(u, v) \text{ is a retained coefficient,} \\ 0 & - \qquad \text{otherwise.} \end{cases} \qquad (1.3)$$

The values of the elements $m_0(u, v)$ are selected in accordance with the requirement the retained coefficients $\hat{s}_0(u, v) = m_0(u, v)s_0(u, v)$ to be these with maximum energy, calculated for all image blocks. The transform $[S_0(2^n)]$ of the block $[B(2^n)]$ is defined through direct 2D OT:

$$[S_0(2^n)] = [T_0(2^n)][B(2^n)][T_0(2^n)], \qquad (1.4)$$

where $[T_0(2^n)]$ is a matrix of size $2^n \times 2^n$ for the decomposition level $p = 0$, used to perform the selected 2D OT, which could be DFT, DCT, WHT, KLT, etc.

The remaining coefficients in the decomposition presented by Eq. 1.1 are the approximating difference matrices $[\hat{E}_{p-1}(2^{n-p})]$ for levels $p = 1, 2, ..., r$. They comprise the sub-matrices $[\hat{E}_{p-1}^{k_p}(2^{n-p})]$ of size $2^{n-p} \times 2^{n-p}$ for $k_p = 1, 2, ..., 4^p$, obtained through quadtree division of the matrix $[\hat{E}_{p-1}(2^{n-p})]$. Each sub-matrix $[\hat{E}_{p-1}^{k_p}(2^{n-p})]$ is then defined by the relation:

$$[\hat{E}_{p-1}^{k_p}(2^{n-p})] = [T_p(2^{n-p})]^{-1}[\hat{S}_p^{k_p}(2^{n-p})][T_p(2^{n-p})]^{-1} \text{ for } k_p = 1, 2, ..., 4^p, \quad (1.5)$$

where 4^p is the number of the quadtree branches in the decomposition level p. Here $[T_p(2^{n-p})]^{-1}$ is a matrix of size $2^{n-p} \times 2^{n-p}$ in the level p, used for the inverse 2D OT.

The elements $\hat{s}_p^{k_p}(u, v) = m_p(u, v).s_p^{k_p}(u, v)$ of the matrix $[\hat{S}_p^{k_p}(2^{n-p})]$ are defined by the elements $m_p(u, v)$ of the binary matrix-mask $[M_p(2^{n-p})]$:

$$m_p(u, v) = \begin{cases} 1, & \text{if } s_p^{k_p}(u, v) \text{ is a retained coefficient,} \\ 0 & - \qquad \text{otherwise.} \end{cases} \qquad (1.6)$$

The matrix $[S_p^{k_p}(2^{n-p})]$ is the transform of $[E_{p-1}^{k_p}(2^{n-p})]$ and is defined through direct 2D OT:

$$[S_p^{k_p}(2^{n-p})] = [T_p(2^{n-p})][E_{p-1}^{k_p}(2^{n-p})][T_p(2^{n-p})]. \qquad (1.7)$$

Here $[T_p(2^{n-p})]$ is a matrix of size $2^{n-p} \times 2^{n-p}$ in the decomposition level p, used for the 2D OT of each block $[E_p^{k_p}(2^{n-p})]$ (when $k_p = 1, 2,..., 4^p$), of the difference matrix for same level, defined by the equation:

$$[E_{p-1}(2^n)] = [B(2^n)] - [\hat{B}_0(2^n)] \text{ for } p = 1; \qquad (1.8)$$

$$[E_{p-1}(2^{n-p})] = [E_{p-2}(2^{n-p})] - [\hat{E}_{p-2}(2^{n-p})] \text{ for } p = 2, 3, \ldots, r. \qquad (1.9)$$

In result of the decomposition represented by Eq. 1.1, for each block $[B(2^n)]$, are calculated the following spectrum coefficients:

- all nonzero coefficients of the transform $[\hat{S}_0(2^n)]$ in the decomposition level $p = 0$;
- all nonzero coefficients of the transforms $[\hat{S}_p^{k_p}(2^{n-p})]$ for $k_p = 1, 2, \ldots, 4^p$ in the decomposition levels $p = 1, 2, \ldots, r$.

The spectrum coefficients of same spatial frequency (u, v) from all image blocks are arranged in common data sequences, which correspond to their decomposition level p. The transformation of the 2D data massifs into one-dimensional data sequence is executed, using the recursive Hilbert scan, which preserves very well the correlation between neighboring coefficients.

In order to reduce the decomposition complexity, and in accordance with Eq. 1.1, this could be done recursively, as follows:

$$[B_r'(2^n)] = [B_{r-1}'(2^n)] + [\hat{E}_r(2^n)] \text{ for } r = 1, 2, \ldots, n-1. \qquad (1.10)$$

For the case, when the number of the retained coefficients for each IDP sub-block k_p of size $2^{n-p} \times 2^{n-p}$ is $\sum_{u=0}^{2^{n-p}} \sum_{v=0}^{2^{n-p}} m_p(u, v) = 4$, then their total number for all levels is:

$$N = \sum_{p=0}^{n-1} 4^{p+1} = (4/3)(4^n - 1) \approx (4/3)4^n. \qquad (1.11)$$

In this case the total number of "retained" coefficients is 4/3 times higher than that of the pixels in the block, and hence, the IPD is "overcomplete".

1.3.3 Reduced Inverse Difference Pyramid

For the building of the Reduced IDP (RIDP) [14], the existing relations between the spectrum coefficients from the neighboring IDP levels are used. Let the retained coefficients $s_p^{k_p}(u, v)$ with spatial frequencies (0, 0), (1, 0), (0, 1) and (1, 1) for the sub-block k_p in the IDP level p, be obtained by using the 2D-WHT. Then, except for level $p = 0$, the coefficients (0, 0) from each of the four neighboring sub-blocks in same IDP level are equal to zero, i.e.:

$$s_p^{k_p}(0, 0) = s_p^{k_p+1}(0, 0) = s_p^{k_p+2}(0, 0) = s_p^{k_p+3}(0, 0) = 0 \text{ for } p = 1, 2, \ldots, n-1.$$

$$(1.12)$$

From this, it follows that the coefficients $s_{p+i}^{k_p}(0,0)$ for $i = 0$, 1, 2, 3 could be cut-off, and as a result they should not be saved or transferred. Hence, the total number of the retained coefficients N_R for each sub-block k_p in the decomposition levels $p = 1, 2,\ldots, n-1$ of the RIDP could be reduced by $\frac{1}{4}$, i.e.

$$N_R = 4 + \sum_{p=1}^{n-1} 4^{p+1} - \sum_{p=1}^{n-1} 4^p = 4 + 3 \sum_{p=1}^{n-1} 4^p = 4 + 3\frac{4}{3}(4^{n-1} - 1) = 4^n. \quad (1.13)$$

In this case the total number of the "retained" coefficients for all levels is equal to the number of pixels in the block, and hence, the so calculated RIPD is "complete".

1.3.4 Main Principle for Branched IDP Building

The pyramid BIDP [15, 16] with one or more branches is an extension of the basic IDP. The image representation through the BIDP aims at the enhancement of the image energy concentration in a small number of IDP components. On Fig. 1.2 is shown an example block diagram of the generalized 3-level BIDP. The IDP for each block of size $2^n \times 2^n$ from the original image, called "Main Pyramid", is of 3 levels ($n = 3$, for $p = 0$, 1, 2). The values of the coefficients, calculated for these 3 levels, compose the inverse pyramid, whose sections are of different color each. The coefficients $s(0, 0)$, $s(0, 1)$, $s(1, 0)$ and $s(1, 1)$ in level $p = 0$ from all blocks compose corresponding matrices of size $m \times m$, colored in yellow. These 4 matrices build the "Branch for level 0" of the Main Pyramids. Each is then divided into blocks of size $2^{n-1} \times 2^{n-1}$, on which in similar way are built the corresponding 3-level IPDs ($p = 00, 01, 02$). The retained coefficients $s(0, 1)$, $s(1, 0)$ and $s(1, 1)$ in level $p = 1$ of the Main Pyramids from all blocks build matrices of size $2m \times 2m$ (colored in pink).

Each matrix of size $2m \times 2m$ is divided into blocks of size $2^{n-1} \times 2^{n-1}$, on which in similar way are build corresponding 3-level IDPs ($p = 10, 11, 12$). The retained coefficients, calculated after TOT from the blocks of the Residual Difference in the

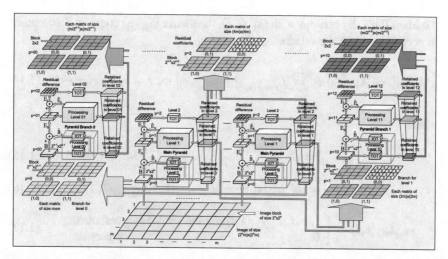

Fig. 1.2 Example of generalized 3-level Branched Inverse Difference Pyramid (BIDP)

last level ($p = 2$) of the Main Pyramids, build matrices of size $4m \times 4m$; from the first level ($p = 00$) of the Pyramid Branch 0—matrices of size ($m/2^{n-1} \times m/2^{n-1}$); and from the first level ($p = 10$) of the "Pyramid Branch 1"—matrices of size ($m/2^{n-2} \times m/2^{n-2}$). In order to reduce the correlation between the elements of the so obtained matrices, on each group of 4 spatially neighboring elements is applied the following transform: the first is substituted by their average value, and each of the remaining 3—by its difference to next elements, scanned counter-clockwise. The coefficients, obtained this way from all levels of the Main and Branch Pyramids are arranged in one-dimensional sequences in accordance with Hilbert scan and after that are quantizated and entropy coded using Adaptive RLC and Huffman. The values of the spectrum coefficients are quantizated only in case that the image coding is lossy. In order to retain the visual quality of the restored images, the quantization values are related to the sensibility of the human vision to errors in different spatial frequencies. To reduce these errors, retaining the compression efficiency, in the consecutive BIDP levels could be used various fast orthogonal transforms: for example, in the zero level could be used DCT, and in the next levels—WHT.

1.3.5 Mathematical Representation for One BIDP Branch

In the general case, the branch g of the BIDP is built on the matrix $[S_{g(u, v)}]$ of size $2^{n-g-1} \times 2^{n-g-1}$, which comprises all spectrum coefficients $s_p^{k_p}(u, v)$ with the same spatial frequency (u, v) from all blocks or sub-blocks k_p in the level $p = g$ of the Main IDPs. By analogy with Eq. (1.1), the matrix $[S_{g(u,v)}]$ could be decomposed in accordance with the relation, given below:

$$[S_{g(u,v)}] = [\tilde{S}_{0,g(u,v)}] + \sum_{s=1}^{r} [\tilde{E}^{k_s}_{s-1,g(u,v)}] + [\tilde{E}_{r,g(u,v)}] \text{ for } r = 1, 2, \ldots, n-1, \quad (1.14)$$

where

$$[\tilde{S}_{0,g(u,v)}] = [T_{0,g(u,v)}]^{-1} [\hat{S}_{0,g(u,v)}] [T_{0,g(u,v)}]^{-1} \text{ for } s = 0; \quad (1.15)$$

$$[\hat{S}_{0,g(u,v)}] = [\hat{s}_{0,g(u,v)}(k, l)] = [m_{0,g(u,v)}(k, l) s_{0,g(u,v)}(k, l)]; \quad (1.16)$$

$$m_{0,g(u,v)}(k, l) = \begin{cases} 1, & \text{if } s_{0,g(u,v)}(k,l) \text{ are the retained coefficients,} \\ 0 & \qquad\qquad\qquad\quad \text{otherwise;} \end{cases} \quad (1.17)$$

$$[S_{0,g(u,v)}] = [s_{0,g(u,v)}(k, l)] = [T_{0,g(u,v)}][S_{g(u,v)}][T_{0,g(u,v)}]; \quad (1.18)$$

$$[E_{s-1,g(u,v)}] = [S_{g(u,v)}] - [\tilde{S}_{0,g(u,v)}] \text{ for } s = 1; \quad (1.19)$$

$$[\tilde{E}^{k_s}_{s-1,g(u,v)}] = [T_{s,g(u,v)}]^{-1} [S^{k_s}_{s,g(u,v)}][T_{s,g(u,v)}]^{-1} \text{ for } s = 2, 3, \ldots, r \text{ and } k_s$$
$$= 1, 2, \ldots, 4^s; \quad (1.20)$$

$$[E_{s-1,g(u,v)}] = [E_{s-2,g(u,v)}] - [\tilde{E}_{s-2,g(u,v)}]. \quad (1.21)$$

All matrices in Eqs. (1.14)–(1.19) are of size $2^{n-g-1} \times 2^{n-g-1}$, and these in Eqs. (1.20) and (1.21)—of size $2^{n-g-s-1} \times 2^{n-g-s-1}$. The decomposition from Eq. (1.14) of the matrix $[S_{g(u,v)}]$ is named Pyramid Branch ($PB_{g(u,v)}$). It is a pyramid, whose initial and final levels are g and r correspondingly ($g < r$). This pyramid represents the branch g of the Main IDPs and contains all coefficients, whose spatial frequency is (u, v).

The maximum number of branches for the levels $p = 0, 1, \ldots, n-1$ of the Main IDPs, built on a sub-block of size $2^{n-p} \times 2^{n-p}$, is defined by the general number of retained spectrum coefficients $M_p = 4^p \sum_{u=0}^{2^{n-p}} \sum_{v=0}^{2^{n-p}} m_p(u, v)$. For the branch g from the level $p = g$ the corresponding pyramid $PB_{g(uv)}$ is of r levels. The number of the coefficients in this branch of the Main IDPs for $p = g, g + 1, \ldots, r$, without cutting-off the coefficients, calculated for the spatial frequency $(0, 0)$, is:

$$N_{g,r} = M_g \sum_{p=g}^{r} 4^p = M_g \left[\sum_{p=0}^{r} 4^p - \sum_{p=0}^{g-1} 4^p \right] = (M_g/3)(4^{r+1} - 4^g). \quad (1.22)$$

In case that the number of the retained spectrum coefficients for each sub-block is set to be $\sum_{u=0}^{2^{n-g}} \sum_{v=0}^{2^{n-g}} m_g(u, v) = 4$, then $M_g = 4^{g+1}$. In this case, from Eq. (1.22) it follows, that the total number of the coefficients in the branch $PB_{g(uv)}$ is $N_{g,r} = (4^{g+1}/3)(4^{r+1} - 4^g)$. Hence, the compression ratio (CR) for $PB_{g(uv)}$ is defined by the relation:

$$CR_{g,r} = \frac{4^{n-g-1}}{N_{g,r,}} = \frac{3}{4} \times \frac{4^{n-g-1}}{4^g(4^{r+1} - 4^g)}, \tag{1.23}$$

where 4^{n-g-1} is the number of the elements in one sub-block of size $2^{n-g-1} \times 2^{n-g-1}$ from $PB_{g(uv)}$.

The compression ratio for the Main IDPs, calculated in accordance with Eq. (1.11), is:

$$CR = \frac{4^n}{N} = \frac{3 \times 4^n}{4(4^n - 1)} \approx \frac{3}{4} \text{ for } 4^n \gg 1. \tag{1.24}$$

From the comparison of the Eqs. (1.23) and (1.24) it follows, that:

$$CR_{g,r} > CR, \text{ if } r \leq n - 3. \tag{1.25}$$

In case that the requirement from Eq. (1.25) for the number of levels r of $PB_{g(u,v)}$ for level g of the Main IDPs is satisfied, the compression ratio for the branch g is higher, than that for each of the basic pyramids. From Eq. (1.25) it follows that the condition $r > 1$ is satisfied, when $n > 4$, i.e., when the image is divided into blocks of minimum size of 16×16 pixels. For this case, to retain the correlation between their pixels high, is necessary the size of the image $(16m) \times (16m)$ to be relatively large. For example, the image should be of size $2k \times 2k$ (for $m = 128$), or larger. Hence, the BIDP decomposition is efficient mainly for images with high resolution.

The correlation between the elements of the blocks of size $2^{n-1} \times 2^{n-1}$ from the initial level $g = 0$ of the Main IDPs is higher than that, between the elements of the sub-blocks of size $2^{n-g-1} \times 2^{n-g-1}$ from the higher levels $g = 1, 2, ..., r$. Because of this, the branching of the BIDP should always start from the level $g = 0$.

1.3.6 Transformation of the Retained Coefficients into Sub-blocks of Size 2 × 2

The aim of the transformation is to reduce the correlation between the retained neighboring spectrum coefficients in the sub-blocks of size 2×2 in each matrix, built by the coefficients of same spatial frequency (u, v) from all blocks (or respectively—from the sub-blocks k_p in the selected level p of the Main IDPs, or

their branches). In order to simplify the presentation, the spectrum coefficients in
the sub-blocks k_p for the level p, are set as follows:

$$A_i = s_p^{k_p+i}(0,0); \ B_i = s_p^{k_p+i}(1,0); \ C_i = s_p^{k_p+i}(0,1); \ D_i = s_p^{k_p+i}(1,1) \text{ for } i = 0, 1, 2, 3.$$
$$(1.26)$$

On Fig. 1.3 are shown matrices of size 2×2, which contain the retained groups
of four spectrum coefficients $s_p^{k_p}(u,v)$, which have same frequencies, (0, 0), (1, 0),
(0, 1) and (1, 1) correspondingly, placed in four neighboring sub-blocks (k_p, $k_p + 1$,
$k_p + 2$, $k_p + 3$) of size $2^{n-p} \times 2^{n-p}$ for the level p of the Main IDPs, or their branches.

In correspondence with the symbols, used in Fig. 1.3, the transformation of the
groups of four coefficients is represented by the relation below [16]:

$$\begin{bmatrix} S_1 \\ S_2 \\ S_3 \\ S_4 \end{bmatrix} = \frac{1}{4} \begin{bmatrix} 1 & 1 & 1 & 1 \\ 0 & 4 & 0 & -4 \\ -4 & 0 & 4 & 0 \\ 0 & 0 & -4 & 4 \end{bmatrix} \begin{bmatrix} P_1 \\ P_2 \\ P_3 \\ P_4 \end{bmatrix}. \tag{1.27}$$

Here P_i, for $i = 1, 2, 3, 4$ represent correspondingly:

- the coefficients A_i, for $i = 1, 2, 3, 4$ with frequencies (0, 0);
- the coefficients B_i, for $i = 1, 2, 3, 4$ with frequencies (1, 0);
- the coefficients C_i, for $i = 1, 2, 3, 4$ with frequencies (0, 1);
- the coefficients D_i, for $i = 1, 2, 3, 4$ with frequencies (1, 1).

Fig. 1.3 Location of the
retained groups of four
spectrum coefficients from 4
neighboring sub-blocks
$k_p + i$ ($i = 0, 1, 2, 3$) of size
$2^{n-p} \times 2^{n-p}$ in the
decomposition level p

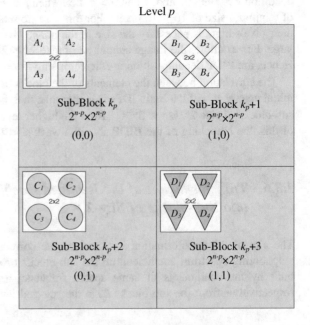

In result of the transform, executed in accordance with Eq. (1.27), each coefficient S_1 has higher value, than the remaining three difference coefficients S_2, S_3, and S_4.

The inverse transform executed in respect of Eq. (1.27) gives total restoration of the initial coefficients P_i, for $i = 1, 2, 3, 4$:

$$\begin{bmatrix} P_1 \\ P_2 \\ P_3 \\ P_4 \end{bmatrix} = \frac{1}{4} \begin{bmatrix} 4 & -1 & -3 & -2 \\ 4 & 3 & 1 & 2 \\ 4 & -1 & 1 & -2 \\ 4 & -1 & 1 & 2 \end{bmatrix} \begin{bmatrix} S_1 \\ S_2 \\ S_3 \\ S_4 \end{bmatrix}, \tag{1.28}$$

Depending on the frequency (0, 0), (1, 0), (0, 1), or (1, 1) of the restored coefficients $P_1 \sim P_4$, they correspond to $A_1 \sim A_4$, $B_1 \sim B_4$, $C_1 \sim C_4$, or $D_1 \sim D_4$. The operation, given in Eq. (1.28) is executed through decoding of the transformed coefficients $S_1 \sim S_4$. The so described features of the coefficients S_1, S_2, S_3, S_4 permit to achieve significant enhancement of their entropy coding efficiency.

The basic quality of the BIDP is that it offers significant decorrelation of the processed image data. As a result, the BIDP permits the following:

- To achieve highly efficient compression with retained visual quality of the restored image (i.e. visually lossless coding), or efficient lossless coding, depending on the application requirements;
- Layered coding and transfer of the image data, in result of which is obtained low transfer bit-rate with gradually increased quality of the decoded image;
- Lower computational complexity than that of the wavelet decompositions [4];
- Easy adaptation of the coder parameters, so that to ensure the needed concordance of the obtained data stream, to the ability of the communication channel;
- Resistance to noises in the communication channel, or due to compression/decompression. The reason for this is the use of TOT in the decoding of each image block;
- Retaining the quality of the decoded image after multiple coding/decoding;

The BIDP could be further developed and modified in accordance to the requirements of various possible applications. One of these applications for processing of groups of similar images, for example, is a sequence of Computer Tomography (CT) images, Multi-Spectral (MS) images, etc.

1.3.7 Experimental Results

The experimental results, given below, were obtained from the investigation of image database, which contained medical images stored in DICOM (*dcm*) format, of various size and kind, grouped in 24 classes. The database was created at the

Table 1.1 Results for the lossless compression of various classes of medical images

Image name size	CTI013 512 × 512 (Group of 14 images)	CTI069 512 × 512 (Group of 275 images)	CTI022 512 × 512 (Group of 14 images)	CTI002 512 × 512 (Group of 14 images)	MGI001 1914 × 2294	NMI001 1024 × 1024	USI 1.2 1020 × 818	USI 1.3 1020 × 818
Image type: CTI/NMI/CRI/MGI/USI								
dcm	545 KB	545 KB	39.6 KB	545 KB	4.19 MB	2.00 MB	2.44 MB	4.19 MB
jp2	64.2 KB	52.2 KB	34.3 KB	3.92 KB	820 KB	103 KB	700 KB	189 KB
tk	63.9 KB	50.5 KB	26.3 KB	2.30 KB	801 KB	71 KB	388 KB	134 KB
Comparison of compression efficiency for the examples above								
dcm/jp2	8.48	10.44	1.15	139.03	5.11	19.41	3.48	22.16
dcm/tk	8.52	10.79	1.51	236.95	5.23	28.17	6.28	31.26

Medical University of Sofia, and comprises the following image kinds: CTI—
computer tomography images; MGI—mammography images; NMI—nuclear magnetic resonance images; CRI—computer radiography images, and USI—ultrasound
images. For the investigation, the DICOM images were first transformed into
non-compressed (*bmp* format), and then they were processed by using various
lossless compression algorithms. A part of the obtained results is given in Table 1.1.

Here are shown the results for the lossless compression of the *bmp* files of still
images, and of image sequences, after their transformation into files of the kind *jp2*
and *tk*. The image file format *jp2* is based on the standard JPEG2000LS, and the *tk*
format—on the algorithms BIDP for single images, combined with the adaptive
run-length lossless coding (ARLE), based on the histogram statistics [17]. For the
execution of the 2D-TOT/IOT in the initial levels of all basic pyramids and their
branches was used the 2D-DCT, and in their higher levels—the 2D-WHT transform. The number of the pyramid levels for the blocks of the smallest treated
images (of size 512×512), is two, and for the larger ones, it is three. The basic IDP
pyramids have one branch only, comprising coefficients with spatial frequency
(0, 0) for their initial levels.

From the analysis of the obtained results, the following conclusions could be
done:

1. The new format *tk* surpasses the *jp2*, especially for images, which contain
   objects, placed on a homogenous background. From the analyzed 24 classes of
   images, 17 are of this kind. Some examples are shown in Table 1.1;
2. Together with the enlargement of the analyzed images, the compression ratio for
   the lossless *tk* compression grows up, compared to that of the *jp2*;
3. The data given in Table 1.1 show that the mean compression ratio for all
   DICOM images after their transformation into the format *tk* is 41:1, while for the
   jp2 this coefficient is 26:1. Hence, the use of the *tk* format for all 24 classes
   ensures compression ratio which is $\approx$40 % higher than that of the *jp2* format.

The experimental results, obtained for the comparison of the coding efficiency
for several kinds of medical images through BIDP and JPEG2000 confirmed the
basic advantages of the new approach for hierarchical pyramid decomposition,
presented here.

1.4 Hierarchical Singular Value Image Decomposition

The SVD is a statistical decomposition for processing, coding and analysis of
images, widely used in the computer vision systems. This decomposition was an
object of vast research, presented in many monographs [18–22] and papers [23–26].
This is optimal image decomposition, because it concentrates significant part of the
image energy in minimum number of components, and the restored image (after
reduction of the low-energy components), has minimum mean square error. One of
the basic problems, which limit, to some degree, the use of the "classic" SVD, is

related to its high computational complexity, which grows up together with the image size.

To overcome this problem, several new approaches are already offered. The first is based on the SVD calculation through iterative methods, which do not require defining the characteristic polynomials of a pair of matrices. In this case, the SVD is executed in two stages: in the first, each matrix is first transformed into triangular form with the QR decomposition, and then—into bidiagonal, through the Householder transforms [27]. In the second stage on the bidiagonal matrix is applied an iterative method, whose iterations stop when the needed accuracy is achieved. For this could be used the iterative method of Jacobi [21], in accordance with which for the calculation of the SVD with bidiagonal matrix is needed the execution of a sequence of orthogonal transforms with rotation matrix of size 2×2. The second approach is based on the relation of the SVD with the Principal Component Analysis (PCA). It could be executed through neural networks [28] of the kind generalized Hebbian or multilayer perceptron networks, which use iterative learning algorithms. The third approach is based on the algorithm, known as Sequential KL/SVD [29]. The basic idea here is as follows: the image matrix is divided into blocks of small size, and on each is applied the SVD, based on the QR decomposition [21]. At first, the SVD is calculated for the first block from the original image (the upper left, for example), and then is used iterative SVD calculation for each of the remaining blocks by using the transform matrices, calculated for the first block (by updating the process). In the flow of the iteration process are deleted the SVD components, which correspond to very small eigen values.

For the acceleration of the SVD calculation several methods are already developed [30–32]. The first, is based on the algorithm, called Randomized SVD [30], a number of matrix rows (or columns) is randomly chosen. After scaling, they are used to build a small matrix, for which is calculated the SVD, and it is later used as an approximation of the original matrix. In [31] is offered the algorithm QUIC-SVD, suitable for matrices of very large size. Through this algorithm is achieved fast sample-based SVD approximation with automatic relative error control. Another approach is based on the sampling mechanism, called the cosine tree, through which is achieved best-rank approximation. The experimental investigation of the QUIC-SVD in [32] presents better results than those, from the MATLAB SVD and the Tygert SVD. The so obtained 6–7 times acceleration compared to the SVD depends on the pre-selected value of the parameter δ which defines the upper limit of the approximation error, with probability $(1 - \delta)$.

Several SVD-based methods developed, are dedicated to enhancement of the image compression efficiency [33–37]. One of them, called Multi-resolution SVD [33], comprises three steps: image transform, through 9/7 biorthogonal wavelets of two levels, decomposition of the SVD-transformed image, by using blocks of size 2×2 up to level six, and at last—the use of the algorithms SPIHT and gzip. In [34] is offered the hybrid KLT-SVD algorithm for efficient image compression. The method K-SVD [35] for facial image compression, is a generalization of the K-means clusterization method, and is used for iterative learning of overcomplete dictionaries for sparse coding. In correspondence with the combined compression algorithm, in [36] is proposed a SVD based sub-band decomposition and multi-resolution representation of digital colour images. In the paper [37] is used the decomposition, called Higher-Order SVD (HOSVD), through which the SVD matrix is transformed into a tensor with application in the image compression.

In this chapter, the general presentation of one new approach for hierarchical decomposition of matrix images is given, based on the multiple application of the SVD on blocks of size 2 × 2 [38]. This decomposition, called Hierarchical SVD (HSVD), has tree-like structure of the kind "binary tree" (full or truncated). The SVD calculation for blocks of size 2 × 2 is based on the adaptive KLT [5, 39]. The HSVD algorithm aims to achieve a decomposition with high computational efficiency, suitable for parallel and recursive processing of the blocks through simple algebraic operations, and offers the possibility for enhancement of the calculations through cutting-off the tree branches, whose eigen values are small or equal to zero.

1.4.1   SVD Algorithm for Matrix Decomposition

In the general case, the decomposition of each image matrix $[X(N)]$ of size $N \times N$ could be executed by using the direct SVD [5], defined by the equation below:

$$[X(N)] = [U(N)][\Lambda(N)]^{1/2}[V(N)]^t = \sum_{s=1}^{N} \sqrt{\lambda_s} \vec{U}_s . \vec{V}_s^t. \tag{1.29}$$

The inverse SVD is respectively:

$$[\Lambda(N)]^{1/2} = [U(N)]^t[X(N)][V(N)]. \tag{1.30}$$

In the relations above, the terms $[U(N)] = [\vec{U}_1, \vec{U}_2, \ldots, \vec{U}_N]$ and $[V(N)] = [\vec{V}_1, \vec{V}_2, \ldots, \vec{V}_N]$ are matrices, composed respectively by the vectors $\vec{U}_s$ and $\vec{V}_s$ for $s = 1, 2, \ldots, N$; $\vec{U}_s$ are the eigenvectors of the matrix $[Y(N)] = [X(N)][X(N)]^t$ (left-singular vectors of the $[X(N)]$), and $\vec{V}_s$—the eigenvectors of the matrix $[Z(N)] = [X(N)]^t[X(N)]$ (right-singular vectors of the $[X(N)]$), for which:

$$[Y(N)]\vec{U}_s = \lambda_s \vec{U}_s, \; [Z(N)]\vec{V}_s = \lambda_s \vec{V}_s; \tag{1.31}$$

$[\Lambda(N)] = diag[\lambda_1, \lambda_2, .., \lambda_N]$ is a diagonal matrix, composed by the eigenvalues λ_s which are identical for the matrices $[Y(N)]$ and $[Z(N)]$.

From Eq. (1.29) it follows that for the description of the decomposition for a matrix of size $N \times N$, $N \times (2N + 1)$ parameters are needed in total, i.e. in the general case the SVD is a decomposition of the kind "overcomplete".

1.4.2 Particular Case of the SVD for Image Block of Size 2 × 2

In this case, the direct SVD for the block $[X]$ of size 2×2 (for $N = 2$) is represented by the relation:

$$[X] = \begin{bmatrix} a & b \\ c & d \end{bmatrix} = [U][\Lambda]^{1/2}[V]^t = \sqrt{\lambda_1}\vec{U}_1\vec{V}_1^t + \sqrt{\lambda_2}\vec{U}_2\vec{V}_2^t = \sum_{s=1}^{2} \sqrt{\lambda_s}\vec{U}_s\vec{V}_s^t$$

(1.32)

or

$$[X] = [C_1] + [C_2],$$ (1.33)

where $[C_1] = \sqrt{\lambda_1}\vec{U}_1\vec{V}_1^t$; $[C_2] = \sqrt{\lambda_2}\vec{U}_2\vec{V}_2^t$; a, b, c, d are the elements of the block $[X]$; λ_1, λ_2 are the eigenvalues of the symmetrical matrices $[Y]$ and $[Z]$, defined by the relations below:

$$[Y] = [X][X]^t = \begin{bmatrix} a & b \\ c & d \end{bmatrix}\begin{bmatrix} a & c \\ b & d \end{bmatrix} = \begin{bmatrix} (a^2+b^2) & (ac+bd) \\ (ac+bd) & (c^2+d^2) \end{bmatrix};$$ (1.34)

$$[Z] = [X]^t[X] = \begin{bmatrix} a & c \\ b & d \end{bmatrix}\begin{bmatrix} a & b \\ c & d \end{bmatrix} = \begin{bmatrix} (a^2+c^2) & (ab+cd) \\ (ab+cd) & (b^2+d^2) \end{bmatrix}.$$ (1.35)

$\vec{U}_1$ and $\vec{U}_2$ are the eigenvectors of the matrix $[Y]$, for which: $[Y]\vec{U}_s = \lambda_s\vec{U}_s$ ($s = 1, 2$);

$\vec{V}_1$ and $\vec{V}_2$ are the eigenvectors of the matrix $[Z]$, for which: $[Z]\vec{V}_s = \lambda_s\vec{V}_s$ ($s = 1, 2$).

$[U] = [\vec{U}_1, \vec{U}_2]$ and $[V]^t = \begin{bmatrix} \vec{V}_1^t \\ \vec{V}_2^t \end{bmatrix}$ are matrices, composed by the eigen vectors $\vec{U}_s$ and $\vec{V}_s$.

In accordance with the solution given in [38] for the case when $N = 2$, the couple direct/inverse SVD for the matrix $[X(2)]$ could be represented as follows:

$$\begin{bmatrix} a & b \\ c & d \end{bmatrix} = \frac{1}{2A}\left\{ \sigma_1\begin{bmatrix} \sqrt{rp} & \sqrt{sp} \\ \sqrt{rq} & \sqrt{sq} \end{bmatrix} + \sigma_2\begin{bmatrix} \sqrt{sq} & -\sqrt{rq} \\ -\sqrt{sp} & \sqrt{rp} \end{bmatrix} \right\} = \sigma_1[T_1] + \sigma_2[T_2]$$
$$= [C_1] + [C_2],$$

(1.36)

$$\begin{bmatrix} \sigma_1 & 0 \\ 0 & \sigma_2 \end{bmatrix} = \frac{1}{2A}\begin{bmatrix} \sqrt{p} & \sqrt{q} \\ -\sqrt{q} & \sqrt{p} \end{bmatrix}\begin{bmatrix} a & b \\ c & d \end{bmatrix}\begin{bmatrix} \sqrt{r} & -\sqrt{s} \\ \sqrt{s} & \sqrt{r} \end{bmatrix} \text{ for } A \neq 0,$$ (1.37)

where

$$A = \sqrt{v^2 + 4\eta^2}, \; \sigma_1 = \sqrt{\frac{\omega + A}{2}}, \; \sigma_2 = \sqrt{\frac{\omega - A}{2}}, \; r = A + v, \; p = A + \mu,$$

$$s = A - v, \; q = A - \mu,$$

(1.38)

$$v = a^2 + c^2 - b^2 - d^2, \; \eta = ab + cd, \; \omega = a^2 + b^2 + c^2 + d^2,$$

$$\mu = a^2 + b^2 - c^2 - d^2.$$

(1.39)

Figure 1.4 shows the algorithm for direct SVD for the block $[X]$ of size 2×2, composed in accordance with the relations (1.36), (1.38) and (1.39). This algorithm is the basic building element—the kernel, used to create the HSVD algorithm.

In accordance with Eq. (1.32) the matrix $[X]$ is transformed into the vector $\vec{X} = [a, b, c, d]^t$, whose components are arranged by using the "Z"-scan. The components of the vector $\vec{X}$ are the input data for the SVD algorithm. After its execution, are obtained the vectors $\vec{C}_1$ and $\vec{C}_2$, from whose components are defined the elements of the matrices $[C_1]$ and $[C_2]$ of size 2×2, by using the "Z"-scan again. In this case however, this scan is used for the inverse transform of all vectors $\vec{C}_1$, $\vec{C}_2$ in the corresponding matrix $[C_1]$, $[C_2]$.

1.4.3 Hierarchical SVD for a Matrix of Size $2^n \times 2^n$

The hierarchical n-level SVD (HSVD) for the image matrix $[X(N)]$ of size $2^n \times 2^n$ pixels $(N = 2^n)$ is executed through multiple applying the SVD on image sub-blocks (sub-matrices) of size 2×2, followed by rearrangement of the so calculated components.

In particular, for the case, when the image matrix $[X(4)]$ is of size $2^2 \times 2^2$ $(N = 2^2 = 4)$, then the number of the hierarchical levels of the HSVD is $n = 2$. The flow graph, which represents the calculation of the HSVD, is shown on Fig. 1.5. In the first level $(r = 1)$ of the HSVD, the matrix $[X(4)]$ is divided into four sub-matrices of size 2×2, as shown in the left part of Fig. 1.5. Here the elements of the sub-matrices on which is applied the $\text{SVD}_{2 \times 2}$ in the first hierarchical level, are colored in same color (yellow, green, blue, and red). The elements of the sub-matrices are:

$$[X(4)] = \begin{bmatrix} [X_1(2)] & [X_2(2)] \\ [X_3(2)] & [X_4(2)] \end{bmatrix} = \begin{bmatrix} \begin{bmatrix} a_1 & b_1 \\ c_1 & d_1 \\ a_3 & b_3 \\ c_3 & d_3 \end{bmatrix} & \begin{bmatrix} a_2 & b_2 \\ c_2 & d_2 \\ a_4 & b_4 \\ c_4 & d_4 \end{bmatrix} \end{bmatrix}.$$

(1.40)

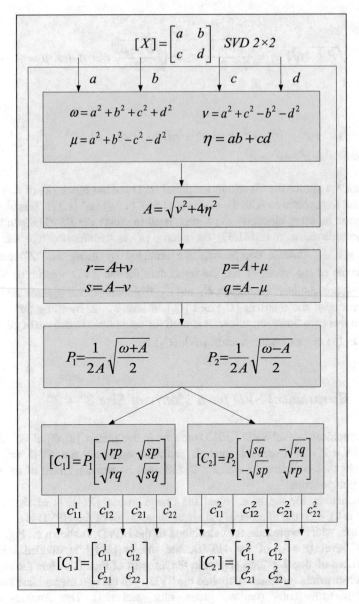

Fig. 1.4 Representation of the SVD algorithm for the matrix [X] of size 2 × 2

On each sub-matrix $[X_k(2)]$ of size 2 × 2 (k = 1, 2, 3, 4), is applied $SVD_{2\times2}$, in accordance with Eqs. (1.36)–(1.39). As a result, it is decomposed into two components:

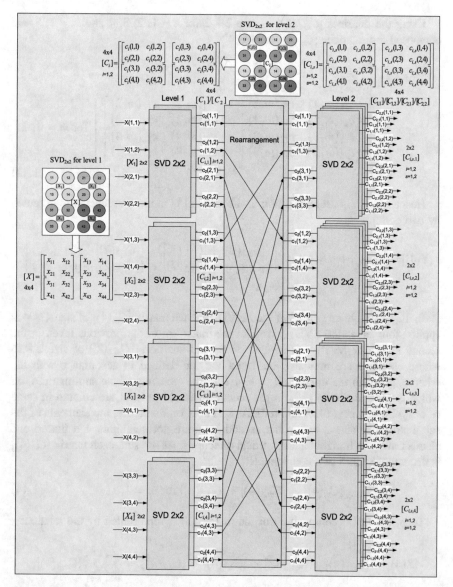

Fig. 1.5 Flowgraph of the HSVD algorithm represented through the vector-radix (2×2) for a matrix of size 4×4

$$[X_k(2)] = \sigma_{1,k}[T_{1,k}(2)] + \sigma_{2,k}[T_{2,k}(2)] = [C_{1,k}(2)] + [C_{2,k}(2)] \text{ for } k = 1, 2, 3, 4; \quad (1.41)$$

where $\sigma_{1,k} = \sqrt{\frac{\omega_{1,k} + A_{1,k}}{2}}$, $\sigma_{2,k} = \sqrt{\frac{\omega_{2,k} - A_{2,k}}{2}}$, $[T_{1,k}(2)] = \vec{U}_{1,k}\vec{V}_{1,k}^t$, $[T_{2,k}(2)] = \vec{U}_{2,k}\vec{V}_{2,k}^t$.

Using the matrices $[C_{m,k}(2)]$ of size 2×2 for $k = 1, 2, 3, 4$ and $m = 1, 2$, are composed the matrices $[C_m(4)]$ of size 4×4:

$$[C_m(4)] = \begin{bmatrix} [C_{m,1}(2)] & [C_{m,2}(2)] \\ [C_{m,3}(2)] & [C_{m,4}(2)] \end{bmatrix}$$

$$= \begin{bmatrix} \begin{bmatrix} c_{11}(m,1) & c_{12}(m,1) \\ c_{13}(m,1) & c_{14}(m,1) \end{bmatrix} & \begin{bmatrix} c_{11}(m,2) & c_{12}(m,2) \\ c_{13}(m,2) & c_{14}(m,2) \end{bmatrix} \\ \begin{bmatrix} c_{11}(m,3) & c_{12}(m,3) \\ c_{13}(m,3) & c_{14}(m,3) \end{bmatrix} & \begin{bmatrix} c_{11}(m,4) & c_{12}(m,4) \\ c_{13}(m,4) & c_{14}(m,4) \end{bmatrix} \end{bmatrix} \text{ for } m = 1, 2.$$

$$(1.42)$$

Hence, the SVD decomposition of the matrix $[X]$ in the first level is represented by two components:

$$[X(4)] = [C_1(4)] + [C_2(4)] = \begin{bmatrix} ([C_{1,1}(2)] + [C_{2,1}(2)]) & ([C_{1,2}(2)] + [C_{2,2}(2)]) \\ ([C_{1,3}(2)] + [C_{2,3}(2)]) & ([C_{1,4}(2)] + [C_{2,4}(2)]) \end{bmatrix}.$$

$$(1.43)$$

In the second level $(r = 2)$ of the HSVD, on each matrix $[C_m(4)]$ of size 4×4 is applied four times the $SVD_{2\times2}$. Unlike the transform in the previous level, in the second level, the $SVD_{2\times2}$ is applied on the sub-matrices $[C_{m,k}(2)]$ of size 2×2, whose elements are mutually interlaced and are defined in accordance with the scheme, given in the upper part of Fig. 1.5. The elements of the sub-matrices, on which is applied the $SVD_{2\times2}$ in the second hierarchical level are colored in same color (yellow, green, blue, and red). As it is seen on the figure, the elements of the sub-matrices of size 2×2 in the second level are not neighbors, but placed one element away in horizontal and vertical directions. As a result, each matrix $[C_m(4)]$ is decomposed into two components:

$$[C_m(4)] = [C_{m,1}(4)] + [C_{m,2}(4)] \text{ for } m = 1, 2. \tag{1.44}$$

Then, the full decomposition of the matrix $[X]$ is represented by the relation:

$$[X(4)] = [C_{1,1}(4)] + [C_{1,2}(4)] + [C_{2,1}(4)] + [C_{2,2}(4)] = \sum_{m=1}^{2} \sum_{s=1}^{2} [C_{m,s}(4)],$$

$$(1.45)$$

Hence, the decomposition of an image of size 4×4 comprises four components in total.

The matrix $[X(8)]$ is of size $2^3 \times 2^3$ ($N = 2^3 = 8$ for $n = 3$), and in this case, the HSVD is executed through multiple calculation of the $SVD_{2\times2}$ on blocks of size 2×2, in all levels (the general number of the decomposition components is eight). In the first and second levels, the $SVD_{2\times2}$ is executed in accordance with the scheme, shown on Fig. 1.5. In the third level, the $SVD_{2\times2}$ is mainly applied on sub-matrices of size 2×2. Their elements are defined in similar way, as shown on Fig. 1.5, but the elements of same color (i.e., which belong to same sub-matrix) are moved three elements away in the horizontal and vertical direction.

The described HSVD algorithm could be generalized for the cases when the image $[X(2^n)]$ is of size $2^n \times 2^n$ pixels. Then the relation (1.45) becomes as shown below:

$$[X(2^n)] = \sum_{p_1=1}^{2} \sum_{p_2=1}^{2} \cdots \sum_{p_n}^{2} [C_{p_1,p_2,\dots,p_n}(2^n)]. \tag{1.46}$$

The maximum number of the HSVD decomposition levels is n, the maximum number of the decomposition components (1.46) is 2^n, and the distance in horizontal and vertical direction between the elements of the blocks of size 2×2 in the level r is correspondingly $(2^{r-1} - 1)$ elements, for $r = 1, 2,\dots, n$.

1.4.4 Computational Complexity of the Hierarchical SVD of Size $2^n \times 2^n$

1.4.4.1 Computational Complexity of the SVD of Size 2×2

The computational complexity could be defined by using the Eq. (1.36), taking into account the number of multiplication and addition operations, needed for the preliminary calculation of the components ω, μ, δ, v, η, A, B, θ_1, θ_2, σ_1, σ_1, defined by the Eqs. (1.38) and (1.39). Then:

- The number of the multiplications, needed for the calculation of Eq. (1.36) is $\Sigma_m = 39$;
- The number of the additions, needed for the calculation of Eq. (1.36) is $\Sigma_s = 15$.

Then the total number of the algebraic operations executed with floating point for SVD of size 2×2 is:

$$O_{\text{SVD}}(2 \times 2) = \Sigma_m + \Sigma_s = 54. \tag{1.47}$$

1.4.4.2 Computational Complexity of the Hierarchical SVD of Size $2^n \times 2^n$

The computational complexity is defined on the basis of $\text{SVD}_{2\times2}$. In this case, the number M of the sub-matrices of size 2×2, which comprise the image of size $2^n \times 2^n$, is $2^{n-1} \times 2^{n-1} = 4^{n-1}$, and the number of the decomposition levels is n.

- The number of $SVD_{2\times 2}$ in the first level is $M_1 = M = 4^{n-1}$;
- The number of $SVD_{2\times 2}$ in the second level is $M_2 = 2 \times M = 2 \times 4^{n-1}$;
- . . . .
- The number of $SVD_{2\times 2}$ in the level n is $M_n = 2^{n-1} \times M = 2^{n-1} \times 4^{n-1}$;

The total number of $SVD_{2\times 2}$ is correspondingly $M_\Sigma = M(1 + 2 + \ldots + 2^{n-1}) = 4^{n-1}(2^n - 1) = 2^{2n-2}(2^n - 1)$. Then the total number of the algebraic operations for the HSVD of size $2^n \times 2^n$ is:

$$O_{HSVD}(2^n \times 2^n) = M_\Sigma \times O_{SVD}(2 \times 2) = 27 \times 2^{2n-1}(2^n - 1). \qquad (1.48)$$

1.4.4.3 Computational Complexity of the SVD of Size $2^n \times 2^n$

For the calculation of the matrices $[Y(N)]$ and $[Z(N)]$ of size $N \times N$ for $N = 2^n$ are needed in total $\Sigma_m = 2^{2n+2}$ multiplications and $\Sigma_s = 2^{n+1}(2^n - 1)$ additions. The total number of the operations is:

$$O_{Y,Z}(N) = 2^{2n+2} + 2^{n+1}(2^n - 1) = 2^{n+1}(3 \times 2^n - 1). \qquad (1.49)$$

In accordance with [40], the number of the operations $O(N)$ for the iterative calculation of all N eigenvalues and the eigen N-component vectors of the matrix of size $N \times N$ for $N = 2^n$ with L iterations, is correspondingly:

$$\begin{aligned} O_{val}(N) &= (1/6)(N - 1)(8N^2 + 17N + 42) \\ &= (1/6)(2^n - 1)(2^{2n+3} + 17 \times 2^n + 42), \end{aligned} \qquad (1.50)$$

$$O_{vec}(N) = N[2N(LN + L + 1) - 1] = 2^n[2^{n+1}(2^n L + L + 1) - 1]. \qquad (1.51)$$

From Eq. (1.31) it follows, that two kinds of eigen vectors $(\vec{U}_s$ and $\vec{V}_s)$ should be calculated, so the number of the needed operations in accordance with Eq. (1.51) should be doubled. From the analysis of the Eq. (1.29) it follows that:

- The number of the needed multiplications for all components is: $\Sigma_m = 2^n(2^{2n} + 2^{2n}) = 2^{3n+1}$;
- The number of the needed additions for all components is: $\Sigma_s = 2^n - 1$.

Then the total number of the needed operations for the calculation of Eq. (1.29) is:

$$O_D(N) = 2^{3n+1} + 2^n - 1 = 2^n(2^{2n+1} + 1) - 1 = 2^n(2^{2n+1} + 1) - 1. \qquad (1.52)$$

Hence, the total number of the algebraic operations, needed for the execution of the SVD of size $2^n \times 2^n$ is:

Table 1.2 Coefficient $\psi_1(n, L)$ of the relative reduction of the computational complexity of the HSVD versus the SVD as a function of n, for $L = 10$

n	2	3	4	5
$\psi_1(n, 10)$	5.94	4.21	3.67	3.44

$$
\begin{aligned}
O_{SVD}(2^n \times 2^n) &= O_{Y,Z}(2^n) + O_{val}(2^n) + 2O_{vec}(2^n) + O_D(2^n) \\
&= 2^{2n+1}[2L(2^n + 1) + 2^{n-1} + 5] + (1/6)(2^{2n+3} + 17 \times 2^n + 42) - 1.
\end{aligned}
\tag{1.53}
$$

1.4.4.4   Relative Computational Complexity of the HSVD

The relative computational complexity of the HSVD could be calculated on the basis of Eqs. (1.53) and (1.48), using the relation below:

$$
\begin{aligned}
\psi_1(n, L) &= \frac{O_{SVD}(2^n \times 2^n)}{O_{HSVD}(2^n \times 2^n)} \\
&= \frac{3 \times 2^{n+1}[2^{n+2}(2^n L + L + 1) + 2^{n+1}(2^n + 3) - 3] + (2^n - 1)(2^{2n+3} + 17 \times 2^n + 42) - 6}{81 \times 2^{2n}(2^n - 1)}.
\end{aligned}
\tag{1.54}
$$

For n = 2, 3, 4, 5 (i.e., for image blocks of size 4 × 4, 8 × 8, 16 × 16 and 32 × 32 pixels), the values of $\psi_1(n, L)$ for $L = 10$ are given in Table 1.2.

For big values of n the relation $\psi_1(n, L)$ does not depend on n and trends towards:

$$
\psi_1(n, L)_{n \to \infty} \Rightarrow 0.1 \times (3L + 1).
\tag{1.55}
$$

Hence, for big values of n, when the number of the iterations $L \geq 4$, the relation $\psi_1(n, L) > 1$, and the computational complexity of the HSVD is lower than that of the SVD. Practically, the value of L is significantly higher than 4. For big values of n the coefficient $\psi_1(n, 10) = 3.1$ and the computational complexity of the HSVD is three times lower than that of the SVD.

1.4.5   Representation of the HSVD Algorithm Through Tree-like Structure

The tree-like structure of the HSVD algorithm of $n = 2$ levels, shown on Fig. 1.6, is built on the basis of the Eq. (1.45), for image block of size 4 × 4. As it could be seen, this is a binary tree. For a block of size 8 × 8, this binary tree should be of $n = 3$ levels.

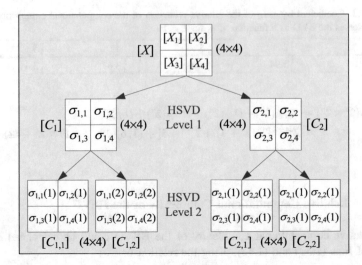

Fig. 1.6 Binary tree, representing the HSVD algorithm for the image matrix [X], of size 4 × 4

Each tree branch has a corresponding eigen value $\lambda_{s,k}$, or resp. $\sigma_{s,k} = \sqrt{\lambda_{s,k}}$ for the level 1, and $\lambda_{s,k}(m)$ or resp. $\sigma_{s,k}(m) = \sqrt{\lambda_{s,k}(m)}$—for the level 2 ($m = 1, 2$). The total number of the tree branches shown on Fig. 1.6, is equal to six. It is possible to cut off some branches, if for them the following conditions are satisfied: $\sigma_{s,k} \vee \sigma_{s,k}(m) = 0$ or $\sigma_{s,k} \leq \Delta_{s,k} \vee \sigma_{s,k}(m) \leq \Delta_{s,k}(m)$, i.e., when they are equal to 0, or are smaller than a small threshold $\Delta_{s,k}$, resp. $\Delta_{s,k}(m)$. To cut down one HSVD component [C_i] in one level, it is necessary all values of σ_i, which participate in this component, to be equal to zero, or very close to it. In result, the decomposition in the corresponding branch could be stopped before the last level n. As a consequence, it follows that the HSVD algorithm is adaptive with respect of the contents of each image block. In this sense, the algorithm HSVD is adaptive and could be easily adjusted to the requirements of each particular application.

From the analysis of the presented HSVD algorithm it follows that its basic advantages compared to the "classic" SVD are:

1. The computational complexity of the full-tree HSVD algorithm (without truncation) for a matrix of size $2^n \times 2^n$, compared to SVD for a matrix of same size, is at least three times lower;
2. The HSVD is executed following the tree-like scheme of n levels, which permits parallel and recursive processing of image blocks of size 2×2 in each level. The corresponding SVD is calculated by using simple algebraic relations;
3. The HSVD algorithm retains the quality of the SVD in respect of the high concentration of the main part of the image energy in the first components of the decomposition. After removal of the low-energy components, the restored matrix has minimum mean square error and is the optimal approximation of the original;

4. The tree-like structure of the HSVD algorithm (a binary tree of n levels) makes more feasible the ability to stop the decomposition earlier in some of the tree branches, for which the corresponding eigen value is zero, or approximately zero. As a result, the computational complexity of the HSVD is additionally reduced, compared to the classic SVD;

5. The HSVD algorithm could be easily generalized for matrices of size different from $2^n \times 2^n$. In these cases each matrix should be divided into blocks of size 8×8, to which is applied the HSVD (i.e., a decomposition of eight components). In case that after the division the blocks at the image borders are incomplete, they should be extended through extrapolation. Such approach is suitable in case, that the number of the decomposition components, which is limited up to 8, is sufficient for the application. If more components are needed, their number could be increased, by dividing the image into blocks of size 16×16, or larger;

6. The HSVD algorithm opens new possibilities for fast image processing in various application areas, as: compression, filtration, segmentation, merging, watermarking, extraction of minimum number of features for pattern recognition, etc.

1.5 Hierarchical Adaptive Principal Component Analysis for Image Sequences

Image sequences are characterized with the huge volumes of visual information and very high spatial and spectral correlation. The decorrelation of this visual information is the first and basic stage of the processing, related to various publication areas, such as: compression and transfer/storage, analysis, objects recognition, etc. For the decorrelation of correlated image sequences, are developed significant number of methods for interframe prediction with movement compensation for temporal decorrelation of moving images and for transform-coding techniques for intra-frame and inter-frame decorrelation. One of the most efficient methods for decorrelation of groups of images is based on the Principal Component Analysis (PCA), known also as Hotelling transform, and Karhunen-Loeve Transform (KLT). This transform is the object of large number of investigations, presented in many scientific monographs [11, 40–47] and papers [12, 48–53]. The KLT is related to the class of linear statistical orthogonal transforms for groups of vectors, obtained, for example, from the pixels of one image, or from a group of matrix images. The PCA has significant role in image analysis and processing, and also in the systems for computer science and pattern recognition. It has a wide variety of application areas: for the creation of optimal models in the image color space [46], for compression of signals and groups of correlated images [41–44, 47], for the creation of objects descriptors in the reduced features' space [50, 51], for image fusion [52] and segmentation [53], image steganography [54], etc.

The PCA has some significant properties: (1) it is an optimal orthogonal transform for a group of vectors, because as a result of the transform, the maximum part of their energy is concentrated in a minimum number of their components; (2) after reduction of the low energy components of the transformed vectors, the corresponding restored vectors have minimum mean square error (MSE); (3) the components of the transformed vectors are not correlated. In particular, in case that the probability distribution of the vectors is Gaussian, their components become decorrelated and independent after PCA. The Independent Components Analysis (ICA) [55] is very close to the PCA in respect of their calculation and properties.

For PCA implementation the pixels of same spatial position in a group of N images compose an N-dimensional vector. The basic difficulty of the PCA implementation is related to the large size of the covariance matrix. For the calculation of its eigenvectors is necessary to calculate the roots of a polynomial of nth degree (characteristic equation) and to solve a linear system of N equations [21, 56]. For large values of N, the computational complexity of the algorithm for calculation of the transform matrix is significantly increased.

One of the basic problems, which limit the use of the PCA, is due to its high computational complexity, which grows up together with the number of the vectors' components. Various approaches are offered to overcome this problem. One of them is based on the PCA calculation through iterative methods, which do not require the definition of the characteristic polynomial of the vectors' covariance matrix. In this case the PCA is executed in 2 stages: in the first, the original image matrix is transformed into a three-diagonal form through QR decomposition [21], and after that—into a bi-diagonal, by using the Householder's transforms [27]. In the second stage, on the bi-diagonal matrix are applied iterative methods, for which the iterations are stopped, after the needed accuracy is achieved. The iterative PCA calculation through the methods of Jacobi and Givens [21, 56], is based on the execution of a sequence of orthogonal transforms with rotational matrices of size 2×2.

One well known approach is based on the PCA calculation by using neural networks [28] of the kind Generalized Hebbian, or Multilayer Perceptron Networks. They both use iterative learning algorithms, for which the number of needed operations can reach several hundreds.

The third approach is based on the Sequential KLT/SVD [29], already commented in the preceding section. In [28, 29] is presented one more approach, based on the recursive calculation of the covariance matrix of the vectors, its eigen values and eigen vectors. In the papers [57, 58] is introduced hierarchical recursive block processing of matrices.

The next approach is based on the so-called Distributed KLT [59, 60], where each vector is divided into sub-vectors and on each is applied Partial KLT. Then is executed global iterative approximation of the KLT, through Conditional KLT, based on side information. This approach was further developed in [61], where is offered one algorithm for adaptive two-stage KLT, combined with JPEG2000, and aimed at the compression of hyper-spectral (HS) images. Similar algorithm for enhanced search is the "Integer Sub-optimal KLT" (Int SKLT) [62], which uses the lifting factorization of matrices. This algorithm is basic for the KLT, executed through a multilevel strategy, also called Divide-and-Conquer (D&C) [63]. In correspondence with this approach, the KLT for a long sequence of images is executed after dividing it into smaller groups, for which the corresponding KLT have lower computational complexity. By applying the KLT on each group, is obtained local decorrelation only. For this reason, the eigen images for the first half of each group in the first decomposition level are used as an input for the next (second) level of the multi-level transform, etc. In the case, when the KLT group contains 2 components only, the corresponding multilevel transform is called Pair-wise Orthogonal Transform (POT) [64]. The experimental results obtained for this transform, when used for HS images, show that it is more efficient than the Wavelet Transform (WT) in respect of Rate-Distortion performance, computational cost, component scalability, and memory requirements.

Another approach is based on the Iterative Thresholding Sparse PCA (ITSPCA) [65] algorithm, aiming at the reduction of the features' space dimension, with minimum dispersion loss.

A fast calculation algorithm (Fast KLT) is known for the particular case, when the images are represented through first order Markov model [66].

In correspondence with the algorithm for PCA randomization [67], on the basis of an accidental choice are selected a certain number of rows (or columns) of the covariance matrix, and on the basis of this approximation, the computational complexity of the KLT is reduced.

In the works [68, 69], are presented hybrid methods for compression of multi-component images through KLT, combined with Wavelets, Adaptive Mixture of Principal Components Model, and JPEG2000.

The analysis of the famous KLT methods shows that: (1) In case of iterative calculations, the number of iterations depends on the covariance matrix of the vectors. In many cases this number is very high, which makes the real-time KLT calculation extremely difficult; (2) In case that the method for multilevel D&C is used, the eigen images from the second half of each group are not transformed in the next levels and as a result, they are not completely decorrelated. Moreover—the selection of the length of each group of images is not optimized.

One of the possible approaches for reducing the computational complexity of PCA for N-dimensional group of images is based on the so-called Hierarchical Adaptive PCA (HAPCA) [70]. Unlike the famous Hierarchical PCA (HPCA) [58], this transform is not related to the image sub-blocks, but to the whole image from one group. For this, the HPCA is implemented through dividing the images into groups of length, defined by their correlation range. Each group is divided into sub-groups of 2 or 3 images each, on which is applied Adaptive PCA (APCA) [71–73], of size 2×2 or 3×3. This transform is performed using equations, which are not based on iterative calculations, and as a result, they have lower computational complexity. To obtain decorrelation for the whole group of images, it is necessary to use APCA of size 2×2 or 3×3, which will be applied in several

consecutive stages (hierarchical levels), with rearranging of the obtained intermediate eigen images after each stage. In result, is obtained a decorrelated group of
eigen images, on which could be applied other combined approaches to obtain
efficient compression through lossless or lossy coding.

1.5.1 Principle for Decorrelation of Image Sequences by Hierarchical Adaptive PCA

The new principle was developed for the transformation of image sequences using
the adaptive PCA (APCA) with transform matrix of size 2×2 or 3×3. The
sequence is divided into groups, whose length is harmonized with their correlation
range. The corresponding algorithm comprises the following steps: (1) correlation
analysis of the image sequence, in result of which is defined the length N of each
group; (2) dividing the processed group into sub-groups of two or three images
each, depending on the length of the group, (3) adding (when necessary) new
interpolated images, which supplements the last sub-group up to two or three
images; (4) defining the number of the hierarchical transform levels on the basis of
the mutual decorrelation, which should be achieved, (5) executing of the HAPCA
algorithm for each group from the image sequence. For this, on each sub-group of
two or three images from the first hierarchical level of HAPCA, is applied
Adaptive PCA (APCA) with matrix of size 2×2 or 3×3. In result, are obtained 2
or 3 eigen images. After that, the eigen images are rearranged so that the first
sub-group of 2 eigen images to comprise the first images from each group, the
second group of 2 or 3 eigen images—the second images from each group, etc. To
each group of intermediate eigen images in the first hierarchical level is applied in
similar way the next APCA with a 2×2 or 3×3 matrix, on each sub-group of 2 or
3 eigen images. In result are obtained the corresponding new intermediate eigen
images in the second hierarchical level. Then the eigen images are rearranged again
so, that the first group of 2 or 3 eigen images contains the first images from each
group before the rearrangement; the second group of 2 or 3 eigen images—the
second image before the rearrangement, etc.

1.5.2 Description of the Hierarchical Adaptive PCA Algorithm

1.5.2.1 Calculation of Eigen Images Through APCA with a 2×2 Matrix

For any 2 digital images of size $S = M \times N$ pixels each, shown on Fig. 1.7, are
calculated the vectors $\vec{C}_s = [C_{1s}, C_{2s}]^t$ for $s = 1, 2, \ldots, S$.

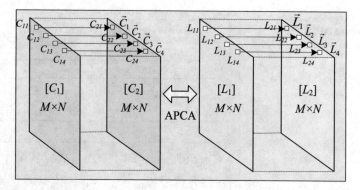

Fig. 1.7 Transformation of the images $[C_1]$, $[C_2]$ into eigen images $[L_1]$, $[L_2]$, through APCA

Each vector is transformed into the corresponding vectors $\vec{L}_s = [L_{1s}, L_{2s}]^t$ through direct APCA using the matrix $[\Phi]$ of size 2×2 in correspondence with the relation:

$$\vec{L}_s = [\Phi](\vec{C}_s - \vec{\mu}) \text{ for } s = 1, 2, \ldots, S. \tag{1.56}$$

where

$$\vec{\mu} = E(\vec{C}_s) = (1/S) \sum_{s=1}^{S} \vec{C}_s = [\bar{C}_1, \bar{C}_2]^t; \quad \bar{C}_1 = E(C_{1s}); \quad \bar{C}_2 = E(C_{2s}); \quad [\Phi] = [\vec{\Phi}_1, \vec{\Phi}_2].$$

On the other hand, the components of the vectors $\vec{\Phi}_1$, $\vec{\Phi}_2$ could be defined using the rotation angle θ of the coordinate system (L_1, L_2) towards the original coordinate system (C_1, C_2), resulting from the APCA execution. Then:

$$\vec{\Phi}_1 = [\cos \theta, \ -\sin \theta]^t; \quad \vec{\Phi}_2 = [\sin \theta, \ \cos \theta]^t, \tag{1.57}$$

where

$$\theta = (1/2) \, arctg[2g_3/(g_1 - g_2)]; \quad g_3 = E(C_{1s}C_{2s}) - (\bar{C}_1)(\bar{C}_2);$$

$$g_1 = E(C_{1s}^2) - (\bar{C}_1)^2; \quad g_2 = E(C_{2s}^2) - (\bar{C}_2)^2.$$

As a result of the transform from Eq. (1.56) on all S vectors, are obtained the corresponding two eigen images $[L_1]$, $[L_2]$, shown in the right part of Fig. 1.7. The transformation from Eq. (1.56) is reversible, and the inverse APCA is represented by the relation:

$$\vec{C}_s = [\Phi]^t \vec{L}_s + \vec{\mu} \text{ for } s = 1, 2, .., S. \tag{1.58}$$

On Fig. 1.8 is shown the algorithm for direct/inverse APCA for a group of two images.

Fig. 1.8 Algorithm for
direct/inverse APCA for a
group of two images

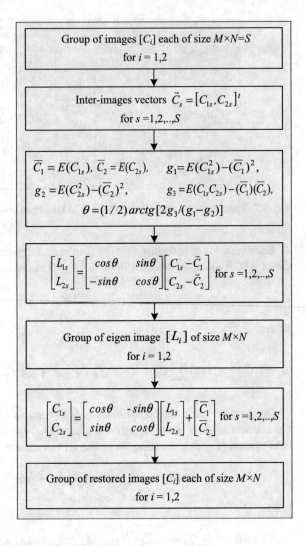

1.5.2.2 Hierarchical APCA Algorithm for a Group of 8 Images

On Fig. 1.9 is shown the 3-level HAPCA algorithm for the case, when the number
of the correlated images in one group (GOI) is $N = 8$; in one sub-group it is
$N_{sg} = 2$; and the number of the sub-groups is $N_g = 4$, i.e. $N = N_{sg} \times N_g$.

As it is shown on Fig. 1.9, on each sub-group of two images from the first
hierarchical level of HAPCA is applied APCA with a 2×2 matrix. In result are
obtained two "eigen" images, colored in yellow and blue correspondingly. After
that, the "eigen" images are rearranged so that the first sub-group of two "eigen"
images comprises the first images from each group, the second group of two
"eigen" images—the second images from each group, etc. For each GOI of 8

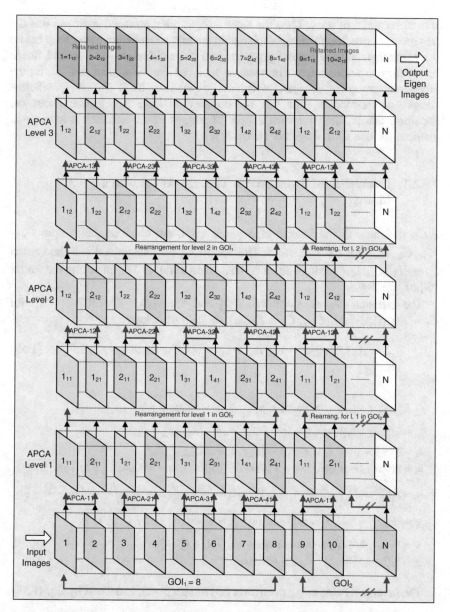

Fig. 1.9 The HAPCA algorithm for direct transform of groups (GOIs) of $N = 8$ images

intermediate eigen images in the first hierarchical level, is applied in similar way the next APCA, with a 2×2 matrix, on each sub-group of two eigen values. In result are obtained two new "eigen" images (i.e. the "eigen" images of the group of two intermediate eigen images), colored in yellow, and blue correspondingly in the second hierarchical level. Then the eigen images are rearranged again, so that the first group of two "eigen" images to contain the first images from each group before the rearrangement; the second group of two "eigen" images—the second image before the rearrangement, etc. In result, is achieved significant decorrelation for the processed group of images, which is a reliable basis for their efficient compression/restoration. For this is necessary to have information about the transform matrix, used for each couple of images in all hierarchical levels—12 matrices for one GOI altogether (when $N = 8$).

1.5.2.3 Calculation of Eigen Images Through APCA with a 3×3 Matrix

From the three digital images of S pixels each, are obtained the vectors $\vec{C}_s = [C_{1s}, C_{2s}, C_{3s}]^t$ for $s = 1, 2, ..., S$. The vectors $\vec{C}_s$ are transformed into the vectors $\vec{L}_s = [L_{1s}, L_{2s}, L_{3s}]^t$ through direct APCA, given in Eq. (1.56), and using the matrix $[\Phi]$ of size 3×3.

The elements Φ_{ij} of the matrix $[\Phi]$ and the vector $\vec{\mu} = [\bar{C}_1, \bar{C}_2, \bar{C}_3]^t$ for $\bar{C}_1 = E(C_{1s}), \bar{C}_2 = E(C_{2s}), \bar{C}_3 = E(C_{3s})$ are defined below:

$$\Phi_{1m} = A_m/P_m; \Phi_{2m} = B_m/P_m; \Phi_{3m} = D_m/P_m \text{ for } m = 1, 2, 3, \tag{1.59}$$

where

$$A_m = (k_3 - \lambda_m)[k_5(k_2 - \lambda_m) - k_4 k_6], B_m = (k_3 - \lambda_m)[k_6(k_1 - \lambda_m) - k_4 k_5],$$

$$D_m = k_6[2k_4 k_5 - k_6(k_1 - \lambda_m)] - k_5^2(k_2 - \lambda_m), P_m = \sqrt{A_m^2 + B_m^2 + D_m^2};$$

$$k_1 = E(C_{1s}^2) - (\bar{C}_1)^2, k_2 = E(C_{2s}^2) - (\bar{C}_2)^2, k_3 = E(C_{3s}^2) - (\bar{C}_3)^2;$$

$$k_4 = E(C_{1s}C_{2s}) - (\bar{C}_1)(\bar{C}_2), k_6 = E(C_{2s}C_{3s}) - (\bar{C}_2)(\bar{C}_3), k_5 = E(C_{1s}C_{3s}) - (\bar{C}_1)(\bar{C}_3);$$

$$\lambda_1 = 2\sqrt{\frac{|p|}{3}}\cos\left(\frac{\varphi}{3}\right) - \frac{a}{3}; \lambda_2 = -2\sqrt{\frac{|p|}{3}}\cos\left(\frac{\varphi + \pi}{3}\right) - \frac{a}{3}; \lambda_3 = -2\sqrt{\frac{|p|}{3}}\cos\left(\frac{\varphi - \pi}{3}\right) - \frac{a}{3};$$

$$q = 2(a/3)^3 - (ab)/3 + c, p = -(a^2/3) + b, \varphi = arccos\left[-q/2 \Big/ \sqrt{(|p|/3)^3}\right];$$

$$a = -(k_1 + k_2 + k_3), b = k_1 k_2 + k_1 k_3 + k_2 k_3 - (k_4^2 + k_5^2 + k_6^2),$$

$$c = k_1 k_6^2 + k_2 k_5^2 + k_3 k_4^2 - (k_1 k_2 k_3 + 2 k_4 k_5 k_6).$$

The inverse APCA, using the matrix $[\Phi]$ of size 3×3, is defined by Eq. (1.58).

1.5.2.4 Hierarchical APCA Algorithm for a Group of 9 Images

In this case, the HAPCA algorithm for a group of nine images $N = 9$ is executed in similar way, as that, shown on Fig. 1.9 for a group of eight images ($N = 8$). Each GOI is divided into $N_g = 3$ sub-groups, each containing $N_{sg} = 3$ images, and the number of the HAPCA decomposition levels is $n = 2$. In the first HAPCA level, in accordance with Eq. (1.56) on the vectors $\vec{C}_s = [C_{1s}, C_{2s}, C_{3s}]^t$ for each of the three sub-groups the APCA is executed. In this case, the elements of the matrix $[\Phi]$ of size 3×3 are defined by Eq. (1.59). In result, for each sub-group the vectors $\vec{L}_s = [L_{1s}, L_{2s}, L_{3s}]^t$ are calculated. After the rearrangement of the vectors components from all sub-groups and their second division into sub-groups of same size ($N_g = 3$), are obtained the corresponding input vectors $\vec{L}_s^1(r) = [L_{1s}^1(r), L_{2s}^1(r), L_{3s}^1(r)]^t$ for the next HAPCA level, etc.

1.5.3 Setting the Number of the Levels and the Structure of the HAPCA Algorithm

1.5.3.1 Number of the HAPCA Levels

The minimum number of levels n_{min} needed for the execution of the HAPCA algorithm for a group of N images could be defined through the analysis of the mutual correlation of the group of transformed N-dimensional vectors, obtained after each hierarchical level. For this, after the execution of the first HAPCA level for the transformed vectors $\vec{L}_s$ for each sub-group (with two or three components), are obtained the N-dimensional vectors $\vec{L}_s^1 = [L_{1s}^1, L_{2s}^1, \ldots, L_{Ns}^1]^t$. After the rearrangement of the components of each vector $\vec{L}_s^1$, it is transformed into the vector $\vec{L}_s^1(r) = [L_{1s}^1(r), L_{2s}^1(r), \ldots, L_{Ns}^1(r)]^t$. The decision to continue with the next (second) HAPCA is based on the analysis of the covariance matrix $[K_L^1(r)]$ of the rearranged vectors $\vec{L}_s^1(r)$ for $s = 1, 2, \ldots, S$, from which could be calculated the achieved decorrelation in the first level. In case that full decorrelation is achieved, the matrix $[K_L^1(r)]$ is diagonal. The HAPCA algorithm could be stopped before the second level even if the decorrelation is not full, provided that the relation below is satisfied:

$$\left\{ \sum_{i=1}^{N} \sum_{j=1}^{N} [k_{i,j}(r)]^2_{|(i \neq j)} \Big/ \sum_{i=1}^{N} \sum_{j=1}^{N} [k_{i,j}(r)]^2_{|(i=j)} \right\} \leq \delta. \tag{1.60}$$

Here $k_{i,j}(r)$ is the element (i, j) of the matrix $[K_L^1(r)]$, and δ is a threshold with preliminary set small value. In case that the condition from Eq. (1.60) is not satisfied, the processing continues with the second HAPCA level. After all calculations are finished, the condition in Eq. (1.60) is checked again, but here $k_{i,j}(r)$ are the elements of the matrix $[K_L^2(r)]$ of the rearranged vectors $\vec{L}_s^2(r)$ in the second level, etc.

1.5.3.2 Structure of the HAPCA Algorithm

The structure of the HAPCA algorithm for one group (GOI) depends on the number of images (N) in it. This number is defined through correlation analysis of the whole image sequence, and most frequently it is in the range from 4, up to 16. In some cases, the number of images in the group is not divisible by the number of the images in a sub-group (N_g), which should be two or three, and then the number of the images N has to be extended by adding m_{int} interpolated images to the GOI. In result, the new value $N_e = N + m_{int}$ becomes divisible by two or three. In Table 1.3, are given the basic parameters of HAPCA for one GOI: N—number of images in the group, n—the number of transform levels, N_{sg}—the number of the sub-groups, N_g—the number of the images in one sub-group, N_e—the number of images in the extended GOI, and m_{int}—the number of the interpolated images in the extended GOI.

The number of the levels n in Table 1.3 is defined through correlation analysis of the whole GOI, and the values of N_{sg} and N_g—on the basis of the requirement for minimum value of the number of interpolated images, m_{int}.

1.5.3.3 Computational Complexity of HAPCA

The computational complexity of the n-levels HAPCA algorithm can be calculated and compared with the classic PCA for a covariance matrix of size $N \times N$ for group of N images with N_{sg} sub-groups for the APCA of size 2×2 or 3×3. In case of classic PCA, this number is $n = N_{sg} = 1$, because there are no hierarchical levels or sub-groups. For this, both algorithms are compared regarding the number of operations O (additions and multiplications) [74] needed for the calculation of the following components:

Table 1.3 Basic parameters of the HAPCA algorithm

N	n	N_{sg}	N_g	$N_e = N_{sg} \times N_g$	$m_{int} = N_e - N$
4	2	2	2	4	0
5	3	3	2	6	1
6	3	3	2	6	0
7	3	4	2	8	1
8	3	4	2	8	0
9	2	3	3	9	0
10	3	4	3	12	2
11	3	4	3	12	1
12	3	4	3	12	0
13	3	5	3	15	2
14	3	5	3	15	1
15	3	5	3	15	0
16	5	8	2	16	0

- Covariance matrices $[K_C]$ of size $N \times N$ for the classic PCA algorithm and for the APCA with size of the transform matrix 2×2 or 3×3 [73]:

$$O_{cov}(N) = (1/2)N(N+1)[N(N-1)+2(N+2)] \text{ for the classic PCA;} \quad (1.61)$$

$$O_{cov}(2) = 30 \text{ for APCA of size } 2 \times 2 \ (N=2); \quad (1.62)$$

$$O_{cov}(3) = 96 \text{ for APCA of size } 3 \times 3 \ (N=3). \quad (1.63)$$

- Calculation of the eigen values of the corresponding $[K_C]$ covariance matrix when the QR decomposition and the Householder transform of $(N-1)$ steps are used for the classic PCA [73]:

$$O_{val}(N) = (N-1)(\frac{4}{3}N^2 + \frac{17}{6}N+7) \text{ for classic PCA;} \quad (1.64)$$

$$O_{val}(2) \approx 12 \text{ for APCA of size } 2 \times 2 \ (N=2); \quad (1.65)$$

$$O_{val}(3) = 55 \text{ for APCA of size } 3 \times 3 \ (N=3). \quad (1.66)$$

- Calculation of the eigen vectors of the corresponding $[K_C]$ covariance matrix in case that iterative algorithm with 4 iterations is used for the classic PCA [73]:

$$O_{vec}(N) = N[2N(4N+5)-1] \text{ for classic PCA;} \quad (1.67)$$

$$O_{vec}(2) = 102 \text{ for APCA of size } 2 \times 2 \ (N=2); \quad (1.68)$$

$$O_{vec}(3) = 303 \text{ for APCA of size } 3 \times 3 \ (N=3). \quad (1.69)$$

- The number of operations needed for the calculation of a group of N eigen images (each of S pixels), obtained in result of direct PCA transform for zero mean vectors, is:

$$O(N,S) = SN(2N-1) \text{ for classic PCA;} \quad (1.70)$$

$$O(2,S) = 6S \text{ for APCA of size } 2 \times 2 \ (N=2); \quad (1.71)$$

$$O(3,S) = 15S \text{ for APCA of size } 3 \times 3 \ (N=3). \quad (1.72)$$

- Using Eqs. (1.61)–(1.72) the total number of operations (*TO*) needed for both algorithms (the classic PCA and the HAPCA-based algorithms with APCA of size 2×2 or 3×3) is:

$$TO_{PCA}(N, S) = \frac{1}{2}N(N+1)[N-1) + 2(N+2)]$$
$$+ (N-1)(\frac{4}{3}N^2 + \frac{17}{6}N+7) + N[2N(4N+5)-1] \quad (1.73)$$
$$+ SN(2N-1);$$

$$TO_{HAPCA-2}(N, S) = nN_{sg}(30+12+102+6S) = nN_g(144+6S); \quad (1.74)$$

$$TO_{HAPCA-3}(N, S) = nN_{sg}(96+55+303+15S) = nN_g(454+15S). \quad (1.75)$$

Having obtained the total number of operations required by the algorithms (1.73) −(1.75), we can compare the computational complexity of both the classic PCA and the proposed algorithms. The reduction of the number of operations needed for these algorithms can be described by the coefficient:

$$\eta_2(N, S) = \frac{TO_{PCA}(N, S)}{TO_{HAPCA-2}(N, S)} = \frac{O_{cov}(N) + O_{val}(N) + O_{vec}(N) + O(N, S)}{nN_{sg}[O_{cov}(2) + O_{val}(2) + O_{vec}(2) + O(2, S)]},$$
$$(1.76)$$

is the ratio of the number of operations for the classic PCA and the proposed HAPCA-2 algorithm (with APCA of size 2 × 2), and:

$$\eta_3(N, S) = \frac{TO_{PCA}(N, S)}{TO_{HAPCA-3}(N, S)} = \frac{O_{cov}(N) + O_{val}(N) + O_{vec}(N) + O(N, S)}{nN_{sg}[O_{cov}(3) + O_{val}(3) + O_{vec}(3) + O(3, S)]},$$
$$(1.77)$$

is the ratio of the number of operations for the classic PCA and the proposed HAPCA-3 algorithm (with APCA of size 3 × 3).

For example, for $N = 8$, $n = 3$ and $N_g = 4$, from Eq. (1.76) is obtained:

$$\eta_2(8, S) = \frac{TO_{PCA}(8, S)}{TO_{HAPCA-2}(8, S)} = \frac{8269 + 120S}{1730 + 72S}. \quad (1.78)$$

For $N = 9$, $n = 2$ and $N_g = 3$, from Eq. (1.77) it follows:

$$\eta_3(9, S) = \frac{TO_{PCA}(9, S)}{TO_{HAPCA-3}(9, S)} = \frac{11987 + 153S}{2724 + 90S}. \quad (1.79)$$

If $S = 2^{18}$, then $\eta_2(8, 2^{18}) = 1.66$ and $\eta_3(9, 2^{18}) = 1.7$, i.e., the coefficient $\eta(S)$ is at least 1.66 times larger than 1 for images of size 512 × 512, or higher (in average, about 2 times). For higher values of N (for example, between 9 and 16), and for big values of S, the coefficient $\eta(S) > 2$.

1.5.4   Experimental Results

The presented experimental results are for sequences of multispectral (MS) images. As an example, was used the test MS sequence "balloons" shown on Fig. 1.10; on Fig. 1.11 is shown the corresponding color image with RGB values, obtained after lighting with neutral daylight. This sequence is from the free-access image database of the Columbia University, USA (http://www1.cs.columbia.edu/CAVE/databases/multispectral/). It contains $N = 15$ MS images of size 512×512 pixels, 16 *bpp*. On

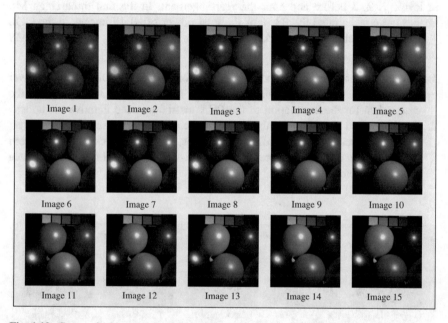

Fig. 1.10 Group of 15 consecutive MS images "balloons"

Fig. 1.11 The color image of "balloons", obtained after lighting with neutral daylight

it was applied the 3-level HAPCA algorithm. The sequence was divided into $N_{sg} =$ 5 sub-groups, each of $N_g = 3$ MS images, and the number of the vectors in each sub-group is $S = 2^{18}$.

On Fig. 1.12 are shown the corresponding eigen MS images, obtained after applying the 3-level HAPCA algorithm on the group of images. As it could be seen from the results shown on Fig. 1.13, the main part of the energy of these 15 images is concentrated on the first eigen MS image, and the energy of the next eigen images decreases rapidly.

The graphics on Fig. 1.13 represent the power distribution of all 15 eigen images in levels 1, 2, 3 before and after the rearrangement. In the first three eigen MS images are concentrated 99, 88 % of the total power of all 15 images in the GOI.

The basic qualities of the HAPCA algorithm for processing of groups of MS images are:

1. Lower computational complexity than PCA for the whole GOI, due to the lower complexity of APCA with matrices of size 2×2 and 3×3 compared to the case, when for the calculation of the PCA matrix are used iterative methods;
2. HAPCA could be used not only for efficient compression of sets of MS images, but also for sequences of medical CT images, video sequences, obtained from stationary TV camera, compression of multi-view images, image fusion, face recognition, etc.;

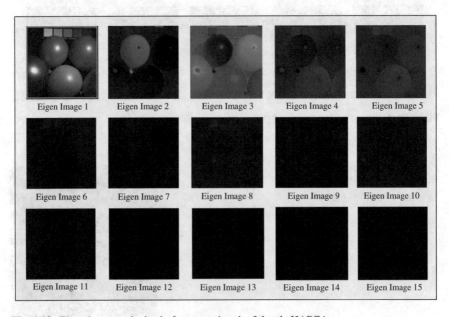

Fig. 1.12 Eigen images, obtained after executing the 3-levels HAPCA

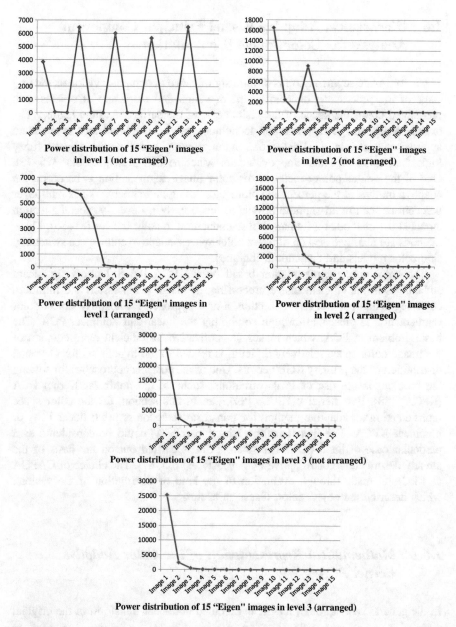

Fig. 1.13 Power distribution of all 15 eigen images in levels 1, 2, 3 before and after the rearrangement

3. There is also a possibility for further development of the HAPCA algorithms, through: the use of Integer PCA for lossless coding of MS images; HAPCA with a matrix of size $N \times N$ (N—a digit, divisible by 2 or 3), but without using numerical methods, etc.

1.6   Hierarchical Adaptive Kernel Principal Component Analysis for Color Image Segmentation

The color image segmentation is of high significance in computer vision as the first stage of the processing, concerning the detection and extraction of objects with predefined color, the shape of the visible part of the surface, and the texture. The existing color image segmentation techniques can be classified into seven main approaches based on: edge detection, region growing, neural network based, fuzzy logic, histogram analysis, Support Vector Machine and principal color [75–79]. One of the contemporary methods for color image segmentation is based on the adaptive models in the perceptual color space, using neural networks as multilayer perceptrons with multi-sigmoid activation function [80]. Recently special attention attracted the methods for human skin segmentation in color images [81–85]. These methods are mainly based on different color spaces, adaptive color space switching, skin color models and detection techniques.

The color space representation based on the PCA [86–88] offers significant advantages in the efficient image processing, as image compression and filtration, color segmentation, etc. In this section, a new approach for adaptive object color segmentation is presented through combining the linear and nonlinear PCA. The basic problem of PCA, which makes its application for efficient representation of the image color space relatively difficult, is related to the hypothesis for Gaussian distribution of the primary RGB vectors. One of the possible approaches for solving the problem is the use of PCA variations, such as: the nonlinear Kernel PCA (KPCA) [88], Fast Iterative KPCA [89], etc. In this section, for the color space representation an adaptive method for transform selection is used: linear PCA or nonlinear KPCA. The first transform (the linear PCA) could be considered as a particular case of the KPCA. The linear PCA is carried out on the basis of the already described Color Adaptive PCA (CAPCA) [85, 87]. The choice of CAPCA or KPCA is made through evaluation of the kind of distribution of the vectors, which describe the object color: Gaussian or not.

1.6.1   Mathematical Representation of the Color Adaptive Kernel PCA

In the general case, through KPCA is executed nonlinear transform of the original centered vectors $\vec{X}_s$ over S pixels ($\vec{X}_s = \sum\limits_{s=1}^{S} \vec{X}_s$) into the high-dimensional space, and then, for the obtained transformed vectors $\Phi(\vec{X}_s)$, the PCA is applied. The aim is, in the new multidimensional space the vertices of the vectors $\Phi(\vec{X}_s)$ to be concentrated in an area, which is accurately enough enveloped by a hyperellipsoid, whose axes are the eigenvectors of the covariance matrix of the vectors $\Phi(\vec{X}_s)$. Figure 1.14 illustrates the idea of the new 3D color space of eigenvectors $\vec{v}_1, \vec{v}_2, \vec{v}_3$ [85].

In particular, it is possible that the vectors $\Phi(\vec{X}_s)$ in the transformed space are represented by their projections on the first eigenvector $\vec{v}_1$ of their covariance matrix, as shown in Fig. 1.15. For the example, shown in this figure, on the eigenvector $\vec{v}_1$ is projected the basic part of the multitude of all transformed vectors $\Phi(\vec{X}_s)$. The original 3D color vectors $\vec{C}_s$ are first centered:

$$\vec{X}_s = \vec{C}_s - \vec{m}_C \text{ for } s = 1, 2, \ldots, S, \tag{1.80}$$

where $\vec{m}_C$ is the mean value of the color vector and then follows some kind of nonlinear transform, which uses the selected nonlinear function $\Phi(.)$. In result, the corresponding N-dimensional vectors, $\Phi(\vec{X}_s)$ ($N \geq 3$) are obtained. The value of N depends on the selected function $\Phi(.)$, used for the nonlinear transform [88].

The covariance matrix $[\tilde{K}_x]$ of the transformed color vectors $\Phi(\vec{X}_s)$ is of size $N \times N$ and can be calculated in accordance with the relation:

$$[\tilde{K}_x] = \frac{1}{S}\sum_{s=1}^{S} \Phi(\vec{X}_s).\Phi(\vec{X}_s)^t = E\{\Phi(\vec{C}_s - \vec{m}_c).\Phi(\vec{C}_s - \vec{m}_c)^t\}, \tag{1.81}$$

where $\Phi(\vec{X}_s) = [\Phi(x_{s1}), \Phi(x_{s2}), .., \Phi(x_{sN})]^t$ for $s = 1, 2,\ldots, S$.

For each eigenvalue $\tilde{\lambda}_i$ and eigenvector $\vec{v}_i = [v_{i1}, v_{i2}, \ldots, v_{iN}]^t$ of the matrix $[\tilde{K}_x]$ the following relation is performed:

$$[\tilde{K}_x]\vec{v}_i = \tilde{\lambda}_i\vec{v}_i \text{ for } i = 1, 2,\ldots, N. \tag{1.82}$$

Fig. 1.14 Plot of skin color samples in the $\vec{v}_1, \vec{v}_2, \vec{v}_3$ eigenvectors space of CAPCA

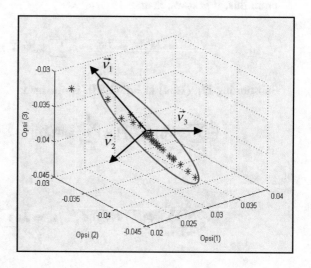

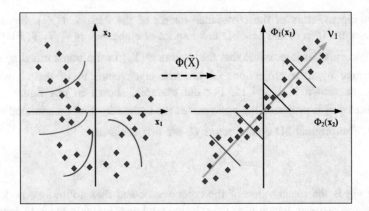

Fig. 1.15 Color space transform with KPCA

After substitution in Eq. (1.82) using Eq. (1.81), is got:

$$[\tilde{K}_x]\,\vec{v}_i = \frac{1}{S}\sum_{s=1}^{S} \Phi(\vec{X}_s)\,\Phi(\vec{X}_s)^t \vec{v}_i = \tilde{\lambda}_i \vec{v}_i \qquad (1.83)$$

In result of the transformation of Eq. (1.83), known as the "kernel trick" [90], for the ith eigenvector is obtained:

$$\vec{v}_i = \frac{1}{S\tilde{\lambda}_i}\sum_{s=1}^{S} [\Phi(\vec{X}_s)^t.\vec{v}_i]\,\Phi(\vec{X}_s) = \sum_{s=1}^{S} \alpha_{si}\Phi(\vec{X}_s), \qquad (1.84)$$

where for $\tilde{\lambda}_i \neq 0$ the coefficient $\alpha_{si} = \frac{\Phi(\vec{X}_s)^t.\vec{v}_i}{S\tilde{\lambda}_i}$.

From this, it follows, that:

$$[\tilde{K}_x]\vec{v}_i = \tilde{\lambda}_i\vec{v}_i = \tilde{\lambda}_i \sum_{s=1}^{S} \alpha_{si}\Phi(\vec{X}_s). \qquad (1.85)$$

Substituting Eq. (1.84) in Eq. (1.83), is obtained:

$$[\frac{1}{S}\sum_{s=1}^{S} \Phi(\vec{X}_s)\Phi(\vec{X}_s)^t] \times [\sum_{l=1}^{S} \alpha_{il}\Phi(\vec{X}_l)] = \tilde{\lambda}_i \sum_{l=1}^{S} \alpha_{li}\Phi(\vec{X}_l)$$

or

$$\frac{1}{S}\sum_{s=1}^{S}\sum_{l=1}^{S} \Phi(\vec{X}_s)\,\Phi(\vec{X}_s)^t\Phi(\vec{X}_l)\alpha_{il} = \tilde{\lambda}_i \sum_{l=1}^{S} \alpha_{li}\Phi(\vec{X}_l),$$

from which follows:

$$\sum_{s=1}^{S} \sum_{l=1}^{S} \Phi(\vec{X}_s) \Phi(\vec{X}_s)^t \Phi(\vec{X}_l) \alpha_{li} = S \tilde{\lambda}_i \sum_{l=1}^{S} \alpha_{li} \Phi(\vec{X}_l). \qquad (1.86)$$

After multiplying the left side of the above equation with the vector $\Phi(\vec{X}_s)^t$, is obtained:

$$\sum_{s=1}^{S} \sum_{l=1}^{S} \Phi(\vec{X}_s)^t \Phi(\vec{X}_s) \Phi(\vec{X}_s)^t \Phi(\vec{X}_l) \alpha_{li} = S \tilde{\lambda}_i \sum_{l=1}^{S} \alpha_{li} \Phi(\vec{X}_l)^t \Phi(\vec{X}_s). \qquad (1.87)$$

The dot product of the vectors $\Phi(\vec{X}_s)$ and $\Phi(\vec{X}_l)$ could be represented through the kernel function $k(\vec{X}_s, \vec{X}_l)$, defined by the relation:

$$k(\vec{X}_s, \vec{X}_l) = \Phi(\vec{X}_s)^t . \Phi(\vec{X}_l) \text{ for } s, l = 1, 2, \ldots, S. \qquad (1.88)$$

Here, the term $k(\vec{X}_s, \vec{X}_l)$ represents the elements (s, l) of the Gram matrix $[K]$ of size $S \times S$, called "kernel matrix". After substituting Eq. (1.88) in Eq. (1.87), is obtained:

$$[K]^2 . \vec{\alpha}_i = S \tilde{\lambda}_i [K] \vec{\alpha}_i. \qquad (1.89)$$

Under the condition, that the matrix $[K]$ is positively defined (i.e. when it eigenvalues are positive) is got a shorter representation than in Eq. (1.89). Then:

$$[K] \vec{\alpha}_i = S \tilde{\lambda}_i \vec{\alpha}_i. \qquad (1.90)$$

From this relation it follows, that $S \tilde{\lambda}_i$ are the eigenvalues of the matrix $[K]$, and $\vec{\alpha}_i = [\alpha_{i1}, \alpha_{i2}, \ldots, \alpha_{iS}]^t$ are the corresponding eigenvectors of same matrix. Taking into account the requirement $\vec{v}_i^t \vec{v}_i = 1$, from Eq. (1.84) is obtained the relation:

$$\sum_{s=1}^{S} \sum_{l=1}^{S} \alpha_{li} \alpha_{si} \Phi(\vec{X}_l)^t . \Phi(\vec{X}_s) = 1 \text{ or } \vec{\alpha}_i^t [K] \vec{\alpha}_i = 1. \qquad (1.91)$$

After substituting Eq. (1.90) in Eq. (1.91) is obtained $S \tilde{\lambda}_i \vec{\alpha}_i^t \vec{\alpha}_i = 1$, from which is defined the square of the module of the vector $\vec{\alpha}_i = [\alpha_{i1}, \alpha_{i2}, \ldots, \alpha_{iS}]^t$:

$$\|\vec{\alpha}_i\|^2 = \vec{\alpha}_i^t . \vec{\alpha}_i = \sum_{s=1}^{S} \alpha_{si}^2 = 1/S \tilde{\lambda}_i. \qquad (1.92)$$

In the general case, the vectors $\Phi(\vec{X}_s)$ in Eq. (1.88) are not centered. In order to apply the PCA on them, they should be centered in advance, and in result are obtained the vectors:

$$\breve{\Phi}(\vec{X}_s) = \Phi(\vec{X}_s) - E\{\Phi(\vec{X}_s)\}, \tag{1.93}$$

where $\vec{m}_{\Phi} = E\{\Phi(\vec{X}_s)\} = \frac{1}{S}\sum_{s=1}^{S}\Phi(\vec{X}_s)$.

The covariance matrix $[\breve{K}]$ of the centered vectors $\breve{\Phi}(\vec{X}_s)$ is of size $S \times S$ and is defined by the relation:

$$[\breve{K}] = \frac{1}{S}\sum_{s=1}^{S}\breve{\Phi}(\vec{X}_s)^t.\breve{\Phi}(\vec{X}_l) = E\left\{\breve{\Phi}(\vec{X}_s)^t.\breve{\Phi}(\vec{X}_l)\right\}. \tag{1.94}$$

The matrix kernel is:

$$\breve{k}(\vec{X}_s, \vec{X}_l) = \breve{\Phi}(\vec{X}_s)^t.\breve{\Phi}(\vec{X}_l) = [\Phi(\vec{X}_s) - \vec{m}_{\Phi}]^t.[\Phi(\vec{X}_l) - \vec{m}_{\Phi}]. \tag{1.95}$$

The relation between the covariance matrices $[\breve{K}]$ and $[K]$ is:

$$[\breve{K}] = [K] - 2[I_{1/s}][K] + [I_{1/s}][K][I_{1/s}], \tag{1.96}$$

where $[I_{1/s}]$ is a matrix of size $S \times S$, whose elements are equal to $1/S$.

The projection of the vector $\Phi(\vec{X}_s)$ on the eigenvector $\vec{v}_i$ in the S-dimensional space is:

$$Pr_{si} = \Phi(\vec{X}_s)^t.\vec{v}_i = \sum_{s=1}^{S}\alpha_{is}\Phi(\vec{X}_i)^t.\Phi(\vec{X}_s) = \sum_{s=1}^{S}\alpha_{is}k(\vec{X}_i, \vec{X}_s) \text{ for } i = 1, 2, 3, \ldots, N. \tag{1.97}$$

Using the projections Pr_{si} of the vector $\Phi(\vec{X}_s)$ on each of the first $k \leq N$ eigenvectors $\vec{v}_i$ (for $i = 1, 2, \ldots, k$), could be taken the decision for the classification of the sth pixel to the dominant color of the selected object, using some of the well-known classifiers, as: SVM, LDA, k-nearest neighbors, neural networks, etc. [89]

To carry out the KPCA, one could use different kinds of kernel functions, such as the polynomial, the Gaussian, the sigmoid, etc. By substituting $\Phi(\vec{X}_s) = \vec{x}$ and $\Phi(\vec{X}_l) = \vec{y}$ the polynomial kernel function of degree d is defined by the relation:

$$k(\vec{x}, \vec{y}) = (\vec{x} \cdot \vec{y})^d. \tag{1.98}$$

For $d = 2$ and if assumed that for the transformation of the 3-component vectors $\vec{X}_s = [x_{s1}, x_{s2}, x_{s3}]^t$ and $\vec{X}_l = [x_{l1}, x_{l2}, x_{l3}]^t$ into N-component is used the nonlinear function $\Phi(.)$, then:

$$\vec{x} = \Phi(\vec{X}_s) = [\Phi_{s1}, \Phi_{s2}, .., \Phi_{sN}]^t, \ \vec{y} = \Phi(\vec{X}_l) = [\Phi_{l1}, \Phi_{l2}, .., \Phi_{lN}]^t \tag{1.99}$$

where the vectors components are defined by the relations:

$$\Phi_{si} = x_{sip_1}^{r_1} x_{sip_2}^{r_2}, \ \Phi_{li} = x_{lip_1}^{r_1} x_{lip_2}^{r_2} \tag{1.100}$$

for r_1, $r_2 = 0, 1, p_1, p_2 = 1, 2, 3, i = 1, 2, ... N$ and $s, l = 1, 2, ..., S$.

In this case the maximum value of N is $N = 9$. In order to reduce the needed calculations, it is suitable to use smaller number of the possible 9 components of the quadratic function $\Phi(.)$.

For example, if assumed $N = 3$ and if only mixed products of the vectors components $\vec{X}_s$ and $\vec{X}_l$ are chosen, then from Eq. (1.100) it follows:

$$\vec{x} = \Phi(\vec{X}_s) = [x_{s1}x_{s2}, x_{s1}x_{s3}, x_{s2}x_{s3}]^t, \ \vec{y} = \Phi(\vec{X}_l) = [x_{l1}x_{l2}, x_{l1}x_{l3}, x_{l2}x_{l3}]^t. \tag{1.101}$$

Then the corresponding kernel function of vectors $\Phi(\vec{X}_s)$ and $\Phi(\vec{X}_l)$ is represented by the polynomial below:

$$k(\vec{x}, \vec{y}) = [\Phi_{s1}, \Phi_{s2}, \Phi_{s3}]^t \cdot [\Phi_{l1}, \Phi_{l2}, \Phi_{l3}] = x_{s1}x_{s2}x_{l1}x_{l2} + x_{s1}x_{s3}x_{l1}x_{l3} + x_{s2}x_{s3}x_{l2}x_{l3}. \tag{1.102}$$

In particular, for $d = 1$, $\Phi(\vec{X}_s) = \vec{X}_s$ and $\Phi(\vec{X}_l) = \vec{X}_l$ the corresponding kernel function is linear:

$$k(\vec{x}, \vec{y}) = [x_{s1}, x_{s2}, x_{s3}]^t \times [x_{l1}, x_{l2}, x_{l3}] = x_{s1}x_{l1} + x_{s2}x_{l2} + x_{s3}x_{l3}. \tag{1.103}$$

From the above, it follows that KPCA is transformed into linear PCA (i.e. PCA is a particular case of KPCA).

1.6.2 Algorithm for Color Image Segmentation by Using HAKPCA

The general algorithm for objects segmentation in the extended color space, based on the Hierarchical Adaptive Kernel PCA (HAKPCA) and the classifier of the reduced vectors, is shown on Fig. 1.16.

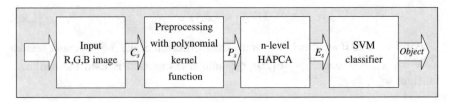

Fig. 1.16 Block diagram of the algorithm for image segmentation in the expanded color space

In the preprocessing block, each color vector $\vec{C}_s = [R_s, G_s, B_s]^t$ is transformed into the corresponding expanded vector $\vec{P}_s$. If the chosen kernel-function is polynomial, and the 3D color space is transformed into a 9-dimensional, then the components p_{is} of vectors $\vec{P}_s$ could be defined as follows:

$$\vec{P}_s = [R_s, G_s, B_s, R_s^2, G_s^2, B_s^2, R_sG_s, B_sG_s, R_sB_s]^t$$
$$= [P_{1s}, P_{2s}, P_{3s}, P_{4s}, P_{5s}, P_{6s}, P_{7s}, P_{8s}, P_{9s}]^t \text{ for } s = 1, 2, \ldots, S.$$

In order to put all components p_{is} in the range [0, 255], for $i = 4, 5, \ldots, 9$: $R_s^2, G_s^2, B_s^2, R_sG_s, B_sG_s, R_sB_s$, are normalized in the range 0–255. The vectors $\vec{P}_s$ are then transformed by the 2-level HAKPCA, whose algorithm is shown in Fig. 1.17. As a result of the transform are obtained the 2-component vectors $\vec{E}_s = [E_{1s}, E_{2s}]^t$, which are used to substitute the input 9-components vectors $\vec{P}_s = [P_{1s}, P_{2s}, P_{3s}, P_{4s}, P_{5s}, P_{6s}, P_{7s}, P_{8s}, P_{9s}]^t$. In this way the performance of the classifier is also simplified, because it has to process the vectors $\vec{E}_s$ in the two-component, instead of the nine-dimensional space.

At its output are separated all pixels in the image, whose corresponding vectors $\vec{E}_s$ are in the area of the cluster, belonging to the object. With this, the color segmentation is finished. In accordance with the algorithm shown in Fig. 1.17, for the 2-level HAKPCA [91], the nine components of each input vector $\vec{P}_s$ are divided into three groups, which contain the three-components vectors

$$\vec{P}_{1s} = [P_{11s}, P_{12s}, P_{13s}]^t, \ \vec{P}_{2s} = [P_{21s}, P_{22s}, P_{23s}]^t, \ \vec{P}_{3s} = [P_{31s}, P_{32s}, P_{33s}]^t,$$

At the first level of HAKPCA, on each group of the three-component vectors $\vec{P}_{ks} = [P_{k1s}, P_{k2s}, P_{k3s}]^t$ for $k = 1, 2, 3$, is performed color APCA with a transform matrix of size 3×3. The so obtained vectors from each group comprise three "eigen" images, shown in Fig. 1.18. These images are rearranged in accordance to the rule:

$$\lambda_1 \geq \lambda_2 \geq \lambda_3 \geq \cdots \geq \lambda_9. \tag{1.104}$$

where $\lambda_i \geq 0$ for $l = 1, 2, \ldots, 9$ are eigen values of the covariance matrices of the three-component vectors $\vec{P}_{ks}$ for each group ($k = 1, 2, 3$) in the first level of

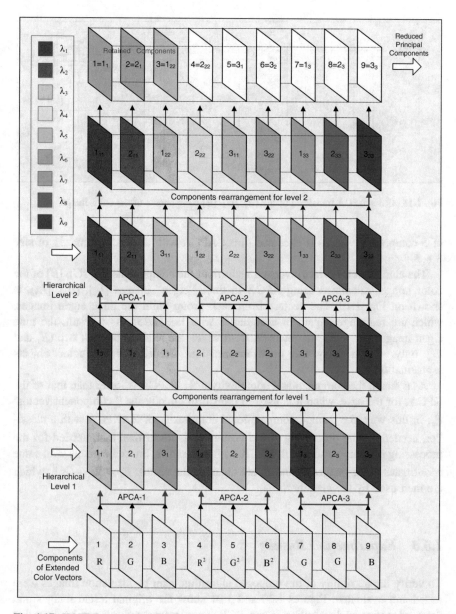

Fig. 1.17 HAKPCA algorithm for direct transform of the extended color image with components $R, G, B, R^2, G^2, B^2, RG, BG, RB$

HAKPCA, arranged as monotonously decreasing sequence of eigen values. After that these components are divided again, this time into 3 groups, of 3 images each.

The vectors, obtained from the pixels with same coordinates in the images from each group, are of 3 components. For the second level of HAKPCA for each group

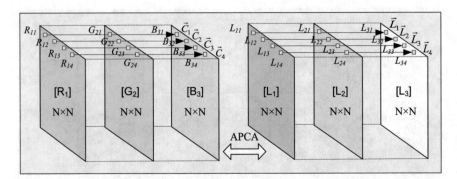

Fig. 1.18 Color APCA transform of the [R], [G], [B] components of the color image

of 3-component vectors is executed direct APCA with a transform matrix of size 3×3.

The algorithm for direct/inverse transform of the components [R], [G], [B] of the color image, calculated using APCA in three "eigen" images [L_1], [L_2], [L_3], is shown on Fig. 1.19. The vectors from each group build the three eigen images, which are rearranged again in accordance with Eq. (1.25). As a result, the nine eigen images $E_1 \sim E_9$ are obtained, from which are retained the first two (E_1 and E_2) only, which carry the main information, needed for the color objects segmentation.

As a result, the computational complexity of HAKPCA is lower than that of the KPCA, for the case, when it is used to transform directly the 9-component vectors $\vec{P}_s$. In this way, the general computational complexity of HAKPCA with a classifier, needed for the processing of the vectors $\vec{P}_s$ is lower than that, needed for the processing of same vectors with KPCA with a classifier. From the pixels with same coordinates in the images E_1 and E_2 are obtained the vectors $\vec{E}_s = [E_{1s}, E_{2s}]^t$, which are then used by the classifier.

1.6.3 Experimental Results

To verify the feasibility of the proposed algorithm, skin pigmentation images were tested and evaluated. Figures 1.20 and 1.21 show the original tested images and their color vectors distribution in the *RGB* space, respectively. It can be seen that their color distributions are considered as non-linear Gaussian ones.

These images are passed through the HAKPCA algorithm (shown on Fig. 1.17). The obtained transformed vectors $\vec{E}_s$ in the new color space E_{1s}, E_{2s}, E_{3s} are plotted in the 3D domain shown in Fig. 1.22. It is easy to notice that the proposed techniques concentrate the energy of the different skin color into very small and close components of transformed vectors.

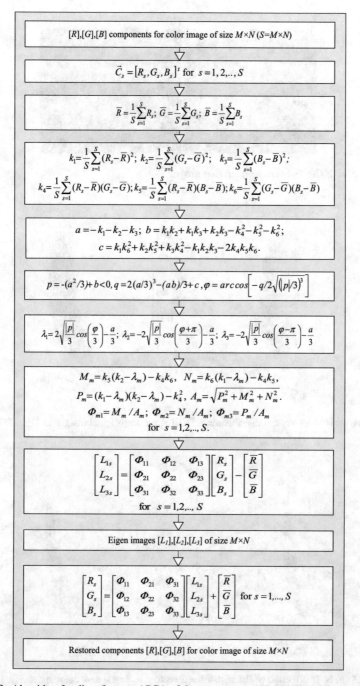

Fig. 1.19 Algorithm for direct/inverse APCA of the components [R], [G], [B] of the color image

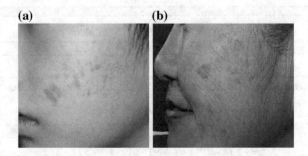

Fig. 1.20 a, b Original skin pigmentation images

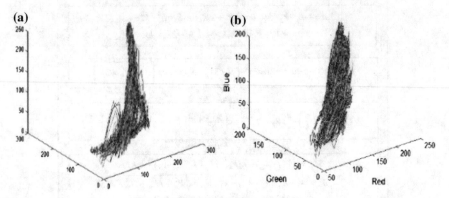

Fig. 1.21 a, b Color vectors distribution in RGB space for original images in Fig. 1.20 a, b

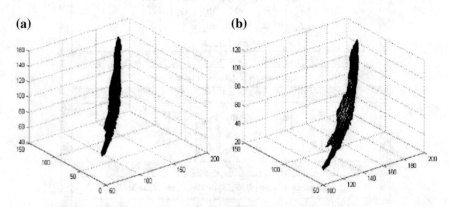

Fig. 1.22 a, b. Distribution of the transformed vectors $\vec{E}_s$ in the new color space E_{1s}, E_{2s}, E_{3s}

Fig. 1.23 Skin color segmentation based on HAKPCA

(a)   **(b)**

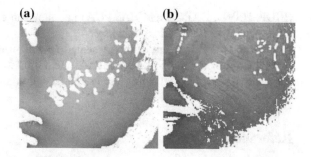

The HAKPCA transformed coefficients are then used to train a classifier. For briefing, fuzzy K-means clustering is used. The segmentation results are shown in Fig. 1.23a, b respectively.

The proposed approach depends mainly on the evaluation of the color vectors distribution. For a non-Gaussian distribution of the vectors, is used HAKPCA. The selected nonlinear transform results in negligible expansion of the original color space, which increases slightly the number of needed calculations. The main advantage of the new approach is that in result of its adaptation in respect to the color vectors distribution, it could be used as universal tool for efficient image processing. One more advantage of HAKPCA towards the KPCA is the lower computational complexity.

On the basis of the presented approach, new algorithm for objects color segmentation was developed, which was distinguished by its high accuracy. This algorithm could be used in the CBIR systems for extraction of objects with preset color, in the computer vision systems for detection and tracking of objects in correspondence to their color under changing surveillance conditions, for automatic control of various manufacturing processes, etc.

1.7 Conclusions

The new approaches for image decomposition, presented in this chapter, are distinguished from the well-known methods by their low computational complexity and high efficiency, which makes them very attractive for future applications in computer vision, video communications, image content protection through digital Watermarking [92], image search in large databases, etc. Besides, the methods for hierarchical decomposition could be also used for the creation of new hybrid algorithms [93, 94] for processing of some specific kinds of images (static, or sequences), of the kind: medical, multi- and hyper spectral, multiview, panoramic satellite photos, etc. Depending on the requirements of the applications, in respect of their efficiency and the computation time needed for the processing, for each image kind could be selected the most suitable from the four decompositions

(BIDP, HSVD, HAPCA and HAKPCA) i.e., they can complement each other and have their own place and significance.

The future development of the presented new algorithms will be focused at the investigation of the possibilities for their integration in the contemporary systems for parallel processing, analysis and image recognition.

References

1. Rao, R., Kim, D., Hwang, J.: Fast Fourier Transform: Algorithms and Applications. Springer, Heidelberg (2010)
2. Gonzalez, R., Woods, R.: Digital Image Processing, 3rd edn. Prentice Hall, NJ (2008)
3. Do, M., Vetterli, M.: Contourlets: a directional multiresolution image representation. In: Proceedings of International Conference on Image Processing, 2002, vol. 1, pp. 357–360 (2002)
4. Kountchev, R., Vl, Todorov, Kountcheva, R.: Linear and non-linear inverse pyramidal image representation: algorithms and applications, Chapter. In: Kountchev, R., Nakamatsu, K. (eds.) Advances in Reasoning-based Image Processing, Analysis and Intelligent Systems, pp. 35–89. Springer, Berlin (2012)
5. Orfanidis, S.: SVD, PCA, KLT, CCA, and all that. In: Optimum Signal Processing, pp. 332–525. Rutgers University (2007)
6. Ahuja, N.: On approaches to polygonal decomposition for hierarchical image representation. Comput. Vis. Graph. Image Proc. **24**, 200–214 (1983)
7. Tadmor, E., Nezzar, S., Vese, L.: Multiscale hierarchical decomposition of images with applications to deblurring, denoising and segmentation. Commun. Math. Sci. **6**, 281–307 (2008)
8. Hidane, M., Lezoray, O., Ta, V., Elmoataz, A.: Nonlocal multiscale hierarchical decomposition on graphs. In: Daniilidis K, Maragos P, Paragios N (eds.) Computer Vision —ECCV 2010, Part IV, pp. 638–650, LNCS 6314. Springer, Berlin (2010)
9. Wu, Q., Xia, T., Yu, Y.: Hierarchical tensor approximation of multidimensional images. In: IEEE International Conference on Image Processing (ICIP'07), San Antonio, TX, 6, IV-49, IV-52 (2007)
10. Cherkashyn, V., He, D., Kountchev, R.: Image decomposition on the basis of an inverse pyramid with 3-layer neural networks. J. Commun. Comput. **6**(11), 1–9 (2009)
11. Mach, T.: Eigenvalue algorithms for symmetric hierarchical matrices. Dissertation, Chemnitz University of Technology, March 13, 175p (2012)
12. Rudin, L., Osher, S., Fatemi, E.: Nonlinear total variation based noise removal algorithms. Physica D **60**, 259–268 (1992)
13. Kountchev, R.: New Approaches in Intelligent Image Processing, 322p. WSEAS Press, Athene (2013)
14. Kountchev, R., Kountcheva, R.: Image representation with reduced spectrum pyramid. In: Tsihrintzis, G., Virvou, M., Howlett, R., Jain, L. (eds.) New Directions in Intelligent Interactive Multimedia, pp. 275–284. Springer, Berlin, Heidelberg (2008)
15. Kountchev, R., Kountcheva, R.: Compression of CT images with modified inverse pyramidal decomposition. In: 12th WSEAS International Conference on Signal Processing, Computational Geometry and Artificial Vision (ISCGAV'12), pp. 74–79, Istanbul, Turkey (2012)
16. Draganov, I., Kountchev, R., Georgieva, V.: Compression of CT images with branched inverse pyramidal decomposition. In: Iantovics, B., Kountchev, R. (eds.) Advanced Intelligent Computational Technologies and Decision Support Systems, vol. 486, VII, pp. 71–81. Springer (2014)

17. Kountchev, R., Todorov, V., Kountcheva, R.: New method for adaptive lossless compression of still images based on the histogram statistics. In: Tsihrintzis, G., Virvou, M., Jain, L., Howlett, R. (eds.) Intelligent Interactive Multimedia Systems and Services, Proceedings of the 4th International Conference on Intelligent Interactive Multimedia Systems and Services (KES-IIMSS 2011), pp. 61−70. Springer (2011)
18. Deprettere, E.: SVD and Signal Processing. Elsevier, New York (1988)
19. Vaccaro, R.: SVD and Signal Processing II. Elsevier, New York (1991)
20. Moonen, M., Moor, B.: SVD and Signal Processing III. Elsevier, New York (1995)
21. Carlen, E.: Calculus++. In: The Symmetric Eigenvalue Problem. Georgia Tech (2003)
22. Andrews, H., Patterson, C.: Singular value decompositions and digital image processing. IEEE Trans. Acoust. Speech Signal Process. ASSP-24, 26–53 (1976)
23. Gerbrands, J.: On the relationships between SVD KLT, and PCA, Pattern Recogn. 14(6), 375–381 (1981)
24. Kalman, D.: A singularly valuable decomposition: the SVD of a matrix. Coll. Math. J 27(1), 2–23 (1996)
25. Rehna, V., Jeyakumar, M.: Singular value decomposition based image coding for achieving additional compression to JPEG images. Int. J. Image Process. Vis. Sci. 1(2), 56–61 (2012)
26. Sadek, R.: SVD based image processing applications: state of the art, contributions and research challenges. Int. J. Adv. Comput. Sci. Appl. 3(7), 26–34 (2012)
27. Householder, A.: The Theory of Matrices in Numerical Analysis. Dover, New York (1975)
28. Diamantaras, K., Kung, S.: Principal Component Neural Networks. Wiley, New York (1996)
29. Levy, A., Lindenbaum, M.: Sequential Karhunen-Loeve basis extraction and its application to images. IEEE Trans. Image Process. 9(8), 1371–1374 (2000)
30. Drinea, E., Drineas, P., Huggins, P.: A randomized singular value decomposition algorithm for image processing applications. In: Manolopoulos, Y., Evripidou, S. (eds.) Proceedings of the 8th Panhellenic Conference on Informatics, pp. 278–288, Nicosia, Cyprus (2001)
31. Holmes, M., Gray, A., Isbell, C.: QUIC-SVD: fast SVD using cosine trees. In: Proceedings of NIPS, pp. 673–680 (2008)
32. Foster, B., Mahadevan, S., Wang, R.: A GPU-based approximate SVD algorithm. In: 9th International Conference on Parallel Processing and Applied Mathematics, 2011, LNCS 7203, pp. 569–578 (2012)
33. Yoshikawa, M., Gong, Y., Ashino, R., Vaillancourt, R.: Case study on SVD multiresolution analysis, CRM-3179, pp. 1–18 (2005)
34. Waldemar, P., Ramstad, T.: Hybrid KLT-SVD image compression. In: IEEE International Conference on Acoustics, Speech, and Signal Processing, pp. 2713–2716. IEEE Computer Society Press, Los Alamitos (1997)
35. Aharon, M., Elad, M., Bruckstein, A.: The K-SVD: an algorithm for designing of overcomplete dictionaries for sparse representation. IEEE Trans. Signal Process. 54, 4311–4322 (2006)
36. Singh, S., Kumar, S.: Singular value decomposition based sub-band decomposition and multi-resolution (SVD-SBD-MRR) representation of digital colour images. Pertanika J Sci. Technol. (JST) 19(2), 229–235 (2011)
37. De Lathauwer, L., De Moor, B., Vandewalle, J.: A multilinear singular value decomposition. SIAM J. Matrix Anal. Appl. 21, 1253–1278 (2000)
38. Kountchev, R., Kountcheva, R.: Hierarchical SVD-based image decomposition with tree structure. Int. J. Reasoning-Based Intell. Syst. (IJRIS) 7(1/2), 114–129 (2015)
39. Kountchev, R., Nakamatsu, K.: One approach for grayscale image decorrelation with adaptive multi-level 2D KLT. In: Graña, M., Toro, C., Posada, J., Howlett, R., Jain, L. (eds.) Advances in Knowledge-Based and Intelligent Information and Engineering Systems, pp. 1303–1312. IOS Press (2012)
40. Fieguth, P.: Statistical image processing and multidimensional modeling. Springer, Science +Business Media (2011)
41. Reed, T.: Digital Image Sequence Processing, Compression, and Analysis. CRC Press (2004)

42. Shi, Y., Sun, H.: Image and Video Compression for Multimedia Engineering: Fundamentals, Algorithms, and Standards, 2nd edn. CRC Press, Taylor & Francis Group, LLC (2008)
43. Mukhopadhyay, J.: Image and Video Processing in the Compressed Domain. CRC Press (2011)
44. Hyvarinen, A., Hurri, J., Hoyer, P.: Natural Image Statistics, A Probabilistic Approach to Early Computational Vision. Springer (2009)
45. Jolliffe, I.: Principal Component Analysis, 2nd edn. Springer-Verlag, NY (2002)
46. Dony, R.: Karhunen-Loeve Transform. In: Rao, K., Yip, P. (eds.) The Transform and Data Compression Handbook. CRC Press, Boca Raton (2001)
47. Fleury, M., Dowton, A., Clark, A.: Karhunen-Loeve Transform—Image Processing. University of Essex, Wivenhoe (1997)
48. Miranda, A., Borgne, Y., Bontempi, G.: New routes from minimal approximation error to principal components, pp. 1−14. Kluwer Academic Publishers (2007)
49. Landqvist, R., Mohammed, A.: Comparative performance analysis of three algorithms for principal component analysis. Radioengineering 15(4), 84–90 (2006)
50. Tipping, M., Bishop, C.: Probabilistic principal component analysis. J. Roy. Stat. Soc. B, 61 (3), 611−622 (1999)
51. Liwicki, S., Tzimiropoulos, G., Zafeiriou, S., Pantic, M.: Euler principal component analysis. Int. J. Comput. Vis. 101(3), 498–518 (2013)
52. Ujwala, P., Uma, M.: Image fusion using hierarchical PCA. In: International Conference on Image Information Processing (ICIIP'11). pp. 1–6 (2011)
53. Abadpour, A., Kasaei, S.: Color PCA eigen images and their application to compression and watermarking. Image Video Comput. 26, 878–890 (2008)
54. Chadha, A., Satam, N., Sood, R., Bade, D.: Image steganography using Karhunen-Loève transform and least bit substitution. Int. J. Comput. Appl. 79(9), 31–37 (2013)
55. Naik, G., Kumar, D.: An overview of independent component analysis and its applications. Informatica 35, 63–81 (2011)
56. Press, W., Teukolsky, S., Vetterling, W.: Numerical Recipes in C, The Art of Scientific Computing, 2nd edn. Cambridge University Press (2001)
57. Erdogmus, D., Rao, Y., Peddaneni, H., Hegde, A., Principe, J.: Recursive PCA using eigenvector matrix perturbation. EURASIP J. Adv. Signal Process. 13, 2034–2041 (2004)
58. Hanafi, M., Kohler, A., Qannari, E.: Shedding new light on hierarchical principal component analysis. J. Chemom. 24(11–12), 703–709 (2010)
59. Amar, A., Leshem, A., Gastpar, M.: Recursive implementation of the distributed Karhunen-Loève transform. IEEE Trans. Signal Process. 58(10), 5320–5330 (2010)
60. Thirumalai, V.: Distributed compressed representation of correlated image sets. Thesis no 5264, Lausanne, EPFL (2012)
61. Saghri, J., Schroeder, S., Tescher, A.: An adaptive two-stage KLT scheme for spectral decorrelation in hyperspectral bandwidth compression. SPIE Proc. 7443, 72–84 (2009)
62. Wongsawat, Y., Oraintara, S., Rao, K.: Integer sub-optimal Karhunen Loeve transform for multi-channel lossless EEG Compression. In: 14th European Signal Processing Conference (EUSIPCO'06), Florence, Italy (2006)
63. Blanes, I., Sagristà, J., Marcellin, M., Rapesta, J.: Divide-and-conquer strategies for hyperspectral image processing. IEEE Signal Process. Mag. 29(3), 71–81 (2012)
64. Blanes, I., Sagristà, J.: Pairwise orthogonal transform for spectral image coding. IEEE Trans. Geosci. Remote Sens. 49(3), 961–972 (2011)
65. Ma, Z.: Sparse principal component analysis and iterative thresholding. Ann. Stat. 41(2), 772–801 (2013)
66. Jain, A.: A fast Karhunen-Loeve transform for a class of random processes. IEEE Trans. Commun. COM-24, 1023–1029 (1976)
67. Rokhlin, V., Szlam, A., Tygert, M.: A randomized algorithm for principal component analysis. SIAM J. Matrix Anal. Appl. 31, 1100–1124 (2009)
68. Kambhatla, N., Haykin, S., Dony, R.: Image compression using KLT, wavelets and an adaptive mixture of principal components model. J. VLSI Signal Process. 18, 287–296 (1998)

69. Bita, I., Barret, M., Pham, D.: On optimal transforms in lossy compression of multi-component images with JPEG2000. Sig. Process. **90**(3), 759–773 (2010)
70. Kountchev, R., Kountcheva, R.: PCA-based adaptive hierarchical transform for correlated image groups. In Proceedings of the International Conference on Telecommunications in Modern Satellite, Cable and Broadcasting Services (TELSIKS'13), pp. 323−332. IEEE, Serbia (2013)
71. Kountchev, R.: Applications of the hierarchical adaptive PCA for processing of medical CT images. Egypt. Comput. Sci. J. **37**(3), 1–25 (2013)
72. Kountchev, R., Kountcheva, R.: Decorrelation of multispectral images, based on hierarchical adaptive PCA. Int. J. WSEAS Trans. Signal Process. **3**(9), 120–137 (2013)
73. Kountchev, R., Kountcheva, R.: Hierarchical adaptive KL-based transform: algorithms and applications. In: Favorskaya, M., Jain, L. (eds.) Computer Vision in Advanced Control Systems: Mathematical Theory, pp. 91−136. Springer, Heidelberg (2015)
74. Arora, S., Barak, B.: Computational complexity: A modern approach. Cambridge University Press (2009)
75. Jie, X., Fei, S.: Natural color image segmentation. In: Proceedings of IEEE IC on Image Processing (ICIP'03), pp. 973–976, Barcelona, Spain (2003)
76. Navon, E., Miller, O., Averabuch, A.: Color image segmentation based on adaptive local thresholds. Image Vis. Comput. **23**, 69–85 (2005)
77. Deshmukh, K., Shinde, G.: An adaptive color image segmentation. Electron. Lett. Comput. Vis. Image Anal. **5**(4), 12–23 (2005)
78. Yu, Z., Wong, H., Wen, G.: A modified support vector machine and its application to image segmentation. Image Vis. Comput. **29**, 29–40 (2011)
79. Wang, X., Wang, T., Bu, J.: Color image segmentation using pixel wise support vector machine classification. Pattern Recogn. **44**, 777–787 (2011)
80. Bhoyar, K., Kakde, O.: Skin color detection model using neural networks and its performance evaluation. J. Comput. Sci. **6**(9), 963–968 (2010)
81. Stern, H., Efros, B.: Adaptive color space switching for face tracking in multi-colored lighting environments. In: Proceedings of the International Conference on Automatic Face and Gesture Recognition, pp. 249–255 (2002)
82. Kakumanu, P., Makrogiannis, S., Bourbakis, N.: A survey of skin-color modeling and detection methods. Pattern Recogn. **40**, 1106–1122 (2007)
83. Ionita, M., Corcoran, P.: Benefits of using decorrelated color information for face segmentation/tracking. Adv. Optical Technol. **2008**(583687) (2008)
84. Hassanpour, R., Shahbahrami, A., Wong, S.: Adaptive Gaussian mixture model for skin color segmentation. World Acad. Sci. Eng. Technol. **41**, 1–6 (2008)
85. Hikal, N., Kountchev, R.: Skin color segmentation using adaptive PCA and modified elliptic boundary model. In: International Proceedings of the IEEE International Conference on Advanced Computer Science and Information Systems (IEEE ICACSIS'11), pp. 407–412, Jakarta, Universitas Indonesia (2011)
86. Abadpour, A., Kasaei, S.: Principal color and its application to color image segmentation. Scientia Iranica **15**(2), 238–245 (2008)
87. Kountchev, R., Kountcheva, R.: Image color space transform with enhanced KLT. In: Nakamatsu, K., Wren, G., Jain, L., Howlett, R. (eds.) New Advances in Intelligent Decision Technologies, pp. 171−182. Springer-Verlag (2009)
88. Gunter, S., Schraudolph, N., Vishwanathan, S.: Fast iterative kernel principal component analysis. J. Mach. Learn. Res. **8**, 1893–1918 (2007)
89. Scholkopf, B., Smola, A., Muller, K.: Kernel principal component analysis. In: Scholkopf, B., Burges, C., Smola, A. (eds.) Advances in Kernel Methods—Support Vector Learning, pp. 327–352. MIT Press, Cambridge, MA (1999)
90. Lee, J., Verleysen, M.: Nonlinear Dimensionality Reduction. Springer (2007)
91. Kountchev, R., Hikal, N., Kountcheva, R.: A new approach for color image segmentation with hierarchical adaptive kernel PCA. In: 18th International Conference on Circuits, Systems,

Communications and Computers (CSCC'14), Proceedings of Advances in Information Science and Applications, pp. 38–43, Santorini Island, I (2014)

92. Kountchev, R., Milanova, M., Todorov, V., Kountcheva, R.: Image watermarking based on pyramid decomposition with CH transform, In: Chan, Y., Talburt, J., Talley, T. (eds.) Data Engineering: Mining, Information, and Intelligence, pp. 353–387. Springer (2010)

93. Draganov, I.: Inverse pyramid decomposition of wavelet spectrum for image compression. Int. J. Reasoning-based Intell. Syst. (IJRIS) 6(1/2), 19–23 (2014)

94. Martino, F., Loia, V., Sessa, S.: A fuzzy hybrid method for image decomposition. Int. J. Reasoning-based Intell. Syst. (IJRIS) 1(1/2), 77–84 (2009)

Chapter 2
Intelligent Digital Signal Processing and Feature Extraction Methods

János Szalai and Ferenc Emil Mózes

Abstract Intelligent systems comprise a large variety of applications, including ones based on signal processing. This field benefits from considerable popularity, especially with recent advances in artificial intelligence, improving existing processing methods and providing robust and scalable solutions to existing and new problems. This chapter builds on well-known signal processing techniques, such as the short-time Fourier and wavelet transform, and introduces the concept of instantaneous frequency along with implementation details. Applications featuring the presented methods are discussed in an attempt to show how intelligent systems and signal processing can work together. Examples that highlight the cooperation between signal analysis and fuzzy c-means clustering, neural networks and support vector machines are being presented.

Keywords Signal processing · Signal-adaptive processing · Frequency domain transforms · Instantaneous frequency · Fuzzy c-means · Fuzzy systems · Support vector machine · Frequency analysis · Time-frequency analysis · Neural networks

2.1 Introduction

Intelligent systems on their own are not always enough to handle complex tasks. In these cases the preprocessing of the signals is a necessary step. Often the preprocessing algorithms not only convert the initial data into a more advantageous format, but they are also capable of realizing feature extraction on the data. This aspect

J. Szalai (✉)
Technical University of Cluj Napoca, 128-130 21 December 1989 Boulevard, 400604 Cluj Napoca, Romania
e-mail: szalai@gmail.com

F.E. Mózes
Petru Maior University of Târgu Mureș, 1 Nicolae Iorga Street, 540088 Târgu Mureș, Romania
e-mail: mozes.ferenc.emil@gmail.com

© Springer International Publishing Switzerland 2016
R. Kountchev and K. Nakamatsu (eds.), *New Approaches in Intelligent Image Analysis*, Intelligent Systems Reference Library 108,
DOI 10.1007/978-3-319-32192-9_2

is equally important as the concept of intelligent systems themselves, for no classification can succeed if the features of the input data are not emphasized correctly beforehand.

The goal of the chapter is to present well-known signal processing methods and the way these can be combined with intelligent systems in order to create powerful feature extraction techniques. In order to achieve this, several case studies are presented to illustrate the power of hybrid systems. The main emphasis is on instantaneous time-frequency analysis, since it is proven to be a powerful method in several technical and scientific areas. The authors' contributions to the computation of the instantaneous frequency and application of the empirical mode decomposition are also presented, highlighting the limitations of the existing methods and showing at the same time a possible approach on T-wave peak detection in electrocardiograms.

Classical signal processing methods have been widely used in different fields of engineering and natural sciences in order to highlight meaningful information underlying in a wide variety of signals. In this chapter we aim to present not only the best known signal processing methods, but also ones that proved to be the most useful. The oldest and most utilized method is the Fourier transform, which has been applied in several domains of scientific data processing, but it has very strong limitations due to the constraints it imposes on the analyzed data. Then the short-time Fourier transform and the wavelet transform are presented, as they provide both temporal and frequency information as opposed to the Fourier transform. These methods form the basis of most applications nowadays, as they offer the possibility of time-frequency analysis of signals. Finally, the Hilbert-Huang transform is presented as a novel signal processing method, which introduces the concept of instantaneous frequency that can be determined for every time point, making it possible to have a deeper look into different phenomena.

The combinations of these methods with intelligent systems are described in the second part of the chapter. Several applications are presented where fuzzy classifiers, support vector machines and artificial neural networks are used for decision making. Interconnecting these intelligent methods with signal processing will result in hybrid intelligent systems capable of solving computationally difficult problems.

2.2 The Fourier Transform

It is a well known fact that periodic functions can be expanded into Fourier series using weighted sums of sines and cosines. In the real world, however, most of the physical phenomena can't be treated as periodical occurrences. For these shapes there exists the Fourier transform which is an integral taken over the whole definition domain. It is thus assumed that the function is represented on the whole real axis. Applying the Fourier transform to real-world signals makes the spectral analysis of phenomena possible, offering more information than it would be available in the time domain.

The definition of the Fourier transform is given by Eq. 2.1 for any integrable function $f : \mathbb{R} \to \mathbb{C}$ [1]

$$F(\omega) = \int_{-\infty}^{\infty} f(t)e^{-2\pi it\omega} dt \qquad (2.1)$$

where t is the time variable and ω is the frequency variable of the Fourier plane.

The Fourier transform describes the original function (or signal) as a bijection, thus the original function can be fully recovered if its Fourier transform is known. This process is achieved by applying the inverse Fourier transform described by Eq. 2.2.

$$f(t) = \int_{-\infty}^{\infty} F(\omega)e^{2\pi it\omega} d\omega \qquad (2.2)$$

The Fourier transform is built upon the Fourier series and in a straightforward manner on the decomposition of functions into sinusoidal basis functions. This can be easily proven using Euler's formula:

$$F(\omega) = \int_{-\infty}^{\infty} f(t)(\cos(-2\pi t\omega) + i\sin(-2\pi t\omega))dt \qquad (2.3)$$

However, most of the practical applications are not dealing with continuous domain signals and functions. Instead they use digitally sampled signals. For these signals an adapted version of the Fourier transform can be used, called the discrete Fourier transform (DFT), defined by Eq. 2.4 for a time series x_n, where n is the sample number, k is the wave number and N is the total number of samples.

$$X_k = \sum_{n=0}^{N-1} x_n e^{\frac{-2\pi ikn}{N}} \qquad (2.4)$$

Naturally, this also has an inverse transform described by Eq. 2.5.

$$x_n = \frac{1}{N} \sum_{k=0}^{N-1} X_k e^{\frac{2\pi ikn}{N}} \qquad (2.5)$$

Whenever possible, the fast Fourier transform (FFT) is used instead of the discrete Fourier transform, mostly based on performance and execution time considerations [2]. The most efficient way to use the FFT is to have signals with number of samples equal to some power of two. This has to do with the way the FFT algorithm is constructed. Further details can be found in [3, 4]. The computational complexity that can be achieved this way is $O(n \log n)$ as opposed to the DFT's $O(n^2)$.

The Fourier transform considers the analyzed signal in its full length, it is not using an analyzing window, thus sometimes identifies false frequency components that in reality aren't present in the signal. This explains why the best results are achieved for full periods of periodic signals.

2.2.1 Application of the Fourier Transform

The FFT is used in various fields, such as multimedia [5], optical [6], seismological [7], spectroscopy [8] or magnetic resonance signal processing [9]. In this part an application in the field of magnetic resonance imaging (MRI) is going to be presented briefly.

Magnetic resonance imaging produces images of the human body by exciting the hydrogen (^{1}H) nuclei with radio frequency pulses and then measuring the radio frequencies emitted by these nuclei as they recover to their initial energy state. Localization of different frequencies is done by modifying the main magnetic field using imaging gradients along the axes of the imaged plane. The frequencies emitted by different nuclei are equal to their precessional frequencies [9]. By measuring the electrical current induced in a receiver coil by the emitted RF frequencies the Fourier-space (or k-space) of the image is constructed using frequencies relative to the position of imaging gradients. The measured spatial frequency spectrum is then transformed to space domain (image domain) using the inverse Fourier transform.

This is an example of using the inverse Fourier transform in more than one dimension. Equations 2.6–2.9 describe the two dimensional continuous Fourier transform, the two dimensional continuous inverse Fourier transform, the two dimensional discrete Fourier transform and respectively, the two dimensional discrete inverse Fourier transform.

$$F(u, v) = \int_{-\infty}^{\infty} \int_{-\infty}^{\infty} f(x, y) e^{-2\pi i(ux + vy)} dx dy \tag{2.6}$$

$$f(x, y) = \int_{-\infty}^{\infty} \int_{-\infty}^{\infty} F(u, v) e^{2\pi i(ux + vy)} du dv \tag{2.7}$$

In Eqs. 2.6 and 2.7 x and y are space variables, u and v are spectral variables.

$$X_{k,l} = \frac{1}{\sqrt{MN}} \sum_{n=0}^{N-1} \sum_{m=0}^{M-1} x_{n,m} e^{-2\pi i \left(\frac{mk}{M} + \frac{nl}{N}\right)} \tag{2.8}$$

$$x_{n,m} = \frac{1}{\sqrt{MN}} \sum_{k=0}^{N-1} \sum_{l=0}^{M-1} X_{k,l} e^{2\pi i \left(\frac{mk}{M} + \frac{nl}{N}\right)} \tag{2.9}$$

In Eqs. 2.8 and 2.9 n and m are discrete space variables, k and l are discrete wave numbers.

Figure 2.1 shows the equivalence between 2D spectral space and 2D image space.

Fig. 2.1 Equivalence between k-space and image space. **a** k-space representation of the Shepp-Logan head phantom. **b** Reconstructed image of the Shepp-Logan head phantom

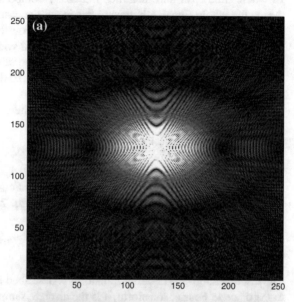

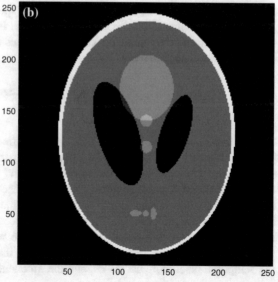

2.3   The Short-Time Fourier Transform

The Fourier transform does not reveal any temporal information about the frequency components present in the signal, making it impossible to locate them in specific applications. The Fourier transform "looks" at the analyzed signal during its whole time span and identifies frequency components as if they were present during the whole signal. Obviously, this is not the case in most of the applications. A way to introduce temporal information in the Fourier transform is to apply it on the signal using a sliding window with a constant width [1]. Sliding this window over the whole length of the signal will offer the possibility to get both spectral and temporal information about the signal. Equation 2.10 defines the short-time Fourier transform, where w is the windowing function, t is the time variable, ω is the frequency variable and τ is the time variable of the spectrogram.

$$F(\omega, \tau) = \int_{-\infty}^{\infty} f(t) e^{-2\pi i \omega t} w(t - \tau) dt \tag{2.10}$$

Similarly to the Fourier transform, the short time Fourier transform is also invertible. The original signal can be recovered using Eq. 2.11.

$$f(t) = \int_{-\infty}^{\infty} \int_{-\infty}^{\infty} F(\omega, \tau) e^{2\pi i \omega t} d\omega d\tau \tag{2.11}$$

The discrete short-time Fourier transform is described by Eq. 2.12, where similarly to the discrete Fourier transform, n is the discrete sample number of the time series x_n and k is the wave number. The time variable of spectrogram is represented by m. Inverting the discrete short-time Fourier transform is not as simple as the continuous one, it is heavily based on knowledge about the window function and the overlap between successive windows.

$$X_{m,k} = \sum_{n=0}^{\infty} x_n w_{n-m} e^{-\frac{2\pi i n k}{N}} \tag{2.12}$$

A very important element of the short time Fourier transform is the windowing function. The way the windowing function and its parameters are chosen, will affect the spectrum produced by the transform. Hann and Hamming windows are often used due to their favorable frequency responses. The most important parameters of a window function are its length and overlap—the resolution of the short time Fourier transform in both the temporal and spectral domain is influenced by these parameters. As we reduce the length of the window, temporal localization gets better in detriment of spectral resolution. If the window width is increased, the temporal resolution decreases but the frequency resolution increases. This phenomenon is strongly related to Heisenberg's uncertainty principle and is expressed by the inequality described with Eq. 2.13 [10].

$$\sigma_t \sigma_f \geq \frac{1}{4\pi} \qquad (2.13)$$

Here, σ_t and σ_f represent the standard deviations in time and frequency, respectively, and their product is bounded. The windowing function that can offer maximal resolution in both domains is the Gaussian window [11]. This limitation on resolution is the major drawback of the transform, making it unsuitable for situations where precise time and frequency information are both essential, e.g. in some electrophysiological applications.

Another drawback of the short time Fourier transform is that it can become numerically unstable [11], i.e. for small perturbations in the initial data set the output would be significantly different. Using analyzing windows is necessary to introduce temporal localization of frequency components of the signal, but there is a drawback to this. The shape of the windowing function will determine the amount of false frequency components in the spectrum and it also influences the amplitude of the spectral components. This is why there exists such a wide range of window functions to accommodate all needs.

2.3.1 Application of the Short-Time Fourier Transform

A wide range of applications exist for the short-time Fourier transform in the audio signal processing domain. However, this transform can be used in image processing.

The short-time Fourier transform can be used to enhance fingerprint images, as it is described in Chikkerur's paper [12]. The authors present an enhancement technique based on frequency domain filtering of the fingerprint image. The short-time Fourier transform is utilized first to get orientation and frequency maps of the original image. Just like the Fourier transform, the short-time Fourier transform can also be extended to higher dimensions. Equation 2.14 presents the 2D form of this transform.

$$X(\tau_1, \tau_2, \omega_1, \omega_2) = \int_{-\infty}^{\infty} \int_{-\infty}^{\infty} I(x,y)\overline{W}(x - \tau_1, y - \tau_2)e^{-j(\omega_1 x + \omega_2 y)}dxdy \qquad (2.14)$$

where $\overline{W}$ is the complex conjugate of a window function, x and y are spatial variables, ω_1 and ω_2 are frequency variables and τ_1 and τ_2 are time variables of the two-dimensional spectrogram. In the case of this application, the window function is a raised cosine. After the short-time Fourier transform was carried out, the whole image can be modeled as a surface wave function:

$$I(x,y) = A(\cos(2\pi r(x\cos\theta + y\sin\theta))) \qquad (2.15)$$

Here r and θ are the frequency and the ridge orientation in the image and A is a constant amplitude. They can be deduced from the short-time Fourier transform and they are considered random variables as they are defined by probability density functions. Thus the joint probability density function is defined by:

$$p(r,\theta) = \frac{|F(r,\theta)|^2}{\int_r \int_\theta |F(r,\theta)|^2} \qquad (2.16)$$

where F represents the Fourier spectrum in polar form.

The marginal density functions for the frequency and the orientation are described by Eqs. 2.17 and 2.18.

$$p(r) = \int_\theta p(r,\theta)d\theta \qquad (2.17)$$

$$p(\theta) = \int_r p(r,\theta)dr \qquad (2.18)$$

Then the ridge orientation is computed as the expected value of the orientation variable:

$$E(\theta) = \frac{1}{2}\tan^{-1}\frac{\int_\theta p(\theta)\,\sin(2\theta)d\theta}{\int_\theta p(\theta)\,\cos(2\theta)d\theta} \qquad (2.19)$$

This average is further smoothened using a Gaussian smoothing kernel.

The ridge frequency image is obtained by calculating the expected value of the frequency variable:

$$E(r) = \int_r p(r)rdr \qquad (2.20)$$

This frequency image is also smoothened, but an isotropic diffusion smoothing is used to avoid errors being propagated from the edges of the image towards the middle of it.

In the next step an energy map is determined, used for differentiating areas which do not contain ridges and thus have very low energy from the short-time Fourier transform, from areas of interest. This energy map is then used as a basis of thresholding in order to get two different regions of the image. Equation 2.21 gives the definition of this energy based region mask.

$$E(x,y) = \log\left(\int_r \int_\theta |F(r,\theta)|^2\right) \qquad (2.21)$$

To further reduce the discontinuities due to the block processing method, a coherence image is produced, which takes into consideration the level of orientation matching around individual points, as described by Eq. 2.22, (x_0, y_0) being the central point and (x_i, y_i) are the points overlapped by the W window. The values of this map will be high when the orientation of a block is similar to neighboring blocks' orientation.

$$C(x_0, y_0) = \frac{\sum_{(i,j) \in W} \cos \left(\theta(x_0, y_0) - \theta(x_i, y_j) \right)}{W \times W} \tag{2.22}$$

The actual image quality enhancement is then produced by applying algorithms described by Sherlock and Monro in [13].

2.4   The Wavelet Transform

The fixed width of the window used in the short time Fourier transform and the limited resolution of both the spectral and temporal domain are key reasons why the STFT cannot be used in many of the applications demanding time-frequency analysis. For example, non-stationary signal analysis depends heavily on determining what frequency components are present in the signal at a certain moment in time as well as on the possibility to search the signal for the occurrences of certain frequency components.

A new transform method was developed to face all these problems, called the wavelet transform. By definition, the wavelet transform is also an integral transform, using windows to slide over the analyzed signal in order to obtain time-frequency information from it [14, 15]. The major difference between this transform and the short time Fourier transform is that the wavelet transform uses windows which have variable width and amplitude, allowing the transform to analyze every bit of the signal. There are many different analyzing windows and they are called wavelets. Every wavelet is generally characterized by two parameters: scale (a) and translation (b). Equation 2.23 describes the continuous wavelet transform where a represents the scaling factor, b is the translation factor, ψ is a continuous function $(\psi \in L^2(\mathbb{R}))$ with $\bar{\psi}$ being its complex conjugate and x is the analyzed signal in the time domain. ψ is also called the mother wavelet.

$$X_{a,b} = \frac{1}{\sqrt{a}} \int_{-\infty}^{\infty} x(t) \bar{\psi} \left(\frac{t - b}{a} \right) dt \tag{2.23}$$

Mother wavelets have zero mean and their square norm is one, as presented by Eqs. 2.24 and 2.25.

$$\int_{-\infty}^{\infty} \psi(t) dt = 0 \tag{2.24}$$

$$\int_{-\infty}^{\infty} |\psi(t)|^2 dt = 1 \tag{2.25}$$

Although applications exist for the continuous wavelet transform (mostly analyzing theoretical aspects, see [16]), most of the problems necessitate a discrete wavelet transform. This can be derived from the continuous transform by quantization and without loosing any redundant information. Equation 2.26 defines the discrete scaling (s) and translation (u) factors used for the definition of wavelet functions.

$$s = 2^{-j}, u = k2^{-j}; j, k \in \mathbb{Z} \tag{2.26}$$

The result of substituting these variables in the integral transform is described by Eq. 2.27.

$$X(j,k) = 2^{j/2} \int_{-\infty}^{\infty} x(t)\psi(2^j t - k) dt \tag{2.27}$$

By discretizing the analyzed signal function, the integral transform becomes a sum as presented by Eq. 2.28.

$$X_{j,k} \approx 2^{j/2} \sum_n x_n \psi(2^j n - k) \tag{2.28}$$

The time-frequency resolution from wavelet decomposition point of view can be represented by the Heisenberg rectangle, where time and frequency is spread proportional to the scaling factor s and $\frac{1}{s}$. With the variation of s the two parameters of the rectangle, i.e. height and width, change accordingly but with a constraint stipulating that the area remains the same, as illustrated in Fig. 2.2. In most cases for a multiscale analysis, a scaling function φ is introduced. The relationship between the two functions φ and ψ is presented below by Eq. 2.29.

$$\left|\hat{\phi}(\omega)\right|^2 = \int_1^{\infty} \left|\hat{\psi}(s\omega)\right|^2 \frac{ds}{s} \tag{2.29}$$

With this notation the frequency analysis of the signal x at the scale of s is computed as follows:

$$X(u,s) = \langle x(t), \varphi_s(t-u) \rangle \tag{2.30}$$

Here $\langle . \rangle$ denotes the inner product of two functions and φ_s is given by:

$$\varphi_s(t) = \frac{1}{\sqrt{s}} \varphi\left(\frac{t}{s}\right) \tag{2.31}$$

In practice, the wavelet transform is computed for a defined number of scales (2^j). The low frequency component $W(u, 2^j)$ is often called the DC component of the

Fig. 2.2 Heisenberg rectangles for different transforms: **a** continuous domain (equal windows in the time direction, no frequency information); **b** Fourier transform (equal windows in the amplitude direction, no time information); **c** short-time Fourier transform (equal windows in both time and frequency direction); **d** Wavelet transform (rectangles with equal area but varying width and height)

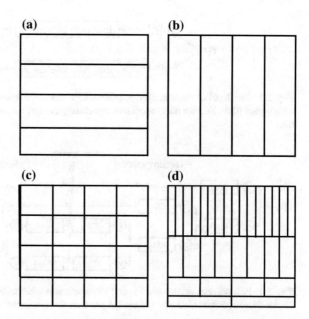

signal. Wavelet coefficients W have a length of $\frac{N}{2^j}$ as the largest depth is bounded by the signal length.

For implementation purposes a set of conjugate mirror filters h and g is constructed using the scaling function φ and the wavelet function ψ. Their definitions are given by Eqs. 2.32 and 2.33.

$$h(n) = \langle \frac{1}{\sqrt{2}} \varphi\left(\frac{t}{2}\right), \varphi(t-n) \rangle \tag{2.32}$$

$$g(n) = \langle \frac{1}{\sqrt{2}} \psi\left(\frac{t}{2}\right), \varphi(t-n) \rangle \tag{2.33}$$

These filter functions have to satisfy the following conditions (here k denotes a filter function):

$$\left|\hat{k}(\omega)\right|^2 + \left|\hat{k}(\omega+\pi)\right|^2 = 2 \tag{2.34}$$

and

$$\hat{k}(0) = 2. \tag{2.35}$$

The discrete orthogonal wavelet decomposition can be calculated by applying the filter functions on the signal recursively. The two functions separate the signal into low and high frequency domains were h is a low-pass filter and g ia a high-pass filter, as presented in Fig. 2.3, while Fig. 2.4 shows the whole decomposition and

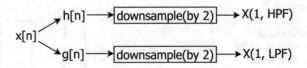

Fig. 2.3 One level of wavelet decomposition. LPF denotes a low-pass filter while HPF stands for a high-pass filter. Further decompositions are usually carried out using the results of the high pass filter

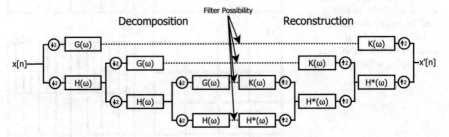

Fig. 2.4 Full wavelet decomposition and reconstruction scheme. Notice the down-sampler blocks on the decomposition side and the up-samplers on the reconstruction side

reconstruction scheme, highlighting the places of eventual filters. A detailed proof of this concept can be found in [15]. Another, more general approach can be achieved by using bi-orthogonal decomposition and reconstruction filters. It allows for a larger room for analyzing, modifying and filtering even multidimensional data.

Having a pair of wavelet function ψ and a reconstruction function χ, the decomposition and reconstruction is a straightforward implementation with H, G and K filters satisfying the Eq. 2.36 [17].

$$\hat{\varphi}(2\omega) = e^{-i\omega s}H(\omega)\hat{\varphi}(\omega)$$
$$\hat{\psi}(2\omega) = e^{-i\omega s}G(\omega)\hat{\psi}(\omega)$$
$$\hat{\chi}(2\omega) = e^{i\omega s}K(\omega)\hat{\chi}(\omega) \tag{2.36}$$
$$|H(\omega)|^2 + G(\omega)K(\omega) = 1$$

As it turns out, filtering a signal is equal to applying a simple scaling function after the transformation. By modifying the wavelet coefficients after the decomposition, we can attenuate or amplify different frequency bands according to the resolution. This can be done even for a defined time-period depending on the scaling function.

2.4.1 Application of the Wavelet Transform

One of the most well-known examples of using the wavelet transform is the JPEG-2000 image compression method. The widely used JPEG image compression format traditionally uses the discrete cosine transform and has limitations both in compression efficiency and the quality of edge preserving. In order to overcome these shortcomings, the JPEG-2000 standard was developed, which uses the discrete wavelet transform in order to decompose the image and then compress it.

As a first step, the image is cut in rectangular tiles and all the following operations are executed on each tile separately. Each tile is transformed using the one dimensional discrete wavelet transform, so 4 results are obtained: one with low-resolution rows and columns, one with low resolution rows and high resolution columns, one with high resolution rows and low resolution columns and one with high resolution rows and columns. This decomposition is then repeated a number of times on the low resolution image block. The decomposition process is called dyadic decomposition. Both lossless and lossy decompositions are possible: when using integer wavelet functions, the result is lossless decomposition, while using real-valued wavelet functions, the result is a lossy decomposition. In both cases the decomposition is done by applying low-pass and high-pass filters built up using wavelet functions. On the reconstruction side quadrature mirror filter pairs [1] of the decomposition side filters are used.

After the decomposition, the obtained coefficient values are quantized, i.e. this is the step where information loss can occur. When using lossless compression, though, there is no quantization, or the quantization step is 1.0. Following quantization, entropy encoding of the bit planes is carried out. After this procedure is done, the image is ready to be stored [18].

2.5   The Hilbert-Huang Transform

2.5.1   Introducing the Instantaneous Frequency

All of the transform methods presented up until now use the classical definition of frequency, i.e. the number of repetitions of a phenomenon in unit time. This means that intra-wave oscillations are totally disregarded when analyzing non-stationary signals. Already in the 1990s it was recognized that there are serious consequences of not taking into consideration these intra-wave changes in frequency.

The basic idea behind the instantaneous frequency is very simple. For example, if we consider a uniform circular motion, then the projection of this movement on one of the x or y axes will result in one of the Fourier series, more accurately a single sine or cosine wave. In this case the instantaneous frequency is constant. The speed of the motion is described as the derivative of the angular position and it is called angular velocity. This angular velocity is exactly the definition of the instantaneous frequency. If the motion is uniform, i.e. has a constant speed, the angular velocity will be constant, thus the frequency will be constant in each time point. However, if the motion is non-uniform, the sinusoid will have time-variable periods. This way, the definition of the instantaneous frequency becomes a generalization of the classical frequency model.

From now on, we can treat every signal as a sum of non-uniform circular motions with varying amplitude where the derivative of the angular position is the instantaneous frequency. Analyzing data from this perspective exceeds the breakdown capability of the Fourier transform as we get an absolute time depending spectral overview.

It was, however, not until recently when the computation of the instantaneous frequency was made easily achievable and Huang [19] contributed significantly to this by developing the Hilbert-Huang transform. Although the new definition of frequency was also known before Huang [20], there were no straightforward algorithms for its computation.

2.5.2 Computing the Instantaneous Frequency

There are many ways of computing the instantaneous frequency of a non-stationary signal. The basic idea is to decompose the original signal into mono-components which have only one oscillation per period and then computing the instantaneous frequency of these components by determining the quadrature of the component and computing the derivative of the phase of the two sub-signals. As we are talking about empirical methods to decompose a signal into mono-components, there is a high risk to end up with elements without any link to the original physical phenomenon. A minimal change in the signal form will result in a totally different decomposition.

The Hilbert-Huang transform is one of the modalities to calculate the instantaneous frequency. It consists of two steps:

1. Determine the mono-components of the original signal. This step uses the empirical mode decomposition algorithm
2. Compute the instantaneous frequency with the Hilbert transform for every mono-component.

The empirical mode decomposition (EMD) is an iterative algorithm that extracts mono-components called intrinsic mode functions (IMF) from the original signal. These IMFs hold two important properties:

1. They have the same number of local maxima and minima, or their number differs at most in one;
2. Their upper and lower envelope averages to zero.

Because it is an iterative algorithm, the exact number of final IMFs is unknown beforehand. The algorithm is described by listing 2.1. The mono-components or intrinsic mode functions can be described by the general equation of a non-uniform, variable-amplitude circular motion:

Algorithm 2.1: Pseudocode of the EMD algorithm

```
1  s ← getSignal();
2  ε ← 10⁻⁵;
3  k ← 1;
4  while !isMonotonic(s) do
5       SDₖ ← 1;
6       hₖ,₀ ← s;
7       i ← 0;
8       while SDₖ > ε do
9            max ← getLocalMaxima(s);
10           min ← getLocalMinima(s);
11           upperEnvelope ← getSpline(maxima);
12           lowerEnvelope ← getSpline(minima);
13           mᵢ ← (upperEnvelope + lowerEnvelope) / 2;
14           hₖ,ᵢ ← hₖ,ᵢ₋₁ − mᵢ;
15           SDₖ ← Σ(hₖ,ᵢ₋₁ − hₖ,ᵢ)² / Σh²ₖ,ᵢ₋₁;
16       cₖ ← hₖ,ᵢ; // the k-th IMF
17       s ← s − cₖ;
18       k ← k + 1;
19  r ← s; // the residue
```

$$c(t) = A(t)\cos(\varphi(t)) \tag{2.37}$$

where $c(t)$ is the IMF, $A(t)$ describes the time-variable amplitude part and $\varphi(t)$ represents the time-variable phase.

Due to their fundamental properties, it is easy to find their quadrature and the Hilbert transform does exactly that. Taking the IMF $c(t)$, the corresponding quadrature $q(t)$ is given by Eq. 2.38.

$$q(t) = \frac{1}{\pi} \int_{-\infty}^{\infty} \frac{c(\tau)}{t - \tau} d\tau \tag{2.38}$$

Then the phase difference between them is computed:

$$\varphi(t) = \arctan \frac{q(t)}{c(t)} \tag{2.39}$$

Then the instantaneous frequency of the IMF will be:

$$\omega(t) = \frac{d\varphi(t)}{dt} \tag{2.40}$$

Although this transform is a powerful tool for calculating the instantaneous frequency of an IMF, there are some limitations imposed by the Bedrosian and Nuttal theorems. According to the Bedrosian theorem [21, 22], the IMFs should satisfy the condition described by Eq. 2.41 in order for the analytical signal provided by the Hilbert transform to make sense.

$$H\{A(t)\cos(\varphi(t))\} = A(t)H\{\cos(\varphi(t))\} \qquad (2.41)$$

This condition states that the IMFs should be band limited, i.e. the spectrum of the time-variable amplitude and the one of the variable phase should not overlap [23]. Failure to conform to this requirement will result in the confusion of amplitude frequencies in the instantaneous frequency of the IMF.

The Nuttall theorem [24] further states that the Hilbert transform of $\cos(\varphi(t))$ is not $\sin(\varphi(t))$ for any arbitrary $\varphi(t)$. Therefore, the result of the Hilbert transform is not necessarily equal to the true quadrature of the IMF. Because there are no further restrictions on what an IMF can be, except the ones presented before, using the Hilbert transform to get the instantaneous frequency of an IMF is considered unsafe. In practice, the instability of the Hilbert transform can be easily spotted in regions where the instantaneous frequency becomes negative or a sudden peak appears in a region where there are no sudden frequency changes in the IMF.

A more accurate way to compute the instantaneous frequency of an IMF is to use the empirical AM-FM decomposition algorithm [23]. This algorithm is also an iterative one, just like the EMD and it is presented in listing 2.2. The aim of this algorithm is to separate the two parts of the IMF, i.e. the amplitude part from the phase part. The model used by this method is that of modulated signals: the amplitude part is considered to be the modulator signal and the phase part the carrier; this is considered the AM modulation in the IMF. The carrier is also modulated in frequency; this constitutes the FM modulation in the IMF. The decomposition results in two signals: $AM(t)$, the amplitude part and $FM(t)$, the frequency part.

Algorithm 2.2: Pseudocode of the empirical AM-FM decomposition

---
1   $s \leftarrow$ getImf();
2   $s_0 \leftarrow s$;
3   $k \leftarrow 1$;
4   **while** $s_k \in [-1; 1]$ **do**
5 | $e \leftarrow$ getSplineEnvelope($|s_{k-1}|$);
6 | $s_k \leftarrow s_{k-1}/e$;
7   $FM \leftarrow s_k$;
8   $AM \leftarrow s/FM$;

The separation process implies successive approximations of the amplitude part by cubic spline functions through the local maxima of the IMF signal. After the two signals are separated, the quadrature of the frequency part is computed:

$$Q(t) = \pm\sqrt{1 - FM^2(t)} \tag{2.42}$$

From here, getting the instantaneous frequency of the IMF matches the Hilbert transform-based method.

It is important to note that this method satisfies the conditions imposed by both the Bedrosian and the Nuttall theorems. However, the method is not free of disadvantages. The cubic spline envelopes often cross the IMF, meaning that they are not always greater or equal than the signal, which is a necessary condition for the algorithm to converge in finite steps. Instead, the method will never converge and a frequency part belonging to the $[-1; 1]$ interval will never be reached. This problem and a possible solution to it is addressed in [25].

First, the intervals where the spline crosses the IMF are identified. These intervals are characterized by an entry and an exit point. The spline is approximated with a line segment over these intervals and the maximum distance between this approximation and the IMF is sought. To do this, the segment and the IMF are rotated to the horizontal axis and the maximum of the IMF is identified. After an inverse geometric transform a new spline is defined, which contains the maximal distance point but not the previous extremum point, thus not intersecting the IMF anymore. These steps are repeated for each interval, until the spline only touches the IMF but does not cross it. The correction mechanism is illustrated in Fig. 2.5 and it is described as a pseudocode in listing 2.3.

Algorithm 2.3: Pseudocode of the empirical AM-FM decomposition

---
1   $s \leftarrow \text{getImf}()$;
2   $s_0 \leftarrow s$;
3   $k \leftarrow 1$;
4   **while** $s_k \notin [-1; 1]$ **do**
5       $e \leftarrow \text{getSplineEnvelope}(|s_{k-1}|)$;
6       $intervalList \leftarrow \text{getCrossingIntervals}(s_{k-1}, e)$;
7   **while** $!\text{isempty}(intervalList)$ **do**
8           $p \leftarrow \text{getMaximumDistancePoint}(s_{k-1}, intervalList_i)$;
9           $e \leftarrow \text{getSplineEnvelope}(|s_{k-1}|, p)$;
10          $intervalList \leftarrow \text{getCrossingIntervals}(s_{k-1}, e)$;
11      $s_k \leftarrow s_{k-1}/e$;
12 $FM \leftarrow s_k$;
13 $AM \leftarrow s/FM$;

Fig. 2.5 A slice created by
the intersection of the spline
and the IMF and the corrected
spline

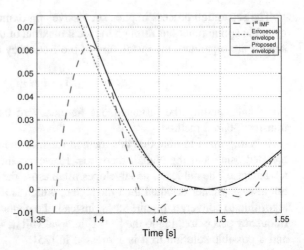

2.5.3 Application of the Hilbert-Huang Transform

2.5.3.1 Detecting the Third and Fourth Heart Sounds in Phonocardiograms

The Hilbert-Huang transform is an ideal tool for analyzing the heart activity since it offers frequency information for every sample point. In the following, a detection of the third and fourth heart sound will be presented through the usage of HHT. The method is described in [26]. These sounds (S3, S4) represent an abnormal activity of the heart indicating failure during diastolic period. Processing sound measurements, i.e. identifying S3, S4 is a noninvasive way to detect early myocardial ischemia. Deficient S3 is related with ventricle activity or problematic blood flow in the rapid filling phase. The fourth heart sound, S4 is located before the first heart sound and is a sign of forceful contractions in an effort to overcome a rigid ventricle. The method localizes the presence of S3, S4 hearts sounds using time-frequency distribution and K-means clustering in electronic stethoscope recordings. The proposed method can be divided into three stages. The first step preprocesses the signal, the second one applies the Hilbert-Huang transform to get a time-magnitude-frequency decomposition followed by a clustering and recognition procedure.

The role of preprocessing is to eliminate high frequency noise due to recording artifacts by a low-pass, finite impulse response filter. Then an envelope is produced using the Hilbert transform noted as $x_{envelope}[n]$ (Eq. 2.43).

$$x_{envelope}(n) = |x(n) + jH\{x(n)\}| \tag{2.43}$$

This envelope is then normalized (Eq. 2.44) to fit the interval $[-1; 1]$ and segmented into systolic (S1-S2) and diastolic (S2-S1) periods.

$$x_{norm} = \frac{x_{envelope}(n)}{\max(x_{envelope}(n))} \qquad (2.44)$$

The detection of these terms is based on the Shannon energy as it shows a high rate of noise rejection property, suppressing low amplitude components. Because of the main components S1 and S2 will stand out, detection and, therefore, segmentation is much easier. Equation 2.45 defines how the Shannon energy is computed.

$$SE(n) = -x_{norm}^2(n)\log(x_{norm}^2(n)) \qquad (2.45)$$

A threshold of 70 % of the maximum Shannon energy is set to differentiate S1 and S2 from other components. Naturally, a higher limit will result in a better filtering but with the risk of missing some S2 sounds. Recognition is built upon the following rules to ensure a correct result:

- If two peaks higher than the threshold are detected within 50 ms, the one with lower energy is eliminated.
- For every interval between the peaks, an interval with shorter length than the previous interval is denoted as a systolic period, while the other one is a diastolic period. The uncertain intervals are annotated.
- For those uncertain intervals, a secondary threshold is set to find S1 or S2 which probably have not been recognized.

As a second step the Hilbert-Huang transformation is applied to the nonlinear signal, extracting the instantaneous frequencies. It is an iterative empirical procedure resulting in a series of intrinsic mode functions and therefore a sum of instantaneous frequencies is calculated for a single time instance (Hilbert-Huang spectrum).

Having a time-frequency map allows for a cluster figure based on pairs of instantaneous frequencies and their amplitudes. Correlating in-scope cluster points with the time scale will reveal the position of S3 and S4 sounds. For the cluster graph the instantaneous frequency with the highest magnitude is selected from the sum of instantaneous frequencies for a particular time instance. Only one with the highest impact factor is selected based on the magnitude. The number of points depends on the resolution of the Hilbert spectrum. These pairs originate from the diastolic interval as it was segmented in the second part of the method. The K-means algorithm was used to divide pairs into different groups. Three separate clusters have been formed depending on the frequency and magnitude distribution. The one of interest is called the abnormal group with the highest frequencies or magnitudes. Usually this group has the fewest of points and their projections on the time axis is periodical. If the points appear right before S1 we found the S4 sound but if they appear periodically after S2 we can identify S3 elements. For those possibly missing components, an iterative method was applied to enhance accuracy.

Records from the Cardiac Auscultatory Recording Database (CARD) of Johns Hopkins University were used to test the proposed method. As a result, 13 signals were processed. The method with iterative recognition, achieved a 90.3 % identification rate for S3, 9.6 % were missed and 9.6 % were marked falsely. For S4, 94.4 % were detected correctly and 5.5 % were missed; 16 % were false positive. Sensitivity for S3 and S4 was 90.4 % and 94.5 %, respectively. Precision was 90.4 and 85.5 %.

Still, the existence of artifacts during diastolic period, such as diastolic murmur or noise produced by the electronic stethoscope, would contribute to misjudgments. Components with low amplitude make the separation of S3 and S4 from background noise difficult. The proposed method aims to detect early heart diseases, such as left ventricle dysfunction, congenital heart failure or myocardial ischemia.

2.5.3.2 Identifying T Waves in ECG Signals

The empirical mode decomposition can also be used in ECG signal analysis. Identifying T waves in an ECG is of special interest, since its morphology can suggest different heart conditions. Finding the positions of these waves is often done manually, thus an automated detector is much needed. In [27] we propose a method that can solve this problem for a wide range of ECG sources.

The general idea behind the method is to identify and remove QRS complexes from the ECG signal, because these are present in most of the IMFs owing to their large amplitude and wide frequency spectrum. A method described in [28] is employed for this task. According to this method, after the ECG is decomposed in IMFs, the Shannon energy of the sum of the first three IMFs is computed, as described by Eqs. 2.46 and 2.47.

$$c(t) = \sum_{k=1}^{3} IMF_i(t) \tag{2.46}$$

$$E_s(t) = -c(t)^2 \log(c(t)^2) \tag{2.47}$$

Then, using a threshold, the Shannon energy (E_s) is filtered to eliminate background noise. The threshold is determined by Eq. 2.48 where N is the number of data points in the Shannon energy.

$$T = \frac{1}{N} \sum_{i=1}^{N} E_s(i) \tag{2.48}$$

The threshold filter substitutes each value less than the threshold with zero and keeps all the other values. Non-zero intervals correspond to the QRS complex intervals of the ECG signal. After identifying each of these intervals, the corresponding places in the original ECG are exchanged with a line segment and the signal is decomposed again using EMD.

In the newly obtained IMF set the 5th, 6th and 7th will contain waves that construct the T wave, thus they are summed. Finally, the peaks in this newly calculated sum appearing right after the positions of QRS complexes are marked as T waves in the original signal. Figure 2.6 presents the steps of the identification process.

Signals from PhysioNet's QT database [29] were used to test the algorithm. This testing resulted in a 95.99 % positive predictability and 99.96 % sensitivity when

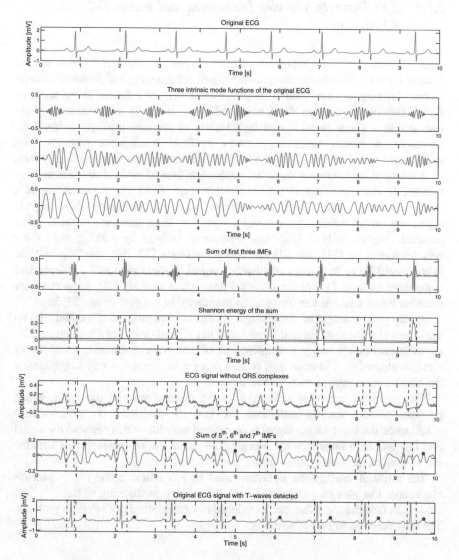

Fig. 2.6 Graphical representation of the T wave identification process

different ECG signals were presented to it. Noise tolerance was also tested by feeding an ECG signal with different amplitudes of noise to the method. An 87.71 % positive predictability was obtained for a signal to noise ratio as low as 12 dB. The corresponding sensitivity was of 100 %.

2.6 Hybrid Signal Processing Systems

2.6.1 The Discrete Wavelet Transform and Fuzzy C-Means Clustering

Medical signal processing represents an essential part of a medical decision making system, however it is able to provide only signal conditioning and feature extraction without offering any information on the diseases that are reflected in the acquired signal. Getting information about any condition is basically a classification problem, and thus it can be addressed by a variety of artificial intelligence methods. Therefore, it makes sense to combine artificial intelligence methods with well-known signal processing methods to assist a medical decision making system.

A classical but also very important application area of medical decision making systems is electrocardiogram (ECG) interpretation. ECG signal processing is a much discussed topic. There are different ways of processing these signals depending on what features are being sought for. Thus, there are methods using classical Fourier analysis [30], time frequency analysis by making use of the wavelet transform [31], the Hilbert-Huang transform [32, 33], the Wigner-Ville distribution [34] or by using a multivariate signal analysis such as the independent component analysis [35] or the principal component analysis [36]. Support vector machine based analysis has also gained popularity in the past years [37, 38].

A very good example of combining a classical signal processing method with an intelligent classification method is presented in [39] and describes a method of ECG signal classification using a combination of the wavelet transform and fuzzy c-means clustering. The main goal of said paper is to describe a way to implement the classification system on a mobile device.

The presented approach makes use of the discrete wavelet transform, which provides discrete wavelet coefficients that describe the signal. To achieve these coefficients, the finite-length signal is convolved with filters that separate the signal into low and high frequency sub-signals. The signal can be reconstructed with the use of quadrature mirror filters from the coefficient sets.

The artificial intelligence algorithm used in this article is the fuzzy c-means clustering. The idea of this method is to do a fuzzy partitioning of the data into classes. In fact, the algorithm can be reduced to a constrained optimization problem. The cost function to be optimized is:

$$J_\beta(z) = \sum_{i=1}^{N} \sum_{j=1}^{c} A_{ij}^{\beta} \|x_i - z_j\|^2 \tag{2.49}$$

In this equation, J_β is the cost function, β is the level of fuzziness, chosen to be 1.5 in this specific application, N is the number of data vectors, c represents the number of clusters, A_{ij} is the grade of membership of data vector i to cluster j, x_i is the ith data vector and z_j is the center of the jth cluster. The grades of membership are computed as Euclidean distances of data vectors from the center of the cluster. The detailed fuzzy c-means algorithm is described in [40].

Figure 2.7 presents the components of the diagnosing system. The ECG signals collected by body sensors are first preprocessed, since they are contaminated by both low and high frequency noise. The baseline drift represents a low frequency component in the signal, while thermal noise is responsible for the high frequency noise. Both artifacts are removed using the discrete wavelet transform, applying the Daubechies-9 wavelet for the signal decomposition. By carefully combining some of the signal components and then reconstructing the signal, the noises can be filtered out. The signals were also normalized to lay between -1 mV and 1 mV. This normalization is also convenient if at a later stage support for different body sensors will be included in the overall system. This step was necessary because in the testing phase the ECG signals from the MIT-BIH database were used, which were recorded by a large variety of devices from all over the world.

The next stage consisted of extracting different features of the signals. In total 6 characteristics were chosen, as follows: the length of a heartbeat period, called the RR period, the length of the QRS complex relative to the RR length, the length of the RS interval relative to the QRS length, the period of the ST segment relative to the RR length, the amplitude of the QRS complex and the amplitude of the T wave. Measuring these parameter is done also by using the DWT and combining components so that the QRS complex and the P-T-U waves could be separated. An ECG time-series was thus characterized by a feature vector having 6 elements corresponding to the measurements of the 6 aforementioned parameters.

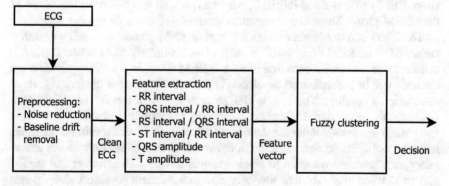

Fig. 2.7 The general architecture of the classification system

In the last stage of the classification algorithm the feature vectors are separated into clusters by the fuzzy c-means algorithm, using a fuzziness number $\beta = 1.5$ and 3 clusters. Two ECG data sets from the MIT-BIH database were used. The first one contained 25 ECG signals while the other one had 23 ECGs. Portions with a length of 10 s were selected from each signal and their feature vectors were fed to the fuzzy c-means clustering algorithm. Feature vectors corresponding to normal ECGs were classified into one cluster and abnormal ones in the other two clusters. There were, however mismatches too. 3 out of 25 ECGs from the first group and 6 out of 23 ECGs from the second group were misclassified.

The diagnostic system was implemented on a mobile device running Windows CE having an ARM9 processor. The wavelet decomposition and fuzzy c-means clustering were implemented in the Matlab environment and then C++ code was generated using Code Generator. According to the article, a nearly real-time execution was made possible on the mobile device.

2.6.2 Automatic Sleep Stage Classification

The correct diagnosis of sleep stages is crucial when it comes to identifying and treating possible sleep apnea, insomnia or narcolepsy conditions. Even today much of this work is done visually by experts, based on polysomnogram (PSG) techniques described by the American Academy of Sleep Medicine (AASM) [41] or by the guidelines defined by Rechtschaffen's and Kales's [42]. Basically, most specialists try to identify 6 repeating sleep stages from electroencephalogram (EEG) measurements combined with electrooculograms (EOG) and electromyogram (EMG) signals to enhance accuracy. The initial stage is characterized by being awake (Awa) followed by S1, transition between wakefulness and sleep. S2 is considered to be the baseline; it may consist of 45–55 % of the entire sleep duration. Stage three and four (S3, S4) represent the recovery mode of the body and mind also known as deep-sleep period followed by the rapid eye movement (REM) stage. From this perspective, two major phases can be distinguished when it comes to sleep, REM and non-REM (NREM), where S1, S2, S3 and S4 are sub-divisions of the NREM phase. These stages repeat themselves 4–5 times during one night.

The EMG activity is almost missing during REM phase. This helps to differentiate S1 from REM using EMG measurements. Similarly EOG is useful when it comes to eye movement detection in S1 and REM sleep. In the following, different methods will be presented, all developed to make the evaluation of sleep signals as automatic as possible. The task to design such a system usually takes into consideration the amount of data and the complexity of classification algorithms. These two parameters greatly influence the overall behavior of the end result. For instance, multichannel EEG devices represent a drawback to patient comfort and in ambulatory environment compared to single channel devices. Therefore recent studies aim to develop methods that use only one measurement to detect sleep stages [43–49]. Another challenge represents the right choice and combination of different feature extraction methods working together with pattern recognition systems. A great variety has been proposed over the years with different and surprising outcomes. For instance Hidden Markov Models (HMM), fuzzy classifiers, different types of artificial neural networks (ANN) using feature vectors based on power spectral densities (PSD), wavelet decompositions or visibility graphs (VG). Still, it is difficult to achieve a higher accuracy like the ones produced by experts using manual techniques. However, in [50] several approaches are proposed for the discussed problem through a particular neural network, financial forecast based method, non-smooth optimization problem and frequency domain analysis. The ANN approach suggests the usage of a time-delay neural network (TDNN) with the property of reducing input data size, an important element when it comes to processing EEG, EMG and EOG simultaneously. A particular output from such a network depends not only on the input but it takes into consideration a range of the previous input values. The benefit of this particularity makes it possible to reduce the input-volume as previous information is already stored in the network. In this study the network was configured with 1 input layer, 3 hidden layers and 1 output layer. Normalized PSG variables were fed as input: direct EEG, EMG and EOG measurements. The output layer consists of 6 nodes corresponding to the 6 sleep stages. As a result, the network was able to obtain a classification accuracy of 76.15 % from thousands of test cases.

A similar model proposed by [51] uses a multi-layer perceptron (MLP), a neural network designed for classification. The network had 5 input nodes and again 6 output nodes for the 6 sleep-stages. Electrode distribution followed the 10–20 system but only the C3-A2 left-right and C4-A1 right-left measurements were used. A well known fact in EEG signal processing is related to the frequency bands that reflect the mind's different states. There are usually 5 frequency bands related to mental activity: δ (<4 Hz), θ (4–7 Hz), α (8–12 Hz), σ (13–16 Hz) and β (>16 Hz). Usually the short time Fourier transform is used to gain an overview in the time-frequency domain. Due to the non-linearly of these signals, the power spectra is used in this case. The proposed method calculates the relative spectral power (RSP) for a window of 30 s, equal to the band spectral power (BSP) divided to the total spectral power (TSP).

$$RSP_i = \frac{BSP_i}{TSP}, i \in \{\delta, \theta, \alpha, \sigma, \beta\} \tag{2.50}$$

After training the network and finding the optimal structure the method was able to recognize sleep states with an accuracy of 76 %. The final conclusion states that Awa, S2, S4 and REM stages are easily recognized, however S1 is confused with S2 and REM, being the hardest stage to identify. By adding the second derivative of the EEG signal to the input vector the overall performance of the system did not improve but remained in the 76–77 % range.

ANN combined with wavelet packet coefficients was suggested in [52] where they used a 3-layer feed forward perceptron with 12 input nodes, 1 hidden layer (with 8 neurons for best results) and 1 output layer (with 4 nodes) as a classifier. The goal of the perceptron was to distinguish between sleep stages Awa, S1 + REM, S2 and SWS also known as deep-sleep stage. Adaptive learning rate has been applied to avoid stagnation on the error surface. This translates to an adaptive learning step solution minimizing the learning period. For testing purposes PhysioBank's EEG Database was used, more precisely Pz-Oz bipolar recordings. Wavelet packets had been chosen as a feature extraction method, allowing for a finer frequency resolution. After fine-tuning the packet tree, the structure shown in Fig. 2.8 was defined as the transformation method supplying feature vectors to the perceptron. The subbands represent the following EEG frequency bands:

1. Delta—0.39–3.13 Hz
2. Theta—3.13–8.46 Hz
3. Alpha—8.46–10.93 Hz
4. Spindle—10.93–15.63 Hz
5. Beta1—15.63–21.88 Hz
6. Beta2—21.88–37.50 Hz

However, the 6 wavelet coefficient groups were further refined to a 5 element classification series as a statistical time-frequency distribution representation. The first element holds the mean quadratic value for each of the 6 bands. The second element is the total energy, followed by multiple elements calculated as the ratio of different energy bands (alpha, delta and theta). The fourth and fifth elements are the mean of the absolute values and the standard deviation of the coefficients in each sub-band. The 5 element series fed to the MLP resulted in a very high classification rate, indicating that the method could discriminate between Awa, S1 + REM, S2 and SWS with a specificity of 94.4 ± 4.5 %, a sensitivity of 84.2 ± 3.9 % and an accuracy of 93.0 ± 4.0 %.

Another feature extraction method uses financial forecasting to predict sleep stages based on the hypothesis that a variable from the sleep measurement depends on the previous values. It relies on conditional probability to foretell the next stage. Raw date is mapped into a 5 symbol series depending on preset derivative thresholds. The new values are tagged as follows: BI (big increase), SI (small

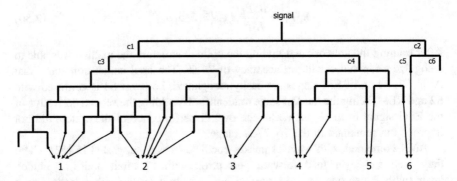

Fig. 2.8 Wavelet packet transform and selected subbands

increase), N (no change), SD (small decrease) and BD (big decrease). After the tagging the method looks for particular sequences like BI, BI, BI, SI, N, SD and SI defining one of the sleep stages. As a result, this approach finds it hard to classify S2 correctly; on the other hand its simplicity is a great advantage with an overall accuracy of above 70 % compared again to expert classification structures.

Non-smoothing optimization is a relatively new extraction method having its roots in the field of signal processing. The main idea behind it is simple, where the algorithm tries to fit a sum of two sine waves on the raw EEG. The deviation from the sine waves represents the information itself. In this case the amplitude is a linear function resulting in a more adequate curve fitting procedure then with the ordinary scalar value. This way sudden changes can be tracked maintaining vital information of the signal. The sum is composed of two sine curves, a low frequency component tracking the baseline wandering and a high frequency component being able to adapt to all kind of shapes.

As a summary to the sleep-stage classification methods we can add that even manual scoring done by experts show differences. Comparisons between results can score below 80 %. This means that the actual automatic sleep stage classification methods are as reliably as the experts.

2.6.3 The Hilbert-Huang Transform and Support Vector Machines

In recent years the car industry has produced a whole new generation of vehicles, more reliable, safer, and with build-in-intelligence to recognize and anticipate different type of engine failures. Wang et al. present in [53] a method of failure detection based on the Hilbert-Huang transform and support vector machines. Based on the failure source, the engine fault diagnosis (EFD) technology defines 4 types of analysis methods: engine performance detection, lubricating oil analysis, vibration based methods, and noise based diagnostic methods. The vibration and noise based EFD methods normally extract the failure features from the vibration or noise signals of a running engine and make decisions based on diagnostic results by using pattern recognition algorithms. Usually such a system is composed of two larger elements. One is responsible for processing the measured signals and supplying useful features, called a feature extractor and the second element is a classifier. The classifier's role is to separate the different engine failures into categories based on feature types.

Noises from a car's engine are non-stationary under working conditions, for this reason the Hilbert-Huang transform presents itself as an ideal candidate supplying time variant mono-components related to circular motions. For the classification algorithm support vector machine (SVM) [54] has been used as a multi-category classifier based on "one against one" voting decision method. Such a model will define an optimal hyper-plane which separates fault types by mapping one group of feature vectors (belonging to a particular engine fault) on one side of the plane and other groups on different sides thus categorizing them geometrically. Basically, for n categories $\frac{n(n-1)}{2}$ SVM models are constructed from a training database for every two differing category. A measurement sample will pass every model and in the end it will be ranked upon the received votes. The final category is the one with the largest number of votes.

For the analysis it is presumed that in fault conditions, particular engine noises will reflect in the amplitude and frequency domain and certain failures will shift the energy quantum from one frequency band to another. Therefore the energy pattern of the intrinsic mode function is considered to be at the heart of the proposed feature extraction method.

The process is performed in 4 steps. First the sound is filtered to remove unwanted noise, and then the IMFs are calculated together with the residual element. Usually the residual and particular IMFs are discarded as they carry almost no energy. After selecting the IMFs, a correlation coefficient is calculated between the original signal and every IMF with Eq. 2.51.

$$\rho_{S,c_i} = \left| \frac{E[(c_i(t) - \mu_{c_i})(S(t) - \mu_S)]}{\sigma_{c_i}\sigma_S} \right| \tag{2.51}$$

Here S represents the original measurement, c_i are the IMFs while μ_{c_i}, σ_{c_i} and μ_S, σ_S are the mean values and the standard deviations of these signals. Every correlation coefficient falls between 0 and 1, where 1 means that the two signals are identical and 0 means that they are totally different. A large value means that the two signals have much in common, giving the method a chance to choose between the IMFs. Usually mono-components with a low correlation coefficient are discarded, therefore only IMFs with relevant information content are processed further. The third step computes the energy moment of the selected IMFs, denoted with E_i.

$$E_i = \sum_{k=1}^{n} \left[(k\Delta t)|c_i(k\Delta t)|^2 \right] \tag{2.52}$$

As a last step a feature vector is constructed from the energy moments.

$$T_E = [E_1, E_2, \ldots, E_7] \tag{2.53}$$

These vectors also describe the energy distribution among the IMFs and reflect their change through time. To further enhance the vector's efficiency to track changes, the marginal spectrum of the IMFS is added to them as the maximum amplitude A_0 and the corresponding instantaneous frequency f_0:

$$T_E = [E_1, E_2, \ldots, E_7, A_0, f_0] \tag{2.54}$$

The differences between features are then used to distinguish between the states of the engine, including failures and normal regimes, making it possible to diagnose engine faults.

The proposed method was tested using a sample car with a fault generator reproducing valves, pistons, rods, crankshaft, timing belt and structure vibrations. A microphone was placed over the engine to record its operation at 2500 rotations per minute. The simulated 7 engine faults are: normal state without faults, a break or short in the circuit of throttle threshold sensor, break or short in the circuit of ignition coil, a break or short in the circuit of the Hall sensor, basic set errors in the throttle control unit, defect in the circuit of the first cylinder injector and defect in the circuit of the third cylinder injector. Every state was recorded 20 times, totaling in 140 signals which were used to train and validate the SVM models through 7 and 9 dimensional feature vectors. An iterative training algorithm has been used to train the SVM models by minimizing an error function. Based on the error function's form, two model types can be distinguished: C-SVM and nu-SVM. However results show that the C-SVM model outperforms the nu-SVM regardless of the dimensions of the feature vectors.

For testing purposes 70 measurements were used to evaluate the proposed method. Measurements included faulty and normal behavior and both 9 and 7 dimensional feature vectors were used to see which model is best suited for EFD. For a total of 56 engine states the 7 dimensional feature vector model classified correctly 80 % of the states, where the one using a 9 dimensional model was able to predict 91.43 % of it right. The difference between the accuracy can be acknowledged to the extra information carried by A_0 and f_0. The outcome is greatly influenced by the model's ability to distinguish between states, however faults 4 and 5 are sometimes confused with state 1 and fault 2. This can be attributed to the fact that faults 2 and 5 represent throttle problems which are hard to distinguish only by engine noise and fault 4 is a failure of the Hall sensor which has almost no effect on the acoustic properties of the engine, easily mistaken with the normal state 1. As a conclusion, the paper states that the noise based HHT SVM method can't be applied to all types of engine failures. Furthermore to enhance precision, a 5 state classifier has been proposed (omitting faults 4 and 5) which resulted in 96 % accuracy. To validate this new model a second set of measurements has been acquired resulting in a test set of 50 states. Again, the model recognized every state with an accuracy of 94–96 %. Thus, the HHT based 9 dimensional feature vectors together with multi-class SVMs for pattern recognition can be used to design noise based EFD with accuracy above 90 %. The approach itself can be extended to be used in different fields of engineering, i.e. machinery diagnostics, speech and image recognition.

2.7 Conclusions

We have presented both classical and new signal processing and artificial intelligence methods in the context of complex digital signal processing and feature extraction. Signal processing methods were revisited in a historical order, from the Fourier transform to the Hilbert-Huang transform, covering the whole interval of frequency and time-frequency analysis.

The Fourier and discrete Fourier transform are the most basic ones still being in use today. They are limited to offering information only on the frequency components but none on their time localization. In contrast, the short-time Fourier transform is capable of offering a basic insight into the time scale of frequency distributions, however, it is limited by the maximum achievable resolution in both time and frequency domains.

The wavelet transform comes to avoid this limitation of resolution by offering variable width analyzing windows, which can adapt to the time scales present in the analyzed signal. Still, it is the Hilbert-Huang transform which introduces the instantaneous frequency and offers information about the signal in each time point.

These methods by themselves are capable to decompose signals in different ways, but are not able to draw conclusions about them. Thus, integrating them in a hybrid system with artificial intelligence methods offers a robust solution to many signal processing problems. We have presented several hybrid systems having various uses: a combination between the wavelet transform and the fuzzy c-means clustering to aid the differentiation of ECG patterns on mobile devices. Another hybrid system consists of using the Hilbert-Huang transform and support vector machines together to perform engine-fault detection. The last presented method combines the power of wavelet packet decomposition and artificial neural networks in order to detect sleep stages, based on EEG, EMG and EOG signals.

In conclusion, the combination of intelligent methods with classical, well-known signal processing transforms yield robust feature extraction systems with applications in different technical and scientific fields, as the presented case studies have shown.

References

1. Kaiser, G.: A Friendly Guide to Wavelets. Birkhäuser (1994)
2. Cooley, J.W., Tukey, J.W.: An algorithm for the machine calculation of complex Fourier series. Math. Comput. **19**, 297–297 (1965)
3. White, S.: A simple FFT butterfly arithmetic unit. IEEE Trans. Circuits Syst. **28**, 352–355 (1981)
4. Johnson, S.G., Frigo, M.: A Modified split-radix FFT with fewer arithmetic operations. IEEE Trans. Signal Process. **55**, 111–119 (2007)
5. Megas, D., Serra-Ruiz, J., Fallahpour, M.: Efficient self-synchronised blind audio watermarking system based on time domain and FFT amplitude modification. Signal Process. **90**, 3078–3092 (2010)

6. Hillerkuss, D., et al.: Simple all-optical FFT scheme enabling Tbit/s real-time signal processing. Opt. Express **18**, 9324–9340 (2010)
7. Zhong, R., Huang, M.: Winkler model for dynamic response of composite caisson–piles foundations: seismic response. Soil Dyn. Earthquake Eng. **66**, 241–251 (2014)
8. Carbonaro, M., Nucara, A.: Secondary structure of food proteins by Fourier transform spectroscopy in the mid-infrared region. Amino Acids **38**, 679–690 (2010)
9. McRobbie, D.W., Moore, E.A., Graves, M.J., Prince, M.R.: MRI from Picture to Proton, 2nd edn. Cambridge University Press (2007)
10. Gabor, D.: Theory of communication. Part 1: the analysis of information. J. Inst. Electr. Eng. Part III: Radio Commun. Eng. **93**, 429–441 (1946)
11. Allen, R.L., Mills, D.: Signal Analysis: Time, Frequency, Scale, and Structure. Wiley, IEEE Press (2004)
12. Chikkerur, S., Cartwright, A.N., Govindaraju, V.: Fingerprint enhancement using STFT analysis. Pattern Recogn. **40**, 198–211 (2007)
13. Sherlock, B.G.: Fingerprint enhancement by directional Fourier filtering. IEEE Proc. Vision, Image, Signal Process. **141**, 87 (1994)
14. Mallat, S. Peyre, G.: A Wavelet Tour of Signal Processing: The Sparse Way, 3rd edn. Academic Press (2009)
15. Daubechies, I.: Ten lectures on wavelets. Soc. Ind. Appl. Math. (1992)
16. Rucka, M., Wilde, K.: Application of continuous wavelet transform in vibration based damage detection method for beams and plates. J. Sound Vib. **297**, 536–550 (2006)
17. Mallat, S., Zhong, S.: Characterization of signals from multi-scale edges. IEEE Pattern Anal. Mach. Intell. **14**, 710–732 (1992)
18. Rabbani, M., Joshi, R.: An overview of the JPEG2000 still image compression standard. Signal Process. Image Commun. 3–48 (2002)
19. Huang, N.E., Shen, Z., Long, S.R., Wu, M.C., Shih, H.H., Zheng, Q., Yen, N.C., Tung, C.C., Liu, H.H.: The empirical mode decomposition and the Hilbert spectrum for nonlinear and non-stationary time series analysis. In: Proceedings of the Royal Society London A, pp. 903–995 (1998)
20. Földvári, R.: Generalized instantaneous amplitude and frequency functions and their application for pitch frequency determination. J. Circuits, Syst. Comput. (1995)
21. Bedrosian, E.: A product theorem for hilbert transforms. Technical report, United States Air Force (1962)
22. Xu, Y., Yan, D.: The Bedrosian identity for the Hilbert transform of product functions. Proc. Am. Math. Soc. **134**, 2719–2728 (2006)
23. Huang, N.E., Wu, Z., Long, S.R., Arnold, K.C., Chen, X., Blank, K.: On instantaneous frequency. Adv. Adapt. Data Anal. **1**, 177–229 (2009)
24. Bedrosian, E., Nuttall, A.H.: On the quadrature approximation to the Hilbert transform of modulated signals. Proc. IEEE **54**, 1458–1459 (1966)
25. Szalai, J., Mozes, F.E.: An improved AM-FM decomposition method for computing the instantaneous frequency of non-stationary signals. In: Proceedings of the 2nd IFAC Workshop on Convergence of Information Technologies and Control Methods with Power Systems, pp. 75–79, May 2013
26. Tseng, Y.L., Ko, P.Y., Jaw, F.S.: Detection of the third and fourth heart sounds using Hilbert-Huang transform. BioMed. Eng. OnLine **11**, 8 (2012)
27. Szalai, J., Mozes, F.E.: T-Wave Detection Using the Empirical Mode Decomposition. Scientific Bulletin of "Petru Maior" University of Tirgu-Mures, **11**, 53–56 (2014)
28. Taouli, B.-R.F., S A.: Detection of QRS complexes in ECG signals based on Empirical Mode Decomposition (2011)
29. Goldberger, A.L., Amaral, L.A.N., Glass, L., Hausdorff, J.M., Ivanov, P.C., Mark, R.G., Mietus, J.E., Moody, G.B., Peng, C.K., Stanley, H.E.: PhysioBank, PhysioToolkit, and PhysioNet: Components of a New Research Resource for Complex Physiologic Signals. Circulation, **101**, e215–e220, circulation Electronic Pages: http://circ.ahajournals.org/cgi/content/full/101/23/e215. (2000) PMID:1085218; doi:10.1161/01.CIR.101.23.e215

30. Sadhukhan, D., Mitra, M.: ECG noise reduction using Fourier coefficient suppression. In: International Conference on Control, Instrumentation, Energy and Communication, pp. 142–146 (2014)
31. Martinez, J.P., Almeida, R., Olmos, S., Rocha, A.P., Laguna, P.: A wavelet-based ECG delineator: evaluation on standard databases. IEEE Trans. Bio-med. Eng. **51**, 570–581 (2004)
32. Huang, Z., Chen, Y., Pan, M.: Time-frequency characterization of atrial fibrillation from surface ECG based on Hilbert-Huang transform. J. Med. Eng. Technol. **31**, 381–389 (2009)
33. Anas, E.M.A., Lee, S.Y., Hasan, M.K.: Exploiting correlation of ECG with certain EMD functions for discrimination of ventricular fibrillation. Comput. Biol. Med. **41**, 110–114 (2011)
34. Chouvarda, I., Maglaveras, N., Boufidou, A., Mohlas, S., Louridas, G.: Wigner-Ville analysis and classification of electrocardiograms during thrombolysis. Med. Biol. Eng. Comput. **41**, 609–617 (2003)
35. Zhu, Y., Shayan, A., Zhang, W., Chen, T.L., Jung, T.-P., Duann, J.-R., Makeig, S., Cheng, C.-K.: Analyzing high-density ECG signals using ICA. IEEE Trans. Bio-med. Eng. **55**, 2528–2537 (2008)
36. Martis, R.J., Acharya, U.R., Mandana, K.M., Ray, A.K., Chakraborty, C.: Application of principal component analysis to ECG signals for automated diagnosis of cardiac health. Expert Syst. Appl. **39**, 11792–11800 (2012)
37. Park, J., Pedrycz, W., Jeon, M.: Ischemia episode detection in ECG using kernel density estimation, support vector machine and feature selection. Biomed. Eng. Online **11**, 30 (2012)
38. Bakul, G., Tiwary, U.S.: Automated risk identification of myocardial infarction using Relative Frequency Band Coefficient (RFBC) features from ECG. Open Biomed. Eng. J. **4**, 217–222 (2010)
39. Tseng, T.-E., Peng, C.-Y., Chang, M.-W., Yen, J.-Y., Lee, C.-K., Huang, T.-S.: Novel approach to fuzzy-wavelet ECG signal analysis for a mobile device. J. Med. Syst. 71–81 (2010)
40. Bezdek, J.C., Ehrlich, R., Full, W.: FCM: the fuzzy c-means clustering algorithm. Comput. Geosci. **10**, 191–203 (1984)
41. Iber, C., Ancoli-Israel, S., Chesson, A., Quan, F.: The AASM Manual for the Scoring of Sleep and Associated Events: Rules, Terminology and Technical Specification. American Academy of Sleep Medicine (2007)
42. Rechtschaffen, A., Kales, A.: A Manual Of Standardized Terminology, Techniques and Scoring Systems for Sleep Stages of Human Subjects. Washington DC Public Health Service (1968)
43. Ronzhina, M., Janousek, O., Kolarova, J., Novakova, J., Honzik, P., Provaznik, I.: Sleep scoring using artificial neural networks. Sleep Med. Rev. **16**, 251–263 (2012)
44. Flexer, A., Gruber, G., Dorffner, G.: A reliable probabilistic sleep stager based on a single EEG signal. Artif. Intell. Med. **33**, 199–207 (2005)
45. Berthomier, C., Drouot, X., Herman-Stoica, M., Berthomier, P., Prado, J., Bokar-Thire, D., Benoit, O., Mattout, J., D'ortho, M.P.: Automatic analysis of single-channel sleep EEG: validation in healthy individuals. Sleep **30**, 1587–1595 (2007)
46. Hsu, Y.L., Yang, Y.T., Wang, J.S., Hsu, C.Y.: Automatic sleep stage recurrent neural classifier using energy features of EEG signals. Neurocomputing **104**, 105–114 (2013)
47. Liang, S.F., Kuo, C.E., Hu, Y.H., Pan, Y.H., Wang, Y.H.: Automatic stage scoring of single-channel sleep EEG by using multiscale entropy and autoregressive models. IEEE Trans. Instrum. Meas. **61**, 1649–1657 (2012)
48. Fraiwan, L., Lweesy, K., Khasawneh, N., Wenz, H., Dickhaus, H.: Automated sleep stage identification system based on time-frequency analysis of a single EEG channel and random forest classifier. Comput. Methods Programs Biomed. **108**, 10–19 (2012)
49. Jo, H.G., Park, J.Y., Lee, C.K., An, S.K., Yoo, S.K.: Genetic fuzzy classifier for sleep stage identification. Comput. Biol. Med. **40**, 629–634 (2010)
50. Sukhorukova, N., et al.: Automatic sleep stage identification: difficulties and possible solutions. In: Proceedings of the 4th Australasian Workshop on Health Informatics and Knowledge Management, pp. 39–44 (2010)

51. Kerkeni, N., Alexandre, F., Bedoui, M.H., Bougrain, L., Dogui, M.: (2005) Neuronal spectral analysis of EEG and expert knowledge integration for automatic classification of sleep stages. CoRR. arXiv:0510083
52. Ebrahimi, F., Mikaeili, M., Estrada, E., Nazeran, H.: Automatic sleep stage classification based on EEG signals by using neural networks and wavelet packet coefficients. In: Conference Proceedings: ... Annual International Conference of the IEEE Engineering in Medicine and Biology Society. IEEE Engineering in Medicine and Biology Society. Annual Conference, 2008, pp. 1151–1154 (2008)
53. Wang, Y.S., Ma, Q.H., Zhu, Q., Liu, X.T., Zhao, L.H.: An intelligent approach for engine fault diagnosis based on Hilbert-Huang transform and support vector machine. Appl. Acoust. 1–9 (2014)
54. Boser, B.E., Guyon, I.M., Vapnik, V.N.: A training algorithm for optimal margin classifiers. In: Proceedings of the Fifth Annual Workshop on Computational Learning Theory—COLT'92, New York, USA, pp. 144–152. ACM Press, July 1992

Chapter 3
Multi-dimensional Data Clustering and Visualization via Echo State Networks

Petia Koprinkova-Hristova

Abstract The chapter summarizes the proposed recently approach for multidimensional data clustering and visualization. It uses a special kind of recurrent networks called Echo state networks (ESN) to generate multiple two-dimensional (2D) projections of the multidimensional original data. For this purpose equilibrium states of all neurons in the ESN are exploited. In order to fit the neurons equilibriums to the data an algorithm for tuning internal weights of the ESN called Intrinsic Plasticity (IP) is applied. Next 2D projections are subjected to selection based on different criteria in dependence on the aim of particular clustering task to be solved. The selected projections are used to cluster and/or to visualize the original data set. Several examples demonstrate possible ways to apply the proposed approach to variety of multidimensional data sets, namely: steel alloys discrimination by their composition; Earth cover classification from hyper spectral satellite images; working regimes classification of an industrial plant using data from multiple measurements; discrimination of patterns of random dot motion on the screen; and clustering and visualization of static and dynamic "sound pictures" taken by multiple randomly placed microphones.

Keywords Echo state network · Clustering · Multi-dimensional data

3.1 Introduction

In spite of numerous developments, clustering and visualization of multidimensional data sets is still a challenging task [15]. There are numerous approaches for solving it including intelligent techniques based on fuzzy logic and neural networks. The present work is focused on application of a special kind of neural

P. Koprinkova-Hristova (✉)
Bulgarian Academy of Sciences, Institute of Information
and Communication Technologies, Acad. G. Bonchev str. bl. 25A,
1113 Sofia, Bulgaria
e-mail: pkoprinkova@bas.bg

© Springer International Publishing Switzerland 2016   93
R. Kountchev and K. Nakamatsu (eds.), *New Approaches in Intelligent
Image Analysis*, Intelligent Systems Reference Library 108,
DOI 10.1007/978-3-319-32192-9_3

networks called Echo state networks (ESN). It is applied in combination with well known fuzzy subtractive clustering procedure. Since the main contribution in this work is in usage of ESN as a feature extracting techniques, the main focus here will be on their role in the overall algorithm for multidimensional data clustering and visualization.

ESN are a representative member of the so called "reservoir computing" approach that is a common name of extensively developing nowadays class of recurrent neural networks (RNN) [26]. The key idea of this approach was to mimic structures in human brain that seem to be composed by randomly connected dynamic non-linear neurons called reservoir whose output is usually linear combination of the current states of all reservoir neurons. Another advantage of such artificial structure is simplified training algorithm since only weights of the connections from the reservoir to the readout neurons are subject to training. Thus instead of gradient descent learning much faster least squares method can be used.

Although the reservoir connections and their weights are randomly generated, in order to prevent improper behavior of such networks Prof. Jaeger, one of the pioneer scientists in this area of research, formulated the rule that reservoir has to have "echo state property" [14]. The basic rule formulation is: the effect of input disturbances should vanish gradually in time that means the dynamic reservoir must be stable. The usual recipe is to generate a reservoir weight matrix with spectral radius below one. However as it was mentioned [26] this condition will not guaranty ESN stable behavior in general so varieties of task-dependent recipes for improvement of reservoir connections were proposed.

Since one of the laws of thermodynamics says that any stable stationary state has a local maximum of entropy [11], it can be expected that maximization of entropy at the ESN reservoir output could increase its stability. This motivated several works proposing ESN reservoir improvement by its entropy maximization [28]. Other authors proposed biologically motivated algorithm called Intrinsic Plasticity (IP) based on mechanisms of changing neural excitability in response to the distribution of the input stimuli [30, 31]. In [16] it was shown that in fact IP training achieves balance between maximization of entropy at the ESN reservoir output and its concentration around the pre-specified mean value increasing at the same time reservoir stability. During investigations in [16] another interesting effect was observed: the reservoir neurons equilibrium states were concentrated in several regions. Then question aroused: is it possible to use this effect for clustering purposes too? This initiated development of the proposed here algorithm for multidimensional data clustering and visualization.

Since ESN are dynamic structures designed initially for time series modeling, using them for static data clustering might seem odd. However the idea for using RNNs in this way is not new. There are examples in the literature like neural systems possessing multi-stable attractors [6] that perform temporal integration aimed at discrimination between multiple alternatives. In other works [1, 13] unsupervised learning procedures that minimize given energy function were proposed aiming at achievement of network equilibrium states that reflect given data structure.

Concerning ESN applications for clustering, there are only few works available. In [32] it was proposed for the first time to use ESN at feature extraction stage of image classification. Their role was to "draw out" silent underlying features of the data to be used further to train a feedforward neural network classifier. In [29] the idea to exploit equilibrium states of the ESN reservoir in order to design multiple-clusters ESN reservoirs was proposed. It was inspired by complex network topologies imitating cortical networks of the mammalian brain. In [25] it was reported that using another kind of IP algorithm in combination with Spike-time Dependent plasticity (STDP) of synaptic weights changes the connectivity matrix of the network in such way that the recurrent connections capture the peculiarities of the input stimuli so that the network activation patterns can be separated by an unsupervised clustering technique.

The idea described in this chapter was motivated initially from stability analysis of ESN and proposed for the first time in [17]. It exploits more or less the same reservoir properties reported by other works but looking from slightly different view point: to consider combinations between steady states of each two neurons in the reservoir as numerous two-dimensional projections of the original multidimensional data fed into the ESN input; next to use these low dimensional projections for data clustering and/or visualization of multidimensional original data. The proposed in [17] initial algorithm was tested and developed further using different data sets [18–23] whose clustering purposes were task dependent. This lead to formulation of new methodology for clustering and visualization of multidimensional data using IP tuning of ESN reservoirs described further in the present work.

The chapter is constructed as follows: next section describes basics of ESN structure, algorithm for IP tuning of reservoir and its effect on the equilibrium states of neurons that motivate described further clustering algorithm; third section contains examples on how the proposed algorithm can be applied to variety of multidimensional data sets, namely steel alloys discrimination by their composition; Earth cover classification from hyper spectral satellite images; working regimes classification of a industrial plant using multiple measurements data; discrimination of patterns of random dot motion; and clustering and visualization of static and dynamic "sound pictures" taken by multiple randomly placed microphones.

3.2 Echo State Networks and Clustering Procedure

3.2.1 Echo State Networks Basics

ESNs are a kind of recurrent neural networks that arise from so called "reservoir computing approaches" [26]. Their basic structure, presented on Fig. 3.1 below, consists of a reservoir of randomly connected dynamic neurons with sigmoid nonlinearities f^{res} (usually hyperbolic tangent):

Fig. 3.1 Echo state network
basic structure

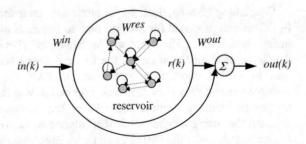

$$r(k) = f^{res}\big(W^{in}in(k) + W^{res}r(k-1)\big) \qquad (3.1)$$

and a linear readout f^{out} (usually identity function) at the output:

$$out(k) = f^{out}\big(W^{out}[\,in(k) \quad r(k)\,]\big) \qquad (3.2)$$

Here k denotes discrete time instant; $in(k)$ is a vector of network inputs, $r(k)$—a vector of the reservoir neurons states and $out(k)$—a vector of network outputs; n_{in}, n_{out} and n_r are the dimensions of the corresponding vectors in, out and r respectively; W^{out} is a trainable $n_{out} \times (n_{in} + n_r)$ matrix; W^{in} and W^{res} are $n_r \times n_{in}$ and $n_r \times n_r$ matrices that are randomly generated and are not trainable. In some applications direct connection from the input to the readout is omitted.

The key idea is that having rich enough reservoirs of nonlinearities will allow to approximate quite complex nonlinear dependence between input and output vectors by tuning only the linear readout weights. Hence the training procedure is simplified to solving in one step Least Squares task [14].

Although this idea seems to work surprisingly well, it appears that initial tuning of reservoir connections to the data that will be fed into the ESN helps to improve its properties. In [30, 31] was proposed a reservoir tuning approach called "intrinsic plasticity" (IP). It is aimed at maximization of information transmission trough the ESN that is equivalent to its output entropy maximization. Motivation of this approach is related to known biological mechanisms that change neural excitability according to the distribution of the input stimuli. The authors proposed a gradient method for adjusting the biases and an additional gain term aimed at achieving the desired distribution of outputs by minimizing the Kullback-Leibler divergence:

$$D_{KL}(p(r), p_d(r)) = \int\limits_{-\infty}^{+\infty} p(r) \ln\!\left(\frac{p(r)}{p_d(r)}\right) dr \qquad (3.3)$$

That is a measure for the difference between the actual $p(r)$ and the desired $p_d(r)$ probability distribution of reservoir neurons output r. Since the commonly used transfer function of neurons is the hyperbolic tangent, the proper target distribution that maximizes the information at the output according to [30] is the Gaussian one with prescribed small variance σ and zero mean μ:

$$p_d(r) = \frac{1}{\sigma\sqrt{2\pi}} \exp\left(-\frac{(r-\mu)^2}{2\sigma^2}\right) \tag{3.4}$$

Hence Eq. (3.3) can be rearranged as follows:

$$D_{KL}(p(r), p_d(r)) = \int\limits_{-\infty}^{+\infty} p(r)\ln p(r)dr - \int\limits_{-\infty}^{+\infty} p(r)\ln p_d(r)dr \tag{3.5}$$

$$= -H(r) + \frac{1}{2\sigma^2}E\left((r-\mu)^2\right) + \ln\frac{1}{\sigma\sqrt{2\pi}}$$

where $H(r)$ is entropy, the last term is constant and the second one determines the deviation of the output from the desired mean value. Thus minimization of (3.5) will lead to compromise between entropy maximization and minimization of distance between μ and r.

In order to achieve those effects two additional reservoir parameters—gain a and bias b (both vectors with n_r size)—are introduced as follows:

$$r(k) = f^{res}\left(diag(a)W^{in}in(k) + diag(a)W^{res}r(k-1) + b\right) \tag{3.6}$$

The IP training is procedure that adjusts vectors a and b using gradient descent.

3.2.2 Effects of IP Tuning Procedure

Theoretical investigations of the effect of the IP improvement procedure on the dynamic properties of reservoir neurons [16] revealed the following two facts (presented graphically on Fig. 3.2 below):

Fig. 3.2 Effect of IP tuning on working region of reservoir neurons

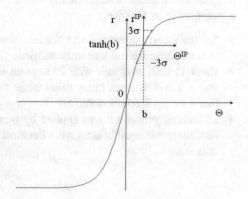

- Due to non-zero bias terms the origin of coordinate system of reservoir outputs with respect to its total input Θ will be moved according to the achieved after IP tuning bias b;
- Since in [30] it was recommended to use Gaussian distribution with zero mean, after IP tuning the reservoir outputs will be squeezed into the interval $[-3\ \sigma, 3\ \sigma]$ of its new coordinate system.

Here capital theta denotes the total input to reservoir neurons, i.e. following the Eqs. (3.1) and (3.6):

$$\Theta = W^{res} r + W^{in} in, \quad \Theta^{IP} = diag(a)\Theta + b \qquad (3.7)$$

Hence we can suppose that overall IP training will lead to some rearrangement of the reservoir neurons equilibrium states [16] r_e accounting for the input data distribution.

In the case of a constant input in_c:

$$r_e = \tanh\left(diag(a)W^{in}in_c + diag(a)W^{res}r_e + b\right) \qquad (3.8)$$

So if $b = 0$ and $in_c = 0$, the equilibrium will be at the origin of the reservoir state space coordinate system. Otherwise it will be moved in dependence on the values of input in_c and bias b vectors. Since the input weights matrix remains constant, the first term in the brackets will be also constant for constant inputs. Thus we can consider it together with the bias term as common bias for a given input vector:

$$r_e = \tanh(diag(a)W^{res}r_e + b_{in_c}), \quad b_{in_c} = diag(a)W^{in}in_c + b \qquad (3.9)$$

Hence the reservoir equilibrium will be different for different input vectors. Moreover, if the input vectors are close in the input space, they will result in close equilibrium points of the reservoir state—a fact that could be exploited for clustering purposes.

The above considerations motivated the experiment with a simple "toy example" described below. It could be extended to multi-dimensional spaces but in order to be able to visualize results clearly our example is three dimensional one. Our experiment is as follows:

- Several clearly separated data clusters (shown on Fig. 3.3 below) were generated in three dimensional unit cube space.
- Random ESN reservoir with 10 neurons was generated and each 3D data point was fed into its input many times while the reservoir achieves corresponding to this data equilibrium state.
- IP training procedure was applied by presenting all generated 3D data and the new reservoir equilibriums were determined for each data point from the input data set.

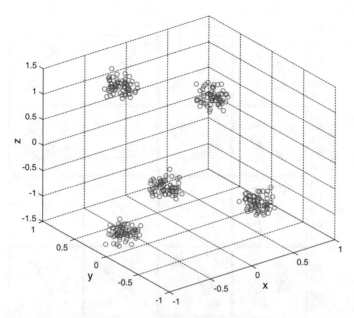

Fig. 3.3 Toy example with five clearly separated clusters of *dots* in 3D space

- Two dimensional plots of all possible combinations between reservoir neurons equilibriums scaled within interval [−1 +1] were generated in order to see if there is some difference before and after IP tuning.

Figure 3.4 presents an example of 2D plots for all possible combinations between several chosen neurons from our ESN reservoir before and after its IP tuning. It is clearly seen that before IP training equilibrium points in two dimensional state spaces are not clearly separated into different clusters. However, after IP training they appeared separable in many of the 2D projections.

The above considerations were transferred further to dynamic input to the ESN [21]. The reservoir output after presenting given time-varying input *in(k)* for all the time steps $k = 0 \div n - 1$ is determined by the following recursive calculation:

$$
\begin{aligned}
r(n) &= f^{res}\big(diag(a)W^{in}in(n-1) + diag(a)W^{res}r(n-1) + b\big) \\
r(n-1) &= f^{res}\big(diag(a)W^{in}in(n-2) + diag(a)W^{res}r(n-2) + b\big) \\
&\cdots \\
r(1) &= f^{res}\big(diag(a)W^{in}in(0) + diag(a)W^{res}r(0) + b\big)
\end{aligned}
\tag{3.10}
$$

Obviously the final reservoir state *r(n)* will depend on the specific characteristics of the time series *in(k)*, $k = 0 \div n - 1$, presented on its input as well as on the "tuned" working region of the reservoir after IP procedure. Hence this property

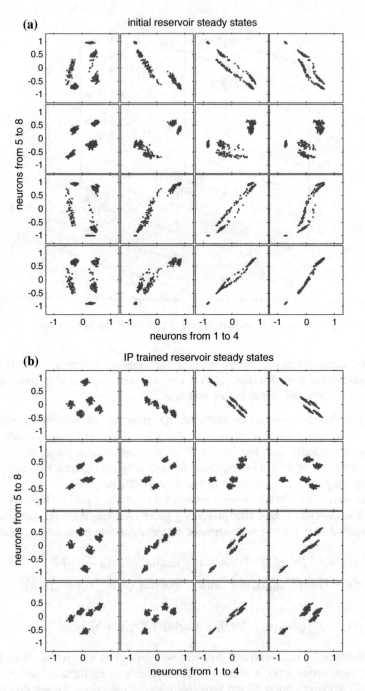

Fig. 3.4 Scaled 2D projections of equilibrium states of several reservoir neurons before (**a**) and after (**b**) IP training

could be exploited for clustering of time series $in(k) = F(k, *)$, $k = 0 \div n - 1$, where $F(k, *)$ could be some nonlinear function of time and other task-specific variables (denoted by *).

3.2.3 Clustering Algorithms

We can consider each couple of neurons equilibriums as a two-dimensional projection of the original multidimensional input space of data taken from different view point. It is obvious that not all possible combinations of two neurons outputs give the same clear picture as can be seen from the figures above. Hence, question arises how to choose a proper 2D projection among the variety of different views?

The initial decision [17] was the following: since the IP training forces reservoir output to distribute according pre-specified Gaussian distribution, we decided to observe the obtained equilibrium states distributions. Left part of Fig. 3.5 shows probability density distributions of equilibrium states of ten chosen neurons of the IP tuned reservoir from our toy example above. As it can be seen, each neuron output distribution is combination of several Gaussian distributions. Stars on the figure mark local maxima on the distribution curves that correspond to the local Gaussian distribution and are found using software from [8]. Hence we can suppose that neurons with bigger number of maxima separate data into bigger number of clusters. So if we choose two dimensional projections formed by neurons with biggest number of probability distribution maxima, we can obtain clearest separation of data. On the left side of Fig. 3.5 is given chosen by this approach 2D projection that obviously separates clearly our initial 3D data into five clusters.

However during testing of this approach [18–23] it was discovered that finding of a proper 2D projection is not always possible. The next step was to test whether two-dimensional density distributions [5, 9] will work better. The chosen in [17] 2D projection according to maximal number of local maxima in 2D density distribution is shown on the right side of Fig. 3.6 below. It is compared with a projection of the same data chosen to have maximal number of clusters on the left side of that figure.

The black dots on these figures represent clusters' centers determined by subtractive clustering procedure [18]. Different classes are marked by different colors. It was observed that using 2D density distributions chooses a 2D projection having equilibrium states along all interval $[-1, 1]$ but grouped into a narrow band. This led to separation of the points into smaller number of classes. In contrast it appears that there are other 2D projections that concentrate the equilibriums in a circle at the middle of this interval having smaller local maxima of the 2D density distribution of points that however reveal bigger number of clusters.

Hence a proper choice of 2D projection has to be task dependent:

- If we don't know in advance the possible number of classes in our multidimensional data and our purpose is to discover as much as possible clusters, a proper strategy will be to cluster the points of all possible 2D projections and to pick up those with maximum clusters revealed;

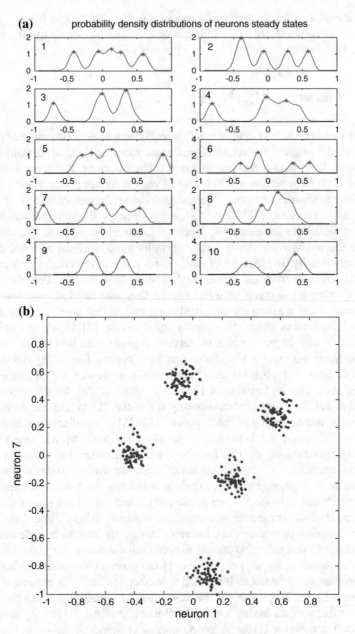

Fig. 3.5 Probability density distributions of all 10 neurons in the reservoir (**a**) with local maxima marked by *stars* and chosen two dimensional projection (**b**)

Fig. 3.6 Chosen 2D projection according to maximum number of clusters (**a**) and maximum number of local maxima in 2D density distribution (**b**) form [18]

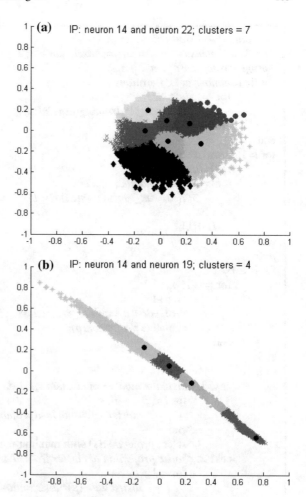

- If we have some idea about the number of classes in our multidimensional data set, then a proper choice can be related to one or two dimensional density distributions or to those projections that reveal the proper number of classes.

In any case, for real multidimensional data the choice of proper 2D projection is not straightforward task because data clusters could not be as easily distinguished as in our firs toy example above. Hence a procedure for clustering of obtained 2D projections is still needed. Since the subtractive clustering procedure [33] was reported as one of the best options in the case of unknown number of clusters [12], we decided to use it at this second step of our clustering approach.

In summary, the proposed here algorithm for 2D visualization and clustering of multidimensional data sets has two task-dependent branches. A description of the algorithm is presented on Fig. 3.7 in Matlab program-like code.

```
in(1:data dimension,1:data size);
n_in=data dimension; n_out=1; n_r=chosen number;
esn=generate_esn(n_in, n_out, n_r);
for it=1:number of IP iterations
        for i=1:data size
                    esn=esn_IP_training(esn, in(:,i));
        end
end
for i=1:data size
        r(0)=0;
        for k=1:chosen number of steps
                    r(k)=sim_esn(esn, in(:,i),r(k-1));
        end
        r_e(i)=r(k);
end
p=0;
for i=1: n_r-1
        for j=i+1: n_r
                    p=p+1;
                    projection(p)=create_projection( r_e(i), r_e(j));
                    visualize projection(p);
        end
end
switch
        case 1: maximum number of clusters needed
                for i=1:p
                            clusters(p)=subclust(projection(p));
                end
                select projection(s) with maximum number of clusters;
        case 2: choose projection to cluster from distribution (1D or 2D)
                for i=1:p
                            distribution(p)=distribution(projection(p));
                end
                select projection(s) having distribution with most maxima;
                clusters(p)=subclust(projection(p));
        case 3: cluster into known number of classes nclass
                for i=1:p
                            clusters(p)=cluster(projection(p), nclass);
                end
                select projection(s) having proper matching of classes;
end
```

Fig. 3.7 Algorithm for multidimensional data clustering and 2D visualization

3.3   Examples

In this section the main results from clustering and visualization of different multidimensional data sets are summarized. They demonstrate variety of practical applications of the proposed above algorithm on wide range of real data. In some of the examples proper preprocessing of original data was done as a first stage of features extraction that depends on their specific characteristics. Since the main focus of the chapter is on ESN in clustering and visualization, the details of this preprocessing are only mentioned. More information about this data-dependent stage can be found in cited references.

3.3.1   Clustering of Steel Alloys in Dependence on Their Composition

The first test of the proposed algorithm [17] was done on a real data set that contains information about 91 steel alloy compositions. Each data point consists of concentrations of the three main alloying elements: carbon (C), silicone (Si) and manganese (Mn) in percents (%). According to knowledge of the experts in the field the steel alloys can be separated into three groups in dependence on concentrations of Si and Mn. Tables 3.1 and 3.2 below summarize the data of the two smaller data groups. The rest of 91 data belong to the third biggest cluster.

The part (a) of Fig. 3.8 presents all 91 data points in the 3D space. The red squares correspond to the data from Table 3.1, the green circles—to the data from Table 3.2 and the blue circles—to the third data cluster.

The part (b) of the Fig. 3.8 presents chosen by our procedure accounting for density distributions (case 2 of the algorithm) of equilibrium states of each neuron in 2D projections obtained after IP training of ESN. The red (squares), blue (dots) and green (circles) marks correspond to the data from the three clusters separated in the (a) part of the figure.

Black stars (a) and squares (b) represent the clusters centers obtained by subtractive fuzzy clustering procedure. Original data are separated into 4 clusters while projected once—in 3 clusters that correspond better to the logical separation of our data set, although the red squares cluster center is moved towards the blue dots cluster due to restricted number of data in the red squares cluster.

Table 3.1 Class one (marked by 5 red squares): Mn ≥ 1.6 %	No	C, %	Si, %	Mn, %
	1	0.35	0.27	1.6
	2	0.45	0.27	1.6
	3	0.305	0.27	1.6
	4	0.36	0.27	1.75
	5	0.4	0.27	1.6

Table 3.2 Class two (marked by 10 green circles): Si ≥ 1.05

No	C, %	Si, %	Mn, %
1	0.355	1.25	0.95
2	0.41	1.4	0.45
3	0.2	1.05	0.95
4	0.315	1.05	0.95
5	0.38	1.2	0.45
6	0.25	1.05	0.95
7	0.31	1.05	0.95
8	0.33	1.2	0.45
9	0.305	1.05	1.15
10	0.34	1.25	0.9

It should be noted that by proper choice of parameters of subtractive clustering algorithm the results could be refined even in case of clustering of original 3D data set. However, since the aim of present investigation was to compare clustering results before and after 2D projection of the original data set, in both cases the same parameters of the subtractive clustering procedure were used.

3.3.2 Clustering and Visualization of Multi-spectral Satellite Images

Multispectral data of eight spectral bands with a spatial resolution of 30 m from Landsat 7 Enhanced Thematic Mapper Plus (ETM +) instrument [24], presented in Table 3.3, are used in this example [18, 19].

Two ESNs with different reservoir size—20 and 100 neurons—were adjusted to these data using IP tuning algorithm. The size of the ESNs input vector is $n_{in} = 8$ following the number of spectral images. Each image has size of 50 × 50 pixels. Hence our data set contains 2500 input vectors of size n_{in} each.

Tables 3.4 and 3.5 represent clustering results obtained by case 1 and case 2 branches of the proposed algorithm. In case 2 2D density distributions of equilibrium states of reservoir neurons were used.

As can be seen, case 1 selects projections with bigger number of clusters in comparison with selections of case 2 for both reservoir sizes. While the case 1 selects only one projection among all possibilities, case 2 gives us four options in the case of bigger reservoir size. In the case of bigger reservoir size obtained clusters number increases, i.e. the preprocessing of multi-dimensional data and creation of many 2-dimensional projections allows us to reveal more detailed classification of the multi-dimensional data set.

The obtained in this way 2D visualization of multispectral data was compared with an orthophoto map of the observed region, GIS classification (CORINE 2000) from Bulgarian Ministry of Regional Development and Public Works [27] and

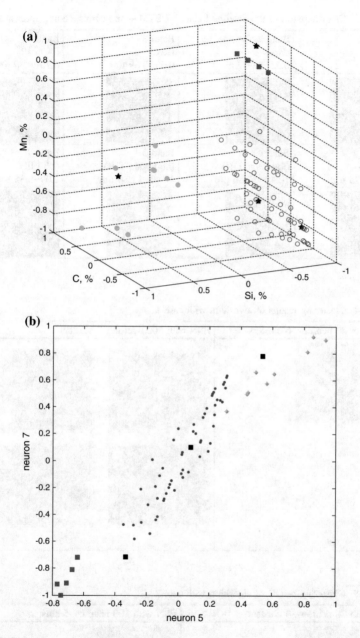

Fig. 3.8 Three-dimensional presentation of 91 steel compositions (**a**) and their separation into three clusters (**b**)

Table 3.3 Characteristics of the Satellite Landsat 7 ETM + and corresponding spectral images

Band 1 (Blue) 0.45−0.52 μm	Band 2 (Green) 0.52−0.60 μm	Band 3 (Red) 0.63−0.69 μm	Band 4 (Near IR) 0.76−0.90 μm
Band 5 (Mid-IR) 1.55−1.75 μm	Band 6 (low gain TIR) 10.40−12.50 μm	Band 6 (high gain TIR) 10.40−12.50 μm	Band 7 (Mid-IR) 2.09−2.35 μm

Table 3.4 Clustering results of algorithm with case 1

ESN with 20 neurons, 6 clusters	ESN with 100 neurons, 7 clusters

Table 3.5 Clustering results of algorithm with case 2

ESN with 20 neurons, 4 clusters	ESN with 100 neurons, 5 clusters

Table 3.6 Comparative maps

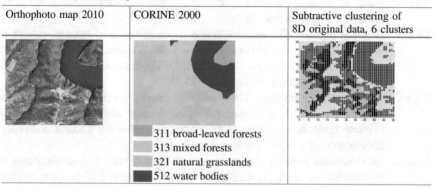

Orthophoto map 2010	CORINE 2000	Subtractive clustering of 8D original data, 6 clusters
	311 broad-leaved forests 313 mixed forests 321 natural grasslands 512 water bodies	

clustering results obtained by subtractive clustering of the original 8-dimensional data set. All comparative maps are given in Table 3.6.

Due to different data scales comparison could not be exact. We can say only that CORINE map separates the observed area into 4 classes—result closer to the one obtained by our algorithm using case 2 and ESN with 20 neurons. In all other cases our algorithm reveals bigger number of clusters. Looking at the orthophoto map with higher spatial resolution of 0.5×0.5 m, although it is hard to correlate exactly each one of the obtained clusters to different types of land covers, we suppose that this classification is able to distinguish more specific features form the multi-spectral data.

3.3.3 Clustering of Working Regimes of an Industrial Plant

The apparatus considered here is a mill fan for fuel preparation in the coal fired power plant Maritsa East 2—a thermal power plant in Bulgaria that is the largest on the Balkan Peninsula. The mill fans are used to mill, dry and feed the coal to the burners of the furnace chamber. The part which suffers the most and should be taken care of is the rotor of the mill fan. Because of the abrasive effect of the coal it wears out and should be repaired by welding to add more metal to the worn out blades. The possibility to predict eventual damages or wear out without switching off the device is significant for providing faultless and reliable work of the equipment avoiding incidents.

In [7, 20] the data archived with 1 min time step by the installed on the site DCS covering the observation periods before and after replacement of the rotor were investigated. The monitored variables are: discharge temperature of the dust-air mixture, vibrations of the nearest to the mill rotor bearing block, load (current) at the system output and the corresponding control actions. Looking at variables tendencies before and after replacement and accounting for expert information available the following major working regimes were distinguished:

Case 0   (stop or manual control regime): the maximum density distribution of control action is below 50 %

Case 1   (starting regime): The maximum density distribution of rotor vibrations is around 2.5 mm, the maximum density distribution of dust-air mixture temperature—around 170−190 °C whiles the maximum density distribution of control action—around 60−75 %

Case 2   (stable working regime): The maximum density distribution of vibrations is around 2.5−3.5 mm, the maximum density distribution of dust-air mixture temperature—around 170−180 °C whiles the maximum density distribution of control action—around 75 %

Case 3   (deterioration regime): The maximum density distribution of vibrations is around 3.5−4.5 mm, the maximum density distribution of dust-air mixture temperature—around 150−170 °C whiles the maximum density distribution of control action—around 75−90 %

Here subject of clustering are 3D data containing the three measured variables. Figure 3.9 presents the processed data and the corresponding to them cluster (case) numbers obtained by proposed algorithm, case 2 using 1D density distributions of reservoir equilibrium states. The revealed regimes are logical since at the beginning we had data before rotor replacement, while after sample number 480 the mill fan

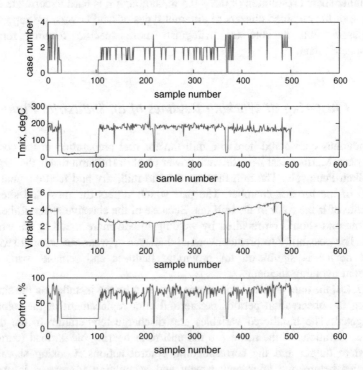

Fig. 3.9 Three dimensional input data and corresponding to them cluster (case) numbers obtained from our clustering procedure

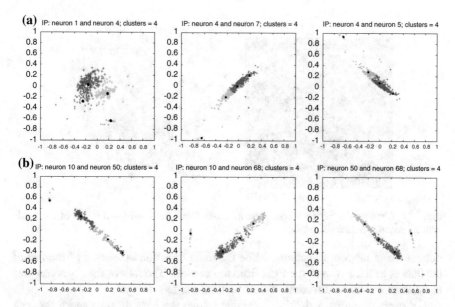

Fig. 3.10 Selected 2D projections from equilibrium states of the ESN reservoir with 10 (**a**) and 124 (**b**) neurons

rotor was replaced and starting regime was initiated. Stopping regimes (from 20 up to 100 sample number, about 130, 220, 280 and 380 sample number) as well as deterioration regime shortly before rotor replacement (after 380 sample number) are also clearly distinguished.

Investigations were done using ESN reservoirs from 10 up to 124 neurons. Some of the selected 2D projections are shown on Fig. 3.10 below. All of them distinguish 4 classes as it was expected by the experts. Hence for this particular data set increasing of ESN reservoir size does not change the number of revealed clusters for all chosen 2D projections. However, it was observed that there are some input vectors that were classified in different clusters by different 2D projections no matter how big or small was ESN reservoir. Hence the choice of a proper 2D projection in this particular case should relay to available experts' assessment.

3.3.4 Clustering of Time Series from Random Dots Motion Patterns

The present data set is received in a study that investigates the sensitivity to motion direction of dynamic stimuli [2, 4]. The stimuli presented to tested people are movies consisting of 25 consecutive frames showing 48 randomly moving dots. An example of a frame is given on the left side of Fig. 3.11. The fused image of all 25 frames of a movie is given on right side of the same figure. A fixed proportion of

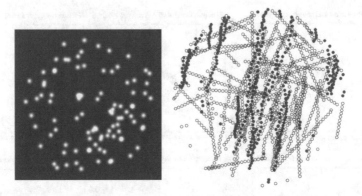

Fig. 3.11 Random dot motion screen presenting one frame (*left*) and fused image of a scenario with 25 consecutive frames (*right*)

dots move in random directions, while the main direction of motion of the rest of the dots is to the left or to the right from the vertical. The task of the observers was to indicate the mean direction of dots motion, i.e. left or right.

For every stimulus with 25 consecutive frames we have 48 time series for each dot. From them we obtain transformed time series consisting of 24 angles of motion direction at each time step. The task to be solved here is to cluster motion patterns of these dynamic data series. Hence instead of reservoir equilibrium states here the reservoir states (as in Eq. (3.10)) achieved after presentation of each input time series are used as features.

In [21] preprocessing of this data was done following the model of human visual perception from [3, 10], namely we used the receptive fields of MST neurons as in [3] to preprocess time series of our motion directions data. In [21] seven receptive fields distributed randomly in the range of moving angles between $-\pi/2$ and $+\pi/2$ were used. Finally, at the output of each one of the seven receptive fields we have a time series consisting of 24 time steps for each one of the experiments, an example of which is presented in Fig. 3.12. These are the dynamic 7-dimensional time series data that was dynamic input to our ESN clustering procedure.

Hence our feature extraction ESN has 7 inputs. Experimental data set contained 3599 successive human trails. Reservoirs with different size starting from 10 neurons up to 100 neurons were tuned and tested. The clustering procedure uses case 1, i.e. it chooses a two dimensional projections with biggest number of clusters. Since increasing of reservoir size led to discrimination of bigger number of clusters and because with these data it is known that real number of clusters should be two or three (the human decision has three classes: left, right and unclear), it become obvious that procedure needs some adjustment to the task peculiarities. Then it was decided to investigate the number of two dimensional projections that cluster the data into two, three, four etc. clusters. In this way an interesting behavior was discovered: the majority of two dimensional projections have only three clusters. Figure 3.13 presents the bar chart containing the number of projections with different number of clusters for 10, 30, 50 and 100 neurons of ESN reservoir.

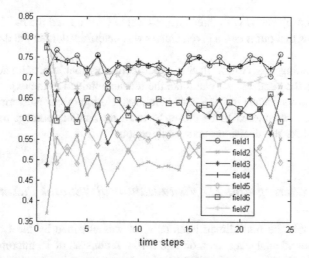

Fig. 3.12 Receptive fields outputs

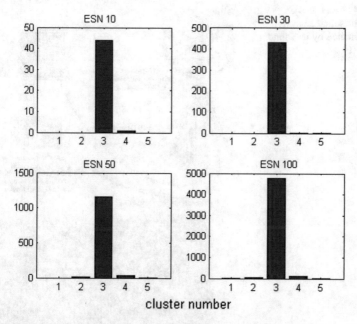

Fig. 3.13 Number of projections with respective number of clusters

The possible explanation is that this reflects the fact that three clusters are closer to human perception. Then a proper idea is to choose among projections with prevailing number of clusters. That is why the next decision was to use this as a "voting" mechanism: for each dynamic data item decisions form all two

dimensional projections with three clusters were collected and the maximal number
of projections that put it into a given cluster was determined; then the data are put to
that cluster.

An interesting observation in this experiment was that approximately 30 % of
data fall into the "undecided" class like the results obtained in the experiments with
human decisions. Hence the proposed here modification of our clustering algorithm
could be also considered to be closer to the way humans percept motion information and take the decision about its direction.

3.3.5   Clustering and 2D Visualization of "Sound Pictures"

In this example the multidimensional data set was obtained by the Brüel & Kjær
system for sound analysis shown on Fig. 3.14a. It consists of 18 microphones array
placed randomly in a wheel grid (called antenna) at which center is mounted a

Fig. 3.14 Brüel & Kjær
system for sound analysis
(**a**) and created by it "sound
picture" (**b**)

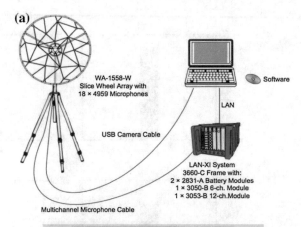

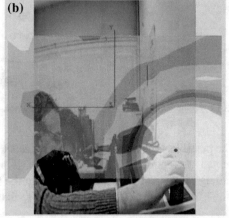

camera. All microphones are connected to a front-end panel and finally both camera and front-end are connected to a computer with software for sensor information processing. The system measures acoustic pressure and visualizes "sound picture" of the observed by camera area as it is shown on Fig. 3.14b.

Considered here multidimensional data set consists of raw measurement data from all 18 microphones of the antenna. A piezo beeper WB 3509 (standard Brüel & Kjær equipment—the red box in the right low corner on the picture on the right side of Fig. 3.14, with frequency of 2.43 kHz was used as sound source. After switching on the beeper the system collects acoustic pressure in Pa for 15.9 ms— period of time predetermined by the system software—from all 18 microphones. The measurements were taken with time step $1.53 * 10^{-5}$ s. The collected data, shown on Fig. 3.15, are periodic signals with variable amplitude and constant frequency of the noise source (the beeper).

The present task needs to consider collected data as "sound picture" so that to be able to map further it with the picture taken from the camera and to determine position of sound sources. Hence it was decided to divide the antenna area (area of stimuli collection) into 16 overlapping square regions shown on Fig. 3.16 (each region is surrounded by a square with different color). The small numbered dots represent microphones positions and the big dot in the center marks camera position. Regions are determined so that each of them contains at least one microphone (e.g. microphone 5 is the only one in upper right region while the maximal number of microphones in region is four and is situated at the center of antenna—region containing microphones 1, 3, 8 and 7).

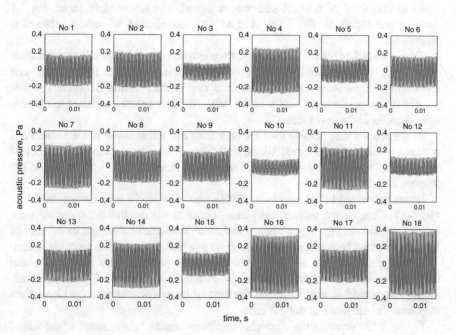

Fig. 3.15 Acoustic pressure data collected from 18 microphones for 15.9 ms

Fig. 3.16 Regions positions
at the antenna area

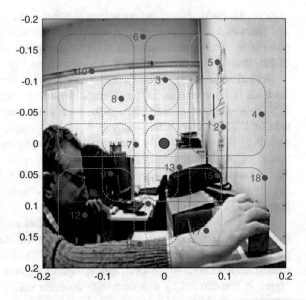

Here preprocessing of raw signals was done in a similar way as it was described
in previous section using the receptive fields of MST neurons. For this purpose the
dynamic range of raw data (that is from −0.4 to 0.4 Pa) was divided into 11
intervals. For each interval a filter with center at the center of interval and variance
equal to one third of interval size was assigned. The obtained in these way 11
features are inputs to the ESN used at the second step of feature extraction
procedure.

Two approaches for the first stage feature extraction were applied: accumulation
of acoustic pressure for all period of measurements (called further 2D) and
accounting for measurements at each time step (called further 3D). Thus it was
possible to prepare a final picture of the "observed" by antenna noise or to observe
time changes in "sound picture".

At the second step of described above feature extraction algorithm we used ESN
reservoirs with different sizes: 10, 30 and 50 neurons. In all cases the number of
inputs of ESN was determined by the number of features, i.e. 11 according to the
number of receptive fields.

Results obtained in 2D case using ESN reservoir with 50 neurons are shown on
Fig. 3.17. Comparison with sound picture obtained via original producer software
on the right side of Fig. 3.14 reveals that the sound source on the right low corner of
the picture could be detected since it belongs to a cluster that differs from the others.

It was observed that acoustic pressure data is periodic with period of about
0.412 ms or approximately 28 time steps. Hence time changes of "sound picture"
during one period as well as for all the time of measurements with 0.412 ms time
step were investigated in the 3D case.

Figure 3.18a presents the clustering results observing "movement" of the sound
wave coming from the noise source through receptive fields for the first period of

Fig. 3.17 Accumulated sound picture clustering results

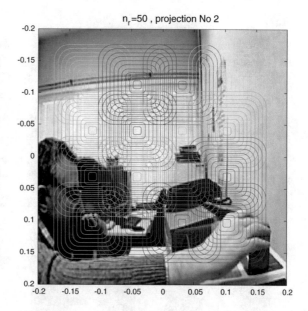

time using ESN with 50 neurons in the reservoir. It can be seen that the last picture (beginning of new period) is the same as the first one, i.e. the clustering reveals periodical characteristics of data.

On Fig. 3.18b time changes of "sound picture" during all time of measurements obtained by the ESN reservoir with 50 neurons are shown. The "unfolded" 3D picture reveals changes in the acoustic pressure amplitude with time. Although all pictures are from the beginning of current period, they gradually change from the beginning of the measurements to their end. This change can be explained by inexact correspondence of sampling frequency and beeper frequency.

In spite of roughness of our sensing fields, the position of the beeper can be exactly estimated in all pictures. In 3D case (accounting for measurements at each time step) number of obtained clusters was 3 or 4 while in 2D case it was about 6 clusters.

3.4 Summary of Results and Discussion

Table 3.7 presents a summary of all five test data sets and their clustering procedure parameters. The size of ESN input vector n_{in} corresponds to the size of features vector that corresponds directly to the size of the raw data set (data sets 1, 2 and 3) or was extracted at the first stage of raw data preprocessing (data sets 4 and 5). Hence there are examples varying from 3-dimensional data sets (1 and 3) through 11-dimensional one in the last example.

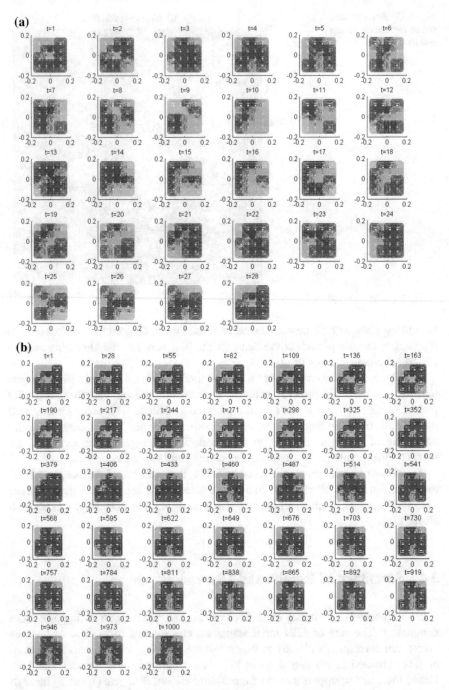

Fig. 3.18 "Movement" of the sound wave for the first period of time (**a**) and obtained "sound picture" during all time of measurements with time step equal to one period (**b**)

Table 3.7 Summary of test data sets and clustering procedure parameters

No	Data set	Features number n_{in}	ESN reservoir size n_r
1	Steel alloys composition	3	10
2	Multi-spectral images	8	20, 100
3	Mill fan	3	10, 124
4	Random dots motion	7	10, 30, 50, 100
5	"Sound pictures"	11	10, 30, 50

The ESN reservoir size is the other important parameter of the proposed clustering procedure. It was varied from 10 up to 124 neurons. Obtained results showed that for small number of features there is no need to use huge reservoirs while with increasing of ESN input vector size the bigger is number of neurons the better are achieved clustering results. This is valid especially for data sets 2 and 5.

It should be noted however that increasing of reservoir size dramatically increases the possible number of 2D projections that is the number of all possible combinations between each two neurons equilibrium states, i.e. $(n_r*(n_r - 1))/2$. Although the increased number of obtained 2D projections increases the chance to obtain several once that separate multidimensional data set as perfect as it is possible, these also increases the computational time of overall procedure. This is mainly due to the slowest or "bottleneck" part of the proposed algorithm, i.e. the subtractive clustering that is an optimization procedure. So increase of ESN reservoir size increases computational costs significantly. This raises a question to be solved: to find a good criterion to chose only most suitable neurons to create smaller number of 2D projections to be clustered.

Another issue that has to be commented is the achieved accuracy of data clustering. Since for all test data sets we had only vague expert information (data sets 1, 3 and 4) or some visual representation (data sets 2 and 5) of a prospective matching of data points to unknown in advance number of clusters there is no way to calculate any numerical criteria about goodness of clustering results. By far only subjective visual or expert assessment of the achieved results is done. Hence the next step of development of the proposed algorithm will be to test it on data sets with clearly defined and known in advance number and position of clusters.

3.5 Conclusions

The described here approach for feature extraction, clustering and 2D visualization of multidimensional data sets showed good performance characteristics with all testing data sets. Although there was no available information about exact clusters number, type and position for all of the exploited in present investigation data sets, the obtained by far results seem logical. Moreover, tests were performed on real data with different nature demonstrating wide range of possible applications. Visual comparison with other information sources about exact data classes demonstrated a lot of similarities, especially in the case of hyper spectral images and sound pictures.

In should be noted that second stage of clustering of the obtained 2D projections from the extracted features was done via well developed method that was proven to be the best in the case of unknown number and position of clusters. In addition this method allows gradual change of membership to several classes using fuzzy sets—a property that was not exploited here. Instead membership was crisp and it was determined by minimal distance to the revealed clusters centers. Further refinement of 2D visualizations could be achieved via determination of fuzzy membership and overlapping classes that will yield more realistic pictures.

Future work will include also exploitation of other well known clustering procedures, especially in case of known number of classes. Testing of the approach on benchmark data sets with exact information about data clusters will be the next step to evaluate numerically the quality of the proposed approach.

Acknowledgments The present work is partially supported by the following projects: AComIn "Advanced Computing for Innovation", grant 316087, funded by the FP7 Capacity Program (Research Potential of Convergence Regions); the projects funded by the Bulgarian Science Fund numbered DDVU02/11, DVU-10-0267/10 and DFNI-I01/8. All real data sets are obtained during work on these projects. The author is grateful to all colleagues and co-authors who contributed by various data sets and comments on the obtained clustering results. Special thanks to Kiril Alexiev, Denica Borisova and Georgi Jelev for their supportive and valuable comments.

References

1. Ackley, D.H., Hinton, G.E., Sejnowski, T.J.: A learning algorithm for Boltzmann machines. Cogn. Sci. **9**, 147–169 (1985)
2. Alexiev, K., Bocheva, N., Stefanov, S.: Assessment of age-related changes in global motion direction discrimination. In: International Conference Automatics and Informatics'11, pp. B277–B280, Sofia, Bulgaria, 3–7 Oct 2011
3. Beardsley, S.A., Ward, R.L., Vaina, L.M.: A neural network model of spiral-planar motion tuning in MSTd. Vision. Res. **43**, 577–595 (2003)
4. Bocheva, N., Bojilov, L.: Neural network model for visual discrimination of complex motion. Comptes rendus de'l Academie bulgare des Sciences **65**(10), 1356–1379 (2012)
5. Botev, Z.I., Grotowski, J.F., Kroese, D.P.: Kernel density estimation via diffusion. Ann. Stat. **38**(5), 2916–2957 (2010)
6. Brody, C.D., Romo, R., Kepecs, A.: Basic mechanisms for graded persistent activity: discrete attractors, continuous attractors, and dynamical representations. Curr. Opin. Neurobiol. **13**, 204–211 (2003)
7. Doukovska, L., Koprinkova-Hristova, P., Beloreshki, S.: Analysis of mill fan system for predictive maintenance. In: International Conference on Automatics and Informatics'11, pp. B-331–B-334, Sofia, Bulgaria, 3–7 Oct 2011
8. Eli Billauer's home page: http://billauer.co.il/peakdet.html. Accessed 2013
9. Fast and accurate state-of-the-art bivariate kernel density estimator by Z. Botev, http://www.mathworks.com/matlabcentral/fileexchange/17204-kernel-density-estimation (updated 2009)

10. Grossberg, S., Pilly, P.K.: Temporal dynamics of decision-making during motion perception in the visual cortex, Technical report BU CAS/CNS TR-2007-001, Feb 2008
11. Haddad, W.M., Chellaboina, V.S., Nersesov, S.G.: Thermodynamics: A Dynamical System Approach, Princeton University Press (2005)
12. Hammouda, K.: A comparative study of data clustering techniques. In: SYDE 625: Tools of Intelligent Systems Design, Course Project, Aug 2000
13. Hinton, G.E., Salakhutdinov, R.: Reducing the dimensionality of data with neural networks. Science **313**(5786), 504–507 (2006)
14. Jaeger, H.: Tutorial on training recurrent neural networks, covering BPPT, RTRL, EKF and the "echo state network" approach, GMD Report 159, German National Research Center for Information Technology (2002)
15. Jain, A.K., Murty, M.N., Flynn, P.J.: Data clustering: a review. ACM Comput. Surv. **31**(3), 264–323 (1999)
16. Koprinkova-Hristova, P., Palm, G.: ESN intrinsic plasticity versus reservoir stability. In: Artificial Neural Networks and Machine Learning—ICANN 2011. Lecture Notes in Computer Science (including subseries Lecture Notes in Artificial Intelligence and Lecture Notes in Bioinformatics), vol. 6791, pp. 69−76 (2011)
17. Koprinkova-Hristova, P., Tontchev, N.: Echo state networks for multi-dimensional data clustering. In: Artificial Neural Networks and Machine Learning—ICANN 2012. Lecture Notes in Computer Science (including subseries Lecture Notes in Artificial Intelligence and Lecture Notes in Bioinformatics), vol. 7552 (PART 1), pp. 571–578 (2012)
18. Koprinkova-Hristova, P., Alexiev, K., Borisova, D., Jelev, G., Atanassov, V.: Recurrent neural networks for automatic clustering of multispectral satellite images, In: Bruzzone, L (ed.) Proceedings of SPIE, Image and Signal Processing for Remote Sensing XIX, vol. 8892, 88920X, 17 Oct 2013. doi:10.1117/12
19. Koprinkova-Hristova, P., Angelova, D., Borisova, D., Jelev, G.: Clustering of spectral images using Echo state networks. In: IEEE International Symposium on Innovations in Intelligent Systems and Applications, IEEE INISTA 2013, Albena, Bulgaria, 19–21 June 2013. doi:10.1109/INISTA.2013.6577633
20. Koprinkova-Hristova, P., Doukovska, L., Kostov, P.: Working regimes classification for predictive maintenance of mill fan systems. In: 2013 IEEE International Symposium on Innovations in Intelligent Systems and Applications, IEEE INISTA 2013, Albena, Bulgaria, 19–21 June 2013. doi:10.1109/INISTA.2013.6577632
21. Koprinkova-Hristova, P., Alexiev, K.: Echo state networks in dynamic data clustering. In: Lecture Notes in Computer Science (including subseries Lecture Notes in Artificial Intelligence and Lecture Notes in Bioinformatics), vol. 8131, pp. 343−350 (2013)
22. Koprinkova-Hristova, P., Alexiev, K.: Sound fields clusterization via neural networks. In: 2014 IEEE International Symposium on Innovations in Intelligent Systems and Applications, INISTA 2014, pp. 368−374, Alberobello, Italy, 23−25 June 2014
23. Koprinkova-Hristova, P., Alexiev, K.: Dynamic sound fields clusterization using neuro-fuzzy approach. In: 16th International Conference, AIMSA 2014, Varna, Bulgaria, 11−13 Sept 2014. Artificial Intelligence: Methodology, Systems, and Applications, Lecture Notes in Computer Science, vol. 8722, pp. 194−205 (2014)
24. Landsat Missions. http://landsat.usgs.gov/. Accessed 2013
25. Lazar, A., Pipa, G., Triesch, J.: Predictive coding in cortical microcircuits. In: Kurkova, V., et al. (eds.) ICANN 2008, Part II, LNCS 5164, pp. 386–395 (2008)
26. Lukosevicius, M., Jaeger, H.: Reservoir computing approaches to recurrent neural network training. Comput. Sci. Rev. **3**, 127–149 (2009)
27. MRRB GIS System security division, http://gis.mrrb.government.bg/pmapper/map_separate-legend.phtml?winsize=medium&language=bg&config=separate-legend. Accessed 2013
28. Ozturk, M., Xu, D., Principe, J.: Analysis and design of echo state networks. Neural Comput. **19**, 111–138 (2007)

29. Peng, X., Guo, J., Lei, M., Peng, Y.: Analog circuit fault diagnosis with echo state networks based on corresponding clusters. In: Liu, et al. (eds.) ISNN 2011, Part I, LNCS 6675, pp. 437–444 (2011)
30. Schrauwen, B., Wandermann, M., Verstraeten, D., Steil, J.J., Stroobandt, D.: Improving reservoirs using intrinsic plasticity. Neurocomputing **71**, 1159–1171 (2008)
31. Steil, J.J.: Online reservoir adaptation by intrinsic plasticity for back-propagation-decoleration and echo state learning. Neural Netw. **20**, 353–364 (2007)
32. Woodward, A., Ikegami, T.: A reservoir computing approach to image classification using coupled echo state and back-propagation neural networks. In: Proceedings of 26th International Conference on Image and Vision Computing, Auckland, New Zealand, pp. 543–458, 29 Nov−1 Dec 2011 (2011)
33. Yager, R., Filev, D.: Generation of fuzzy rules by mountain clustering. J. Intell. Fuzzy Syst. **2** (3), 209–219 (1994)

Chapter 4
Unsupervised Clustering of Natural Images in Automatic Image Annotation Systems

Margarita Favorskaya, Lakhmi C. Jain and Alexander Proskurin

Abstract The chapter is devoted to automatic annotation of natural images joining the strengths of text-based and content-based image retrieval. The Automatic Image Annotation (AIA) is based on the semantic concept models, which are built from large number of patches receiving from a set of images. In this case, image retrieval is implemented by keywords called as Visual Words (VWs) that is similar to text document retrieval. The task involves two main stages: a low-level segmentation based on color, texture, and fractal descriptors (a shape descriptor is less useful due to great variety of visual objects and their projections in natural images) and a high-level clustering of received descriptors into the separated clusters corresponding to the VWs set. The enhanced region descriptor including color, texture (with the high order moments—skewness and kurtosis), and fractal features (fractal dimension and lacunarity) has been proposed. For the VWs generation, the unsupervised clustering is a suitable approach. The Enhanced Self-Organizing Incremental Neural Network (ESOINN) was chosen due to its main benefits as a self-organizing structure and on-line implementation. The preliminary image segmentation permitted to change a sequential order of descriptors entering in the ESOINN as the associated sets. Such approach simplified, accelerated, and decreased the stochastic variations of the ESOINN. Our experiments demonstrate acceptable results of the VWs clustering for a non-large natural image sets. Precision value of clustering achieved up to 85–90 %. Our approach show better precision values and execution time as compared with fuzzy

M. Favorskaya · A. Proskurin
Institute of Informatics and Telecommunications, Siberian State Aerospace University, 31 Krasnoyarsky Rabochy, Krasnoyarsk 660037, Russian Federation
e-mail: favorskaya@sibsau.ru

A. Proskurin
e-mail: proskurin.av.wof@gmail.com

L.C. Jain (✉)
Bournemouth University, Fern Barrow, Poole, UK
e-mail: Lakhmi.Jain@canberra.edu.au

L.C. Jain
University of Canberra, Canberra ACT 2601, Australia

© Springer International Publishing Switzerland 2016
R. Kountchev and K. Nakamatsu (eds.), *New Approaches in Intelligent Image Analysis*, Intelligent Systems Reference Library 108,
DOI 10.1007/978-3-319-32192-9_4

c-means algorithm and classic ESOINN. Also issues of parallel implementation of unsupervised segmentation in OpenMP and Intel Cilk Plus environments were considered for processing of HD-quality images. Execution time has been increased on 26–32 % using the parallel computations.

Keywords Unsupervised clustering · Visual words · Self-organizing incremental neural network · Automatic image annotation · Image features · Image segmentation

4.1 Introduction

Nowadays, the image browsing and retrieval are the embedded WWW tools, which are available for many users in anytime and anywhere. However, the retrieval systems require the development of efficient software tools that is caused by the increasing visual data growth. The image retrieval systems have three frameworks: text-based (since 1970s), content-based (since 1980s), and automatic annotation (since 2000s). In Text-Based Image Retrieval (TBIR) systems, the images are manually annotated by text descriptors [1]. This leads to inaccuracy and duration of a user work. The Content-Based Image Retrieval (CBIR) is free from such disadvantages. The queries into Content-Based Retrieval Systems (CBRS) can be different, for example, a retrieval of features (color, shape, spatial location), abstract objects, real objects, or listed events [2–4]. Image retrieval in Automatic Image Annotation (AIA) assumes that the images can be retrieved in the same manner as text documents. The basic idea of the AIA is to implement the unsupervised learning based on semantic concept models extracted from large number of image samples [5–9]. The images can be retrieved by keywords called as Visual Words (VWs). The AIA systems join the advantages of both TBIR and CBIR systems and additionally solve the task of automatic image annotation using semantic labels.

This chapter is devoted to the retrieval of abstract objects, which is based on the VWs extraction from a non-large set of images. The task involves two main stages: a low-level segmentation based on color, texture, and fractal descriptors (a shape descriptor is less useful due to great variety of visual objects and their projections in natural images [10]) and a high-level clustering of received descriptors into the separated clusters corresponding to the VWs set. Sometimes it is useful to divide images in two global categories: natural or urban scenes, and according to such classification tune an unsupervised procedure of extraction of low-level features. The goal is to develop fast and accurate methods, which are suitable for natural images annotation in the AIA framework.

For the VWs detection, the unsupervised clustering is a suitable approach. The Enhanced Self-Organizing Incremental Neural Network (ESOINN) was chosen due to its main benefits in unsupervised clustering and on-line implementation. This network can be trained adaptively and store the previous data with any increasing volume of input information.

The chapter is organized as follows. A brief literature review is provided by Sect. 4.2. The main statements, methods, and algorithms of unsupervised segmentation and unsupervised clustering are detailed in Sects. 4.3 and 4.4, respectively. Section 4.5 presents a discussion of experimental results of precision and computational speed involving experiments with images from the dataset IAPR TC-12 Benchmark [11]. Conclusion and remarks of future development are drawn in Sect. 4.6.

4.2   Related Work

The literature review includes two main issues: the analysis of unsupervised image segmentation for extraction of low-level features (Sect. 4.2.1) and the overview of unsupervised image clustering (Sect. 4.2.2) in order to receive the high-level semantic descriptors.

4.2.1   Unsupervised Segmentation of Natural Images

Any image, especially natural, involves a set of regions with different textures and colors. During the last decade, many heuristic segmentation methods and algorithms have been designed, which can be concerned to three main approaches:

- Region-based approach including grid-based method, when an image is roughly divided into blocks [12], threshold-based methods of gradient gray-scales image [13], contour-based methods evolving a curve around an object [14], methods of morphological watersheds with preliminary image pyramid building in order to detect the centers of crystallization [15–18], region-based methods including a region growing approach [19–21].
- Model-based approach involving graph-based methods, among which a normalized graph cut [22], statistical models using Bayesian model, Markov chain, Expectation Maximization (EM) algorithm, and others [23–25], auto-regressive models [26, 27], clustering algorithms like k-means, which are used to classify pixels into different classes [28].
- Structured-based approach using Haralick structural methods for texture segmentation [29], image segmentation by clustering of spatial patterns [30].

A majority of known CBIR and AIA systems use a region-based segmentation as the close technique for a human vision. Let us notice that for the AIA systems, the unsupervised segmentation is strongly recommended approach.

The unsupervised color image segmentation method based on the estimation of Maximum A Posteriori (MAP) on the Markov Random Fields (MRFs) was proposed by Hou et al. [31]. This method works under the assumption that there are n pixels of m ($m \ll n$) colors in the image I, and any two colors, fore and back, are perceptually distinguishable from each other. The authors used the energy functions approximately in the non-iteration style. A new binary segmentation algorithm based on the slightly tuned Lanczos eigensolver was designed.

The effective unsupervised color image segmentation algorithm, which uses the multi-scale edge information and spatial color content, was represented by Celik and Tjahjadi [32]. The multi-scale edge information is extracted using Dual-Tree Complex Wavelet Transform (DT-CWT). The segmentation of homogeneous regions was obtained using a region growing followed by a region merging in the Lightness and A and B (LAB) color space in the research [33]. The authors proposed the edge-preserving smoothing filter, which removes a noise and retains a contrast between regions. The authors show that their approach provides better boundaries of objects than JSEG and mean-shift algorithms. However, the unsupervised color image segmentation works non-well in the textured images. Sometimes the color image segmentation needs in a priory information and has high computational cost. The use of statistical pattern recognition and Artificial Neural Networks (ANN) with multi-layer perceptron topology was suggested by Haykin [34] in order to segment and make a clustering of images into the pre-determined classes.

Also some hybrid methods exist, for example, 2D autoregressive modeling and the Stochastic Expectation-Maximization (SEM) algorithm. The last one was developed by Cariou and Chehdi [27]. The proposed texture segmentation method has three steps. First, 2D causal non-symmetric half-plane autoregressive modeling of the textured image is realized. Second, the parameters of identifiable mixed distributions and the corrected number of classes are calculated using the SEM algorithm. The second step is finalized by coarse, block-like image pre-segmentation. Third, the original image ought to refine the pixel-based segmentation applying the Maximizer of Posterior Marginals (MPM). Using this hierarchical model in a Bayesian framework, the authors obtained a reliable segmentation by means of Gibbs sampling. This approach provided good segmentation/classification results above 90 % of correct classification with maximum value 99.26 %. The disadvantage is the complicated mathematical calculations.

The well-known method of J-image SEGmentation (JSEG) is concerned to the unsupervised segmentation based on a color-texture model. In the pioneer research of Deng and Manjunath [35], the given color-texture patterns and the estimations of their homogeneity were used. First, the image colors are quantized to several representative classes in the color space without considering the spatial distributions of the colors. Then pixel values are replaced by their corresponding color class labels to form a class-map of the image (J-image). The received class-map can be represented as a special type of homogeneous color-texture regions. Second, a spatial segmentation is executed into this class-map without considering the corresponding pixel color similarity. This work became the basis of following modifications and improvements.

An improved version, combining the classical JSEG algorithm with a local fractal estimator, permits to improve the boundary detection [36]. A model of the texture features using a mixture of Gaussian distributions, which components can be degenerate or nearly-degenerate, was developed by Yang et al. [37]. The authors show the efficiency of their simple agglomerative clustering algorithm derived from a lossy data compression approach. Using 2D texture filter banks or simple fixed-size windows, the algorithm effectively segments an image minimizing the overall coding length of the feature vectors.

Statistical Region Merging (SRM) algorithm based on perceptual growing and region merging was proposed by Nock and Nielsen [38]. An unsupervised GSEG algorithm [21] is based on color-edge detection, dynamic region growth, and multi-resolution region merging procedure. A Partion-based SEGmentation (PSEG) algorithm uses a hierarchical approach, according to which the spatially connected regions group together based on the mean vectors and covariance matrices of a multi-band image [39]. Also the authors introduced the inner and the external measures based on Gaussian distribution, which estimate the goodness for each portion in the hierarchy.

The approach for color–texture segmentation based on graph cut techniques finds optimal color–texture segmentation by regarding it as a minimum cut problem in a weighted graph [40]. A texture descriptor called as texton was introduced to efficiently represent texture attributes of the given image, which is derived from the complex Gabor filtered images estimated in various directions and scales. In the research [40], the texton feature is defined as a magnitude of textons rather than a histogram of textons, which makes it highly effective to apply the graph cut techniques. The problem of color-texture segmentation is formulated in terms of energy $E(\cdot)$ minimization with graph cuts by Eq. 4.1, where A is the data and smoothness constraint, Θ denotes the mixture model parameters, $\lambda > 0$ specifies a balance between a data term $U(A, \Theta)$ and a prior term $V(A)$.

$$E(A, \Theta) = \lambda \cdot U(A, \Theta) + V(A) \qquad (4.1)$$

The segmentation energy should be minimized with respect to the labeling A and the model parameter Θ. This method provides better precision and recall results in comparison with JSEG algorithm. The following essential extension of multilayer graph cut approach using multivariate mixed Student's t-distribution and regional credibility merging one can find in [41].

The Blobworld segmentation is widely used method. It is closed to the JSEG algorithm. The pixel clustering is executed in a color-texture-position feature space. First, a common distribution of these features is modeled by a Gaussian mixture. Second, the EM algorithm estimates the parameters of received model. The pixel-cluster membership produces a resulting coarse segmentation of the objects. Vogel et al. proposed the adapting version of Blobworld algorithm, which was called BlobContours segmentation [42]. The idea of displaying the intermediate segmented images as the layers lays in the basis of BlobContours segmentation. Each EM iteration is displayed as a layer, and the user can examine, which layer is the best one. The flood-fill algorithm calculates the average true color for each region instead of using the connected component algorithm from the original Blobworld. However, the blob-approach has a restricted application in the unsupervised segmentation.

Many authors use the measures of similarity to estimate and compare the experimental results. Some of such measures one can find in [43].

4.2.2 Unsupervised Clustering of Images

The good decision for extraction of high-level semantic features is the use of semi-supervised or unsupervised machine learning techniques. The goal of supervised learning is to determine the output values, and the goal of unsupervised learning is to re-distribute the input data into classes and describe these classes. Support Vector Machine (SVM) classification, Bayesian classification, and decision tree technique are concerned to supervised methods, which form the high-level results from the low-level features. In this research, the unsupervised methods are considered for clustering of low-level features into the VWs representation. The traditional k-means clustering, fuzzy c-means, and clustering based on Self-Organizing Neural Network (SONN) including their modifications are often applied approaches in the CBRS. Let us discuss some approaches for such clustering.

The color moments and Block Truncation Coding (BTC) were used in [44] to extract features as the inputs of k-means clustering algorithm. The basis of color moments (mean, standard deviation, and skewness) uses assumption that a distribution of color in an image can be interpreted as a probability distribution. An image is split into R, G, and B components separately, the average values of each component are determined. Then the features are calculated as a set of color moments for R, G, and B values, which are higher and lower the corresponding averages. Such heuristic algorithm can not provide a high accuracy of clustering because color features are computed in a whole image. The closed approaches one can find in [45, 46].

The application of Radial-Based Function Neural Network (RBFNN) for semantic clustering was proposed by Rao et al. [47]. The authors applied the hierarchical clustering algorithm to group the images into classes based only on the color RGB-content; however, the result of received accuracy is absent in research [47].

Self-Organizing Fuzzy Neural Network (SOFNN) can be concerned to a special type of the SONN. The first group of Fuzzy Neural Network (FNN) with the self-tuning capabilities requires the initial rules prior to train. The second group of the FNN is able to automatically create the fuzzy rules from the training data set. In opposite of a traditional clustering, when classes are disjointed, a fuzzy clustering suggests so called soft clustering scheme. In this case, each pattern is associated with every class by a membership function, in other words each class is a fuzzy set of all patterns. A traditional clustering can be obtained from a fuzzy clustering using a threshold of a membership value. The most popular fuzzy clustering algorithm is a Fuzzy C-Means (FCM) algorithm. It is better than the k-means algorithm avoiding local minimums. The design of membership functions is the most important problem in a fuzzy clustering because they determine the similarity decomposition and the centroids of classes. An incremental clustering is based on the assumption that it is possible to consider instances one at a time and assign them to the existing classes.

In research [48], the SOFNN was proposed as extended RBFNN, which is a functional equivalent to Takagi-Sugeno-Kang fuzzy systems. First, a self-organizing clustering approach is used to form the structure and obtain the initial values of parameters in a network. Second, a hierarchical on-line self-organizing learning paradigm is employed to adjust the parameters and the structure of the SOFNN. The algorithm of incremental learning was developed, which is capable to generate automatically fuzzy rules according to a simple error criterion based on the differences between calculated and desired output values.

Tung and Quek suggested a Generic Self-Organizing Fuzzy Neural Network (GenSOFNN), which overcomes the drawbacks of fuzzy neural network approach connecting with the necessity of prior knowledge such as a number of classes [49]. The proposed GenSOFNN did not require a pre-definition of the fuzzy rules. The authors show that its training cycle takes place in a single pass of the training data and demonstrated the on-line applications of the GenSoFNNs.

Three new learning algorithms for Takagi-Sugeno-Kang fuzzy system based on a training error and a genetic algorithm were proposed by Malek et al. [50]. First two algorithms involve two stages. In the first stage, the initial structure of the FNN was created by estimating the optimum points of training data in input-output space using k-nearest neighbor algorithm and c-means methods, respectively. This stage keeps adding new neurons based on an error-based algorithm. In the second stage, the redundant neurons were recognized and removed using a genetic algorithm. Third algorithm built the FNN by a single stage using a modified version of error algorithm. These algorithms were evaluated using two examples: by function of two nonlinear inputs and identification of nonlinear dynamic system.

Fuzzy clustering can be applied with other techniques, for example, invariant moments as the invariant shape features [51]. One of the connected problems is a semantic gap removal, which appears between low-level and high-level features because the images, which are identical in a spatial domain, can be non-identical in a semantic domain [52].

For our experiments, two ways were chosen: without preliminary segmentation of natural images and with preliminary segmentation, description of the last one is located in next Sect. 4.3.

4.3   Preliminary Unsupervised Image Segmentation

Segmentation of natural images is a complicated task due to a great set of regions with various colors and textures. The basic JSEG algorithm [35] with some modification was applied in order to obtain good segmentation results. The segmentation task can be interpreted as an optimization task for search of such division of image, which possesses the predetermined properties according to some functional.

The authors of research [35] referred this functional as J-functional, which estimates a quality of segmentation based on a color distribution. However, the direct optimization of J-functional is a high resource task. The JSEG algorithm uses a greedy algorithm of optimization, which searches local optimums in each of iterations, calculating J-functional in a neighborhood of each pixel.

Two independent steps including color quantization and spatial segmentation are used in this method. In order to extract only a few representative colors, the colors in image are coarsely quantized. For natural images, 10–20 colors are enough for good segmentation. Each pixel is replaced by corresponding color class label. The image of labels is called a class-map, which can be interpreted as a special texture. Each point belongs to a color class in a class-map. In natural images, such classes usually have the overlapping distributions. Under such assumptions, the authors of research [35] consider Z as the set of all N data points in a class-map, $z = (x, y)$, where x, y are spatial coordinates, $z \in Z$ with the mean m provided by Eq. 4.2, on the one hand,

$$m = \frac{1}{N} \sum_{z \in Z} z \tag{4.2}$$

and, on the other hand, suppose that Z is classified into classes Z_i, $i = 1, \ldots, C$ with mean m_i of the N_i points in class Z_i as it is written in Eq. 4.3.

$$m_i = \frac{1}{N_i} \sum_{z \in Z_i} z \tag{4.3}$$

The total variance S_T of class-map points is determined by Eq. 4.4.

$$S_T = \sum_{z \in Z} \|z - m\|^2 \tag{4.4}$$

The total variance of points S_W belonging to the same class is defined by Eq. 4.5.

$$S_W = \sum_{i=1}^{C} S_i = \sum_{i=1}^{C} \sum_{z \in Z} \|z - m_i\|^2 \tag{4.5}$$

Then J-functional can be calculated by Eq. 4.6.

$$J = (S_T - S_W)/S_W \tag{4.6}$$

If value of J is large, then the color classes are more separated from each other, and points inside a class are strongly connected between themselves. The average $\bar{J}$ can be defined by Eq. 4.7, where J_k is J-functional over region k, M_k is a number of points in region k, N is a total number of points in a class-map.

$$\bar{J} = \frac{1}{N} \sum_k M_k J_k \qquad (4.7)$$

A better segmentation means a lower value of $\bar{J}$. Equation 4.6 is a criterion of minimization of segmentation. However, the global optimization of $\bar{J}$ is impossible because any image can be segmented by various ways. Instead of this, J-image is generated, where pixel values correspond to local J-values, which are calculated over small window centered in the pixels. The local J-values become large near a region boundary. The J-image can be represented as a 3D map containing valleys and hills, which are correspond to the region insides and the region boundaries. The size of local window is a multi-scale parameter. A window with small sizes (9×9 pixels) is useful to detect edges, and a window with large sizes is used for boundary detection.

A spatial segmentation as a second step of the JSEG algorithm is based on a region-growing method. The pixels of J-image with minimal functional values are accepted as the seeds. A growing process is realized by jointing the neighbor pixels to the seeds. As a result, an initial segmentation is received sometimes with small over-segmented regions. To avoid this artifact, such regions are merged based on color similarity. The agglomerative procedure is applied: the distance values between two neighbor regions are calculated according to a color histogram, the pairs of regions with minimal distance value is merged, then a color histogram is recalculated and the distance values are update. The procedure is repeated until the predetermined maximum distance value between regions will not be achieved.

Our improvement of JSEG results deals with decreasing the original image in four times (the upper level of image pyramid with Gaussian blurring), application of JSEG algorithm to small-sized image, and following stretching transformation of J-image to the initial sizes of original image. A convolution of the transformed J-image with original image provides a final segmentation. The segmentation results for images 38019, 38225, 38755, 39986, and 38756 are represented in Fig. 4.1. The test images were taken from DB IAPR TC-12 Benchmark [11]. The size of original images is 480×360 pixels. The size of decreased images was 240×180 pixels.

Our approach provides better visual segmentation results without considering the non-significant for the AIA small regions in original images.

4.4 Feature Extraction Using Parallel Computations

In this research, the JSEG algorithm was chosen as a pre-segmentation stage and realized in the designed software tool. Consider the main color, texture, and fractal features extraction (Sects. 4.4.1, 4.4.2 and 4.4.3, respectively) from a pre-segmented

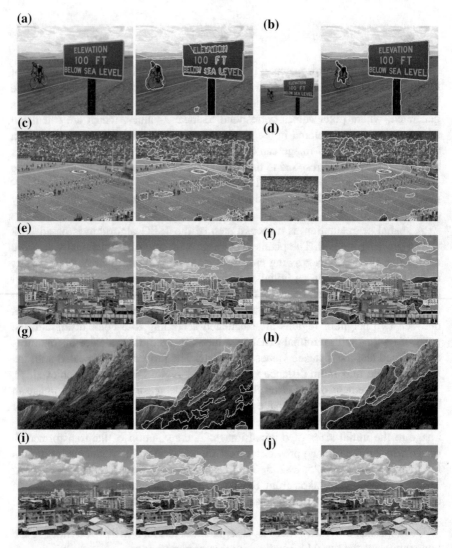

Fig. 4.1 Visual results of JSEG algorithm: **a, c, e, g, i** original images and JSEG results; **b, d, f, h, j** the resized in four times original images and JSEG results

image in order to create a common image descriptor as a set of region features. The enhanced region descriptor is built in Sect. 4.4.4. Section 4.4.5 provides a description of parallel computations of features.

4.4.1 Color Features Representation

A number of important color features, which are extracted from images or regions, have been proposed in literature including Color Histogram (CH) [53], Color Moments (CM) [54], Color Coherence Vector (CCV) [55], Color Correlogram [56], among others. Notice that MPEG-7 standard restricts a number of color features including Dominant Color Descriptor (DCD), Color Layout Descriptor (CLD), Color Structure Descriptor (CSD), and Scalable Color Descriptor (SCD) [57].

The color moments such as mean, standard deviation, and skewness are the simplest and popular features. They are applied to each component of color spaces mentioned below:

- Red, Green, Blue (RGB).
- Lightness and A and B are the color-opponent dimensions based on nonlinearly compressed CIE (International Commission on Illumination; usually abbreviated CIE for its French name, Commission internationale de l'éclairage) XYZ coordinates (LAB).
- Lightness, Uniform chromaticity scale, Valence (LUV). CIE LUV and CIE LAB were adopted simultaneously by the CIE.
- Hue, Saturation, Value (HSV) or Hue, Saturation, Lightness (HSL).
- Hue, Min, Max, Difference (HMMD).

In current research, the color features (mean, standard deviation, and skewness for each color component) are extracted as the low-level features of each region in HSV-color space. According to the theory of moments, normalized mean μ_c, normalized standard deviation σ_c, and normalized skewness θ_c (values of these parameters are normalized relative to a pixel amount into current region) are calculated for each HSV-component by Eqs. 4.8–4.10, where p_i^c is a pixel value of corresponding color component, *NP* is a number of pixels in a current region. Let us remember that preliminary image segmentation was executed using the JSEG algorithm.

$$\mu_c = \frac{1}{NP} \sum_{i=1}^{NP} p_i^c \tag{4.8}$$

$$\sigma_c = \frac{1}{NP} \left(\sum_{i=1}^{NP} \left(p_i^c - \mu_c \right)^2 \right)^{\frac{1}{2}} \tag{4.9}$$

$$\theta_c = \frac{1}{NP} \sum_{i=1}^{NP} \left(p_i^c - \mu_c \right)^3 \tag{4.10}$$

As a result, nine color features are received and included as $FC_0, ..., FC_8$ components in feature vector, describing a current region.

4.4.2 Calculation of Texture Features

The calculation techniques for texture features are very different. Often statistical texture features are based on moments or local statistical measures such as the six Tamura texture features [58]. The Tamura features include coarseness, directionality, regularity, contrast, line-likeness, and roughness. First three characteristics are more significant, and second three ones are the secondary parameters. The MPEG-7 standard has employed regularity, directionality, and coarseness as the texture browsing descriptor [57]. Unfortunately, the Tamura and the MPEG-7 texture descriptors are non-invariant to a scale.

Also it is possible to calculate the statistical features using a Gray-Level Co-occurrence Matrix (GLCM) [59]. The GLCM provides information about the positions of pixels having similar gray level values. Each element of such matrix contains a number of all pairs of pixels separated by displacement vector **d**, which includes gray levels i and j. Haralick et al. [60] suggested a set of 14 textural features extracted from a co-occurrence matrix. Homogeneity, contrast, and entropy are the main parameters, which are calculated from the GLCM. However, the experiments show that these parameters do not make essential contribution into improvement of CBIR accuracy but increase a computational cost.

The spectral characteristics based on 2D wavelet transform and a Gabor transform have high cost for the CBIR and the AIA. The advantage of Gabor transform is an invariance to a scale. Galloway [61] introduced five original features of run-length statistics, which were built using the analysis of image gray levels. At present, run-length statistics have a historical meaning.

Let z be a random value of intensity, $h(z_i)$ is its histogram, $i = 0, 1, 2, ..., Q–1$, Q is a number of brightness levels. Statistical features into a current image region such as normalized average AV, normalized dispersion DS, normalized homogeneity HM, normalized smoothness SM and improved normalized smoothness ISM, normalized entropy EN and improved normalized entropy IEN, normalized skewness SK, and normalized kurtosis KR are provided by Eqs. 4.11–4.19, where SR is a region area, μ_3 and μ_4 are moments of 3rd and 4th orders, σ^3 and σ^4 are standard deviation in 3rd and 4th degrees. All these values are normalized relative to a region area SR.

$$AV = \frac{1}{SR} \sum_{i=0}^{Q-1} z_i h(z_i) \tag{4.11}$$

$$DS = \frac{1}{SR} \sum_{i=0}^{Q-1} (z_i - AV)^2 h(z_i) \tag{4.12}$$

$$HM = \frac{1}{SR} \sum_{i=0}^{Q-1} h^2(z_i) \tag{4.13}$$

$$SM = \frac{1}{SR}\left(1 - \frac{1}{1 + DS\big/(Q-1)^2}\right) \tag{4.14}$$

$$ISM = \begin{cases} -\log SM, & \text{if } SM > 0 \\ 10, & \text{if } SM = 0 \end{cases} \tag{4.15}$$

$$EN = -\frac{1}{SR}\sum_{i=0}^{Q-1} h(z_i)\log_2 h(z_i) \tag{4.16}$$

$$IEN = EN\big/\log_2 Q \quad Q > 1 \tag{4.17}$$

$$SK = \frac{1}{SR}\frac{\mu_3}{\sigma^3} = \frac{1}{SR}\cdot\sum_{i=1}^{Q-1}\left(\left(\frac{z_i - AV}{\sqrt{DS}}\right)^3\cdot h(z_i)\right) \tag{4.18}$$

$$KR = \frac{1}{SR}\frac{\mu_4}{\sigma^4} - 3 = \frac{1}{SR}\cdot\sum_{i=1}^{Q-1}\left(\left(\frac{z_z - AV}{\sqrt{DS}}\right)^4\cdot h(z_i)\right) - 3 \tag{4.19}$$

If parameter $SM = 0$, then its value is forcibly maintained into $NSM = 10$ (small empirical value, differing from 0). Normalized entropy NEN indicates some equalization effect in dark and bright areas of image [62, 63].

Thus, seven texture features (AV, DS, HM, ISM, IEN, SK, and KR) are used as the FT_9, ..., FT_{15} components of feature vector, describing a current region.

4.4.3 Fractal Features Extraction

It is well-known, that natural texture surfaces are the spatial isotropic fractals and their 2D intensity function are also fractals. Connected domain A in a topological n-space is self-similarity, when domain A includes N separated non-overlapping and self-similarity copies, and each of copies is reduced by a coefficient r along all coordinate axes. Fractal dimension FD of connected domain A is determined by Eq. 4.20.

$$FD = \log N\big/\log(1/r) \tag{4.20}$$

Usually fractal surfaces demonstrate a statistical self-similarity, when each of N copies is identical to an original surface by all statistical features. However, to determine a dimension of fractal texture region using Eq. 4.20 is difficult and sometimes impossible. In research [64], two ways for definition of fractal dimension FD were investigated using a cube cover and based on the probability estimations. Let us calculate a measure of domain A on a set R^n. Suppose that domain A is covered by n-ary cube with sizes L_{max}. If domain A is a reduced copy by coefficient r, then $N = r^{-FD}$ sub-cubes exist. Therefore, a number of cubes with sizes $L = r \cdot L_{max}$, which are necessary to cover a whole domain, is determined by Eq. 4.21.

$$N(L) = 1/r^{FD} = [L_{max}/L]^{FD} \qquad (4.21)$$

A simple procedure to determine fractal dimension FD by Eq. 4.21 involves a cover of connected domain A by a grid from n cubes with a side length L and calculation a number of non-empty K cubes. Then fractal dimension FD is determined from a line slope of $\{\log L; -\log N(L)\}$ in R^n space.

Another way to determine fractal dimension FD uses a probability approach. Let $P(m, L)$ be a probability that m points into a cube with length side L are located near a random point of connected domain A. Let total number of points into connected domain A be equal M (in our case, a connected domain A is an image). If a grid from cubes with a length side L is imposed in an image, then a number of cubes, including m points, is determined as $(M/m) \cdot P(m, L)$ and will be proportional to a power dependence L^{-FD}.

However, various fractal structures with a similar fractal dimension FD can have very different textures. The term "lacunarity" was introduced by Mandelbrot [65] to describe such fractals. Mandelbrot proposed some procedures to define a lacunarity FL, the most known of which has a view of Eq. 4.22, where M is a weight of fractal structure, $\langle M \rangle$ is an estimated weight.

$$FL = \left\langle \left(M/\langle M \rangle - 1\right)^2 \right\rangle \qquad (4.22)$$

Lacunarity FL demonstrates the difference between a weight of fractal structure and an estimated weight. This feature is a statistical characteristic of the second order and changes in a following manner. Lacunarity has low value for a fine-grained textures and high value for a coarse-grained textures. Weight of fractal structure M is a function of parameter L (Eq. 4.23), where k is a proportional coefficient [65].

$$M(L) = kL^{FD} \qquad (4.23)$$

Also lacunarity FL can be estimated based on a probability approach. Probability $P(m, L)$ includes data for average distortion of weight in fractal structure. Therefore, lacunarity FL can be calculated by Eq. 4.24.

$$FL(L) = \frac{M^2(L) - |M(L)|^2}{|M(L)|^2} = \frac{\displaystyle\sum_{m=1}^{N} m^2 P(m, L) - \left|\sum_{m=1}^{N} m\, P(m, L)\right|^2}{\left|\displaystyle\sum_{m=1}^{N} m\, P(m, L)\right|^2} \qquad (4.24)$$

Lacunarity estimating by Eq. 4.24 is well for textures with large area but it is non-useful for small area image regions. Let us simplify Eq. 4.24 by introduction of function $C(L)$ provided by Eq. 4.25, where $M_D(L)$ is an average density of weight into a cube with a length side L, $N_P(L)$ is a quotient of division the number cubes with a length side L, which are necessary for a full cover of fractal structure, on the number of points into this fractal structure.

$$C(L) = \frac{M_D(L) - N_P(L)}{M_D(L) + N_P(L)} \tag{4.25}$$

If the smallest texton is less then L, then a weight of fractal structure will distributed uniformly into each cube. In this case, values $M_D(L)$ and $N_P(L)$ have close values, and $C(L) \to 0$. If the smallest texton is large than L, then $C(L) \to 1$. If L value increases, then $C(L) \to 1$ for all fractal structures. Therefore, function $C(L)$ will include the data about textons in both cases.

Two fractal features FD and FL are two components FF_{16} and FF_{17} of feature vector describing a current region.

4.4.4 Enhanced Region Descriptor

Using parameters from Sects. 4.4.1–4.4.3, one can construct a region vector $\mathbf{RF} = \{FC_0, \dots, FC_8, FT_9, \dots, FT_{15}, FF_{16}, FF_{17}\}$, which later will be transformed to Region Descriptor $\mathbf{RD}_{ij}$. Values of $\mathbf{RD}_{ij}$ are normalized to the intervals of input values of neural network, where i is a counter of regions in an image j, j is a counter of images in an image set. As a result, an image descriptor $\mathbf{ID}_j = \{\mathbf{RD}_{1j}, \dots, \mathbf{RD}_{ij}, \dots\}$ and a set descriptor $\mathbf{SD} = \{\mathbf{ID}_1, \dots, \mathbf{ID}_j, \dots\}$ will be constructed. The extended region descriptor is our contribution in the unsupervised clustering for image annotation problem. For simplicity, denote a set of Region Descriptor $\{\mathbf{RD}_{ij}\}$ as a weight input vector $\mathbf{W}_x$, because region descriptors enter to the inputs of classical ESOINN randomly.

A transition from low-level features to high-level semantics is usually tracked by reducing the "semantic gap", which includes four categories:

- Object ontology to define high-level concepts.
- Introduction a relevant feedback into retrieval loop for continuous learning of users' intention.
- Generation semantic templates to support high-level image retrieval.
- Supervised or unsupervised learning methods to associate low-level features with query concepts.

Our choice deals with the last one due to high possibilities of self-organizing approach. Let us remark that a redundancy is eliminated during segmentation stage in order to avoid an over-segmentation of natural images.

4.4.5 Parallel Computations of Features

The parallelizing of program code includes the types mentioned below:

- Parallelizing of data means a multiple execution of the same algorithm with various input data. Data are divided into fragments, and each fragment is processed by an allocated computer core.
- Functional parallelizing is a parallel execution of sets of operations by functional feature. Simple example of such functional decomposition is a decomposition of task into subtasks such as input of initial data, processing, output of results, visualization of results, etc. Functional parallelizing is achieved using sequential or sequential-parallel "conveyor" between subtasks. Each subtask provides a parallelizing of data inside.
- Algorithmic parallelizing finds such fragments in algorithm, which can be executed in parallel. Synthesis of parallel algorithms based on algorithmic parallelizing is called an algorithmic decomposition. During algorithmic decomposition, it would like to divide a task into large and rarely connecting branches with homogeneous distribution of data processing along the branches. Main distinction between algorithmic and functional parallelizing is in following. Functional parallelizing merges only functional close operators from algorithm, and algorithmic parallelizing does not consider a functional similarity of operators.

For implementation of parallel algorithms, some standards are available, among which OpenMP standard [66] is used for parallelizing of program code in languages C, C++, and Fortran. Also the extension of language C++ with parallel possibilities called as Intel Cilk Plus [67] is developed.

In OpenMP standard, a paralleling is executed explicitly by insert the special directives and by call the additional functions in a program code. The standard OpenMP realizes the parallel computations in the multi-thread mode, when the "main" thread creates a set of sub-threads, and a current task is distributed between the sub-threads. First, a program is executed in "sequential" area with single "main" thread (process). Second, several sub-threads are generated in "parallel" area, and the program code is distributed between them. Third, all sub-threads except the "main" thread are finalized, and again a "sequential" area is continued. The standard OpenMP supports the embedding of parallel areas.

The Intel Cilk Plus environment is a dynamic thread scheduler, including a set of keywords. Keywords inform a compiler about the application of scheme scheduling. A parallel Cilk-program creates a task queue. The "executors" capture the tasks, and free thread performs a current task. In the Intel Cilk Plus environment, the semantics of sequential program is supported. However, a program can be executed in sequential or parallel modes due to available resources. Use of extended index notation is an essential difference in comparison with OpenMP standard that provides a paralleling of vector instructions of processor.

The enhanced region descriptor involves color, texture, and fractal features calculated in a neighborhood of considered pixel. Calculation of color and textural features (Eqs. 4.8–4.19) can be implemented in a parallel mode. Fractal features requires a separate non-parallel computation. The color and texture features are computed by two steps. First, stochastic data acquisition is accomplished in a neighborhood of current pixel: the normalized means of color channels are calculated using Eq. 4.8, and local histogram is built based on texture features. Second, color and texture features (Eqs. 4.9–4.19) are calculated directly. Two basic cycles are implemented in parallel mode. There are an external cycle for image with sizes $(w/k_w) \times (h/k_h)$, where w and h are width and height of image, respectively, k_w and k_h are width and height of image segment, respectively, and the internal cycles for segment with sizes $k_w \times k_h$.

For parallel computation of texture features, a whole image is divided into segments, and a processing of segments is distributed between cores of processor. A way of image partitioning in vertical/horizontal bands or rectangle blocks determines a structure of parallel procedure. In order to increase a computational cost, a parallelizing of external cycles in whole image is required. For this purpose, a processor directive "#pragma omp parallel for" in the case of OpenMP standard and a keyword "cilk_for" in the case of the Intel Cilk Plus environment can be applied. The calculations of color and texture features do not connected. Therefore, the additional parallel areas in random access memory can be determined for color and texture features separately.

4.5   Clustering of Visual Words by Enhanced SOINN

As a result of features extraction (Sect. 4.3), any image can be represented as a set of regions with corresponding region vectors $\mathbf{RF} = \{FC_0, \ldots, FC_8, FT_9, \ldots, FT_{15}, FF_{16}, FF_{17}\}$ as a collection of color, texture, and fractal features. Direct comparison of feature sets in a metric space is not preferable due to segmentation errors and noises. Therefore, a clustering methodology is a single way to receive good results. In literature review (Sect. 4.2), it was shown that the unsupervised clustering is more suitable for the VWs detection, and among unsupervised clustering methods the SOINN was chosen.

The clustering procedure groups the regions of all annotated images into subsets (VWs) in such manner that the regions with similar features are grouped together, while the regions with different features belong to the different classes. Formally, a clustering structure $\mathbf{S}$ is represented as a set of subsets $\mathbf{C} = \{C_1, \ldots, C_K\}$, Eq. 4.26. Consequently, any element in $\mathbf{S}$ belongs to one and only one subset.

$$\mathbf{S} = \bigcup_{k=1}^{K} C_i \quad \text{and} \quad C_i \cap C_j = 0 \quad \text{for} \quad i \neq j \qquad (4.26)$$

The ESOINN proposed by Furao et al. [68] is applied as the useful unsupervised clustering technique in many applications: robots navigation [69, 70] microarray data analysis [71], multi-agent systems [72], among others. The basic concepts of ESOINN are discussed shortly in Sect. 4.5.1, and algorithm of ESOINN is presented in Sect. 4.5.2.

4.5.1 Basic Concepts of ESOINN

The ESOINN was developed to overcome the main disadvantages of the two-layer SOINN as mentioned below:

- The separated training of the first layer and the second layer.
- The second layer is unsuitable for on-line incremental training: the changing of training results in the first layer causes the re-training of the second layer.
- The necessity of user-determined parameters, if a within-class insertion appears.
- The SOINN cannot separate a set with the high-density overlapping areas.

The ESOINN is adapted using a single-layer network structure. To build an edge between nodes, the ESOINN adds a condition to judge, and after some training iterations it separates nodes to the different subclasses deleting edges, which lie in the overlapping areas. The ESOINN achieves the within-class insertion slightly but it is more suitable for on-line or even life-long training tasks than two-layer SOINN.

A single layer of ESOINN is continuously adapted according to the input data structure defining a number and a topology of classes. When an input vector enters, the ESOINN finds two nearest nodes as the winner and the second winner by the predetermined metric. Using a threshold criterion of similarity (the maximum distances between vectors owing to the same cluster), the network judges: an input vector belongs to the winner or the second winner cluster or not. A distribution of input data is unknown, and a threshold criterion is updated adaptively for each separate node. A threshold criterion for node T_i is calculated by Eq. 4.27, where N_i is a set of neighbor nodes, $\mathbf{W}_i$ and $\mathbf{W}_j$ are the weight vectors of nodes i and j, respectively.

$$T_i = \max_{j \in N_i} \|\mathbf{W}_i - \mathbf{W}_j\| \tag{4.27}$$

If a node i has not the connected neighbor nodes, then a threshold criterion Eq. 4.27 is transformed in Eq. 4.28, which is defined as a minimum distance between nodes, where N is a set of all network nodes.

$$T_i = \min_{j \in N \setminus \{i\}} \|\mathbf{W}_i - \mathbf{W}_j\| \tag{4.28}$$

An input vector is inserted as the first node of new class, if distance between an input vector and the winner or the second winner is more than a threshold value between the winner and the second winner. If an input vector belongs to the cluster of the winner or the second winner, then an edge between the winner and the second winner is created with 0 "age", and the "age" of all edges linked to the winner is increased by 1.

Then a density p_i of the winner is updated by Eq. 4.29, where $\overline{d_i}$ is a mean value of distances between node j and its neighbor nodes.

$$p_i = 1 \Big/ \left(1 + \overline{d_i}\right)^2 \tag{4.29}$$

If a mean value of distances between node j and its neighbor nodes is large, then a number of nodes and a density p_i of node i will have small values, and vice versa. For each iteration λ, only a density of winner-node is calculated. The accumulated density h_i of winner-node is provided by Eq. 4.30, where n is a total number of iterations (calculated as $n = LT/\lambda$, LT is a total number of input vectors), K is a number of iterations, when a density value for node i exceeds 0.

$$h_i = \frac{1}{K} \cdot \sum_{l=1}^{n} \sum_{k=1}^{\lambda} p_i \tag{4.30}$$

After re-calculation of density, a counter of wins M_i (for a winner-node) is increased by 1. The change of weight vectors of the winner ΔW_i and its neighboring nodes ΔW_j ($j \in N_i$) are determined by Eqs. 4.31 and 4.32, where $\mathbf{W}_x$ is a weight input vector.

$$\Delta W_i = \frac{1}{M_i} \cdot \left(\|\mathbf{W}_x\| - \|\mathbf{W}_i\|\right) \tag{4.31}$$

$$\Delta W_j = \frac{1}{100 \cdot M_i} \cdot \left(\|\mathbf{W}_x\| - \|\mathbf{W}_j\|\right) \tag{4.32}$$

Then all edges, the "age" of which is higher a threshold value age_{max}, are removed. If number of input vectors does not achieved λ iterations, then a following input vector is submitted. Otherwise, the overlaps between classes are detected and removed by including additional subclasses.

One can find the detailed description of algorithms for separation a composite class into subclasses, a building the edges between nodes, and a classifying nodes to the different classes in research [69].

4.5.2   Algorithm of ESOINN Functioning

The algorithm of the ESOINN functioning includes the steps mentioned below.

Step 1. Set a minimal number of the predetermined segments into a set of images. Number of the pre-determined segments is defined using the JSEG algorithm.

This is our first proposed distinction in comparison with the ESOINN, which initialization is started always from two random nodes. (Let the ESOINN with our color-texture-fractal descriptor be called as the dESOINN.)

Step 2. Input a weight input vector $\mathbf{W}_x$. The second proposed distinction connects with the order of ranked vectors $\{\mathbf{W}_x\}$ at the inputs of the dESOINN. Vectors $\{\mathbf{W}_x\}$ is sorted according to the results of previous segmentation. Therefore, the winner and the second winner are defined at the first steps, and at the following steps the dESOINN is trained using the remaining samples from current segment. Sometimes an input vector $\mathbf{W}_x$ cannot be associated with the current winner owing to coarse segmentation errors. Such input vector ought to be rejected from a sample. This approach reinforces the current winner and makes the stochastic dESOINN more stable. Then the input vectors $\{\mathbf{W}_x\}$ concerning to another segment are clustered by the dESOINN. The proposed approach is especially useful for clustering of non-large set of natural images.

Step 3. Define the nearest node (the winner) a_1 and the second nearest node (the second winner) a_2. If a distance between the input vector $\mathbf{W}_x$ and nodes a_1 or a_2 exceeds threshold values T_i calculating by Eqs. 4.27 and 4.28, then the input vector is considered as a new node and added to a node set. Go to Step 2.

Step 4. Increment the "age" of all edges connecting with a node a_1 by 1.

Step 5. Define the edge creation necessary between nodes a_1 and a_2.

Step 6. Recalculate the accumulated density h_i of the winner-node by Eq. 4.30.

Step 7. Increment the counter of wins M_i by 1.

Step 8. Calculate the weight vectors of the winner ΔW_i and its neighboring nodes ΔW_j using Eqs. 4.31 and 4.32.

Step 9. Remove the edges, "age" of which has more value than a pre-determined parameter age_{max}.

Step 10. If a number of the input vectors $\mathbf{W}_x$ is multiple to a parameter λ, then it is required to update the subclasses for each node and remove the "noisy" nodes using Eqs. 4.33 and 4.34, where N is a number of nodes in a node set, c_1 and c_2 are the empirical coefficients. Equations 4.33 and 4.34 are used, if a node has two or one neighbors, respectively

$$h_i < c_1 \cdot \sum_{j=1}^{N_a} h_j / N \qquad (4.33)$$

$$h_i < c_2 \cdot \sum_{j=1}^{N_a} h_j / N \qquad (4.34)$$

In experiments, the non-large sets of images were used. Therefore, the additional condition Eq. 4.35 was introduced to remove the single nodes, where c_3 is an empirical coefficient.

$$h_i < c_3 \cdot \sum_{j=1}^{N_a} h_j / N \qquad (4.35)$$

Step 11. If a clustering process is finished, then it is needed to determine a number of classes, the output sample vectors for each class, and stop the algorithm.

Step 12. Go to Step 2, if the ESOINN continues to work.

4.6   Experimental Results

For experiments, 120 images from the dataset IAPR TC-12 Benchmark [11] were selected as the 10 sets including 12 images in each set. An example set (12 images and their segmented prototypes by JSEG algorithm with removal small size fragments, Set NN 01) are presented in Fig. 4.2.

The first type of experiments was directed to obtain the precision estimations. The average values of color-texture-fractal features for segments from a set of images, representing in Fig. 4.2, are summarized in Table 4.1. As a result, five clusters (VWs) were determined by the dESOINN as it is show in Fig. 4.3.

As one can see, the clusters from Fig. 4.3 represent three types of objects—"Houses", "Sky", and "Water". However, the dESOINN divided the cluster "Houses" into three clusters. This decision reflects the differences in values of color and texture features (Table 4.1).

Three algorithms were compared: fuzzy c-means, the ESOINN, and the dESOINN. The last one begins its work by use a predetermined minimum number of clusters provided by the JSEG algorithm. For initialization, the following parameters were applied: $\lambda = 50$, $age_{max} = 5$, $c_1 = 0.01$, $c_2 = 0.3$, and $c_3 = 1.05$.

The average precision of algorithms PRC was calculated by Eq. 4.36, where C is a total number of clusters, NTP_i is a number of true positive examples (true detected regions) and NFP_i is a number of false positive examples (false detected regions) in cluster i relatively the expert estimations.

$$PRC = \frac{1}{C} \sum_{i=1}^{C} \frac{NTP_i}{NTP_i + NFP_i} \qquad (4.36)$$

All calculations had been repeated 100 times, and then a precision was averaged out. The parameter "Number of clusters" was chosen as the most frequent value during clustering. The generalized estimations of precision and execution time for fuzzy c-means, ESOINN, and dESOINN algorithms are summarized in Table 4.2. For experiments, PC Acer JM50-HR with a single processor core, Intel Core i5-2430 M 2,4 GHz, RAM Kingston 1333 MHz (PC3-10700) DDR3 8 GB, VC NVIDIA GeForce GT 540 M, 1 GB, SSD Smartbuy, 128 GB was used.

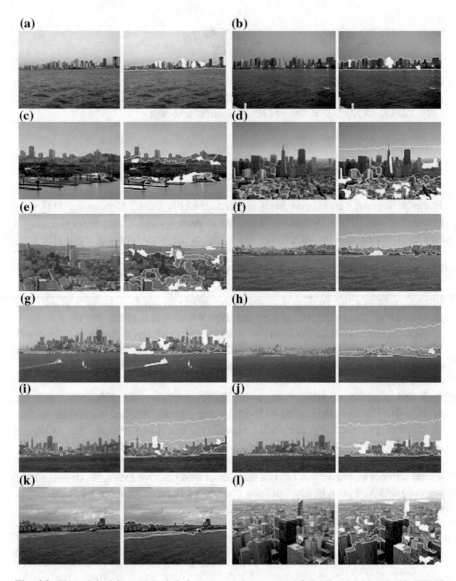

Fig. 4.2 The original images and their segmented prototypes from the database IAPR TC-12 Benchmark: **a** image 2954, **b** image 2956, **c** image 38056, **d** image 38060, **e** image 38063, **f** image 38097, **g** image 38129, **h** image 38183, **i** image 38273, **j** image 38277, **k** image 39458, **l** image 40417

The data from Table 4.2 shows that a precision of VWs using the ESOINN and the dESOINN is better than received by fuzzy c-means algorithm. Also the determined clusters are close for human perception. A creation of VWs by ESOINN or dESOINN is slowly in 3–5 times against fuzzy c-means algorithm for small

Table 4.1 The average values of color-texture-fractal features of Set NN 01

Feature	Cluster 0	Cluster 1	Cluster 2	Cluster 3	Cluster 4
FC_0 (mean H)	0.552435	0.568377	0.458251	0.0596694	0.285296
FC_1 (mean S)	0.324153	0.534038	0.160444	0.0120252	0.185824
FC_2 (mean V)	0.84274	0.444355	0.366143	0.952904	0.492806
FC_3 (st_deviation H)	0.00832996	0.00378337	0.0365661	0.0267154	0.0854395
FC_4 (st_deviation S)	0.00842481	0.030075	0.0393522	0.00138649	0.0738243
FC_5 (st_deviation V)	0.00783369	0.0274488	0.0433435	0.000830239	0.120126
FC_6 (skewness H)	0.00011649	-7.84582e-07	-0.000495041	0.00225364	-2.34724e-05
FC_7 (skewness S)	-2.24345e-06	-5.33866e-05	0.000101857	2.29497e-08	0.000930306
FC_8 (skewness V)	-1.35649e-06	4.79419e-05	9.35369e-05	-1.85642e-10	0.00132605
FT_9 (mean)	0.84274	0.444355	0.366143	0.952904	0.492806
FT_{10} (variance)	0.000128232	0.00102089	0.00232532	1.8876e-06	0.0162171
FT_{11} (homogeneity)	0.243134	0.0658959	0.0429404	0.844896	0.0175021
FT_{12} (im_smoothness)	3.892003	2.992619	2.635473	5.724094	1.801059
FT_{13} (im_entripy)	0.301775	0.54859	0.624985	0.0471946	0.77565
FT_{14} (skewness)	-0.0365511	0.227385	0.078331	0.270309	0.603937
FT_{15} (kurtosis)	0.242881	0.928843	0.838573	1.24215	1.0476
FF_{16} (fractal_dim)	2.43106	2.34751	2.51404	2.21268	2.88331
FF_{17} (lacunarity)	0.000203711	0.00578365	0.0225221	2.29768e-06	0.0867445

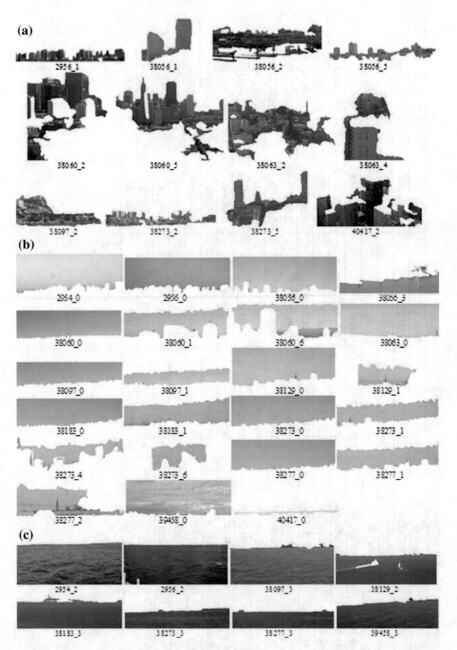

Fig. 4.3 Five clusters of VWs determined by dESOINN, Set NN 01: **a** cluster 0, which includes 12 segments 2956_1, 38056_1, 38056_2, 38056_5, 38060_2, 38060_5, 38063_2, 38063_4, 38097_2, 38273_2, 38273_5, 40417_2, **b** cluster 1, which includes 23 segments 2954_0, 2956_0, 38056_0, 38056_3, 38060_0, 38060_1, 38060_6, 38063_0, 38097_0, 38097_1, 38129_0, 38129_1, 38183_0, 38183_1, 38273_0, 38273_1, 38273_4, 38273_6, 38277_0, 38277_1, 38277_2, 39458_0, 40417_0, **c** cluster 2, which includes 8 segments 2954_2, 2956_2, 38097_3, 38129_2, 38183_3, 38273_3, 38277_3, 39458_3, **d** cluster 3, which includes 6 segments 38063_1, 38097_4, 38097_5, 38129_3, 38183_2, 38183_4, **e** cluster 4, which includes 11 segments 2954_1, 38056_4, 38060_3, 38060_7, 38060_7, 38063_3, 38063_5, 38277_4, 39458_1, 39458_2, 40417_1

(d)

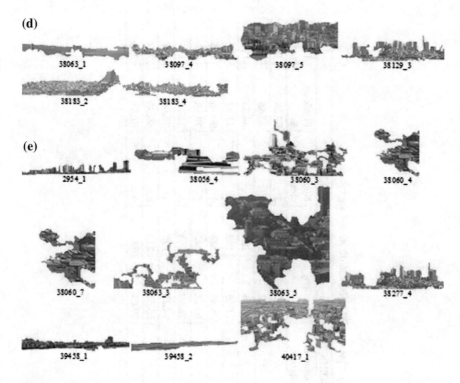

(e)

◀ **Fig. 4.3** (continued)

number of samples. However, the unsupervised clustering with large sets of images will be promising. The main benefits of ESOINN algorithm are a possibility of the unsupervised clustering and the on-line implementation. This network can be trained by novel data adaptively, and it stores the previous data with any increasing volume of input information.

The additional experiments provided a comparison of precision and execution time results with large number of annotated images. Seven sets with 50, 100, 150, 200, 250, 300, and 350 images from the dataset IAPR TC-12 Benchmark [11] were selected and tested by basic ESOINN, DBSCAN algorithm [73], X-Means algorithm [74], and dESOINN. The experiments were implemented using the same PC Acer JM50-HR (Table 4.3).

For the ESOINN initiation, the same parameters were tuned as in the main experiment. The calculations had been repeated 100 times, and then the precision values were averaged. The often received values are chosen as the parameter "Number of clusters".

The plots in Fig. 4.4 show the generalized precision and time dependences for these four algorithms. One can see that the dESOINN algorithm provides better results for large number of annotated images.

Table 4.2 The comparative results of precision and execution time for fuzzy *c*-means, ESOINN, and dESOINN algorithms

Set NN	VWs, determined by expert	Fuzzy *c*-means			ESOINN			dESOINN		
		Number of clusters	Precision, %	Time, ms	Number of clusters	Precision, %	Time, ms	Number of clusters	Precision, %	Time, ms
00	5	5	65.53	79	5	65.13	193	5	**67.28**	186
01	3	6	81.04	98	6	83.22	195	5	**84.66**	182
02	5	5	63.90	68	5	65.54	194	5	**66.63**	193
03	4	4	68.41	56	4	71.16	182	4	**71.47**	182
04	3	5	82.59	91	4	88.84	222	4	**90.58**	215
05	5	6	69.07	162	6	69.13	242	6	**70.68**	236
06	4	4	72.23	69	4	**73.89**	183	5	73.71	175
07	4	4	70.93	65	5	**74.27**	275	4	73.84	251
08	6	8	67.16	162	8	65.50	223	8	**69.52**	223
09	6	6	73.79	114	6	79.07	244	6	**81.83**	232

Table 4.3 The comparative results of precision and execution time for ESOINN, DBSCAN, X-Means, and dESOINN algorithms

Number of images	VWs, determined by expert	ESOINN			DBSCAN			X-Means			dESOINN		
		Number of clusters	Precision, %	Time, ms	Number of clusters	Precision, %	Time, ms	Number of clusters	Precision, %	Time, ms	Number of clusters	Precision, %	Time, ms
50	45	50	64.39	1896	46	74.7	18447	60	78.83	1887	57	90.91	1346
100	67	67	62.37	6620	80	75.2	73824	74	80.24	5406	71	86.5	4004
150	91	102	61.74	12687	118	72.43	164172	88	78.18	17544	91	84.78	7446
200	111	123	61.51	19631	127	71.08	294536	116	75.05	27336	109	85.87	12083
250	136	149	62.69	35671	146	70.5	465530	134	73.52	35598	138	83.02	18753
300	155	161	61.12	48397	144	69.24	657850	165	75.6	42738	156	83.61	26034
350	165	170	59.87	60404	174	67.59	878639	175	72.92	56049	165	82.38	33843

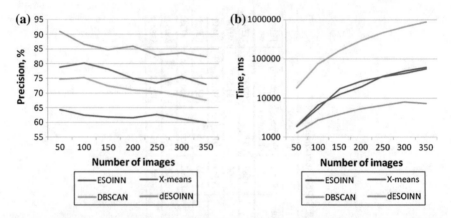

Fig. 4.4 Generalized dependences for ESOINN, DBSCAN, X-Means, and dESOINN algorithms: **a** precision, **b** execution time

The second type of experiments was devoted to increase a computational speed by parallel processing. A version of non-optimized algorithm was realized in Microsoft Visual C++ 2010 package. Versions of optimized algorithm were implemented in OpenMP and Intel Cilk Plus environments. The test dataset contained six sets, each set included 10 images. These six sets were formed from images with different resolutions, i.e. 1920 × 1080 pixels, 2560 × 1600 pixels, 2800 × 2100 pixels, 3840 × 2160 pixels, 3646 × 2735 pixels, and 4096 × 3072 pixels. A test parallel processing was executed in personal computer and servers with different configurations.

For mentioned above six sets of images, the mean values of processing time were estimated using various paralleling algorithms. Four samples, including 40 chosen randomly images, were formed. Each randomly chosen image was processed 25 times. Such methodology permits to decrease the influence of external factors such as activity of background task of operating system and available free hardware resources. Experiments show that a computational speed increases on 26–32 % using parallelizing algorithms. Hardware characteristics influence on computational speed directly. For example, a processing time using Intel Core i5-3450 (3.1 GHz) was in 2–2.2 times less than a processing time using Intel Core 2 Quad Q6600 (2.4 GHz).

A processing time of non-optimized algorithm was defined as 1 in order to calculate a relative speedup factor for optimized algorithms. The speedup factors were calculated for all six sets of images using 2, 3, and 4 threads. Average values of speedup factors for parallelizing algorithms are represented in Fig. 4.5.

Also additional tests were executed in order to compare OpenMP and Intel Cilk Plus environments. The results are drawn in Table 4.4. One can see the advantage of Intel Cilk Plus environment.

Fig. 4.5 Mean values of speedup factors

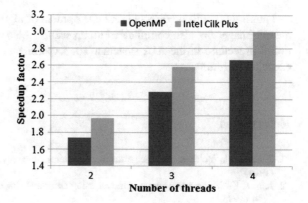

Table 4.4 Values of speedup factors

Image sizes	Number of threads								
	OpenMP			Intel Cilk Plus			Differences in values of speedup factors, %		
	2	3	4	2	3	4	2	3	4
2 MB	1.75	2.29	2.68	1.93	2.51	2.92	10.27	9.79	9.04
4 MB	1.74	2.30	2.64	1.96	2.58	2.99	12.90	12.11	13.38
6 MB	1.73	2.27	2.65	2.00	2.61	3.01	15.40	14.77	13.34
8 MB	1.74	2.29	2.69	1.94	2.58	2.99	11.05	12.35	11.18
10 MB	1.72	2.28	2.66	1.99	2.61	3.02	15.92	14.58	13.58
12 Mb	1.73	2.27	2.66	1.98	2.62	3.05	14.17	15.35	14.58
Mean value	1.74	2.28	2.67	1.97	2.58	3.00	13.29	13.16	12.52

4.7   Conclusion and Future Development

In this chapter, the AIA issues were investigated by the VWs extraction from the restricted image sets. The enhanced feature set describing an image region was suggested, which includes color, texture, and fractal features. Three algorithms were compared: fuzzy c-means, the ESOINN, and color-texture-fractal descriptor ESOINN (dESOINN) in the precision estimation and the execution time. The precision of VWs using the ESOINN and the dESOINN is higher then fuzzy c-means algorithm provides. The experiments demonstrate that the unsupervised clustering of large sets of images based on the neural network approach is promising than other algorithms.

Parallelizing algorithms permit to increase essentially an image processing including images of HD-quality. A computational speed was increased on 26–32 % using algorithms with parallel implementation. Also experiments show that Intel Cilk Plus environment provides the speedup values on 9–14 % higher in comparison with OpenMP environment due to effective balance of loading.

Future work will be directed on development of algorithms, annotating not only natural images but also complicated urban scenes. Also an image categorization is a useful procedure in the AIA systems. The experiments with other datasets will be executed in future.

References

1. Tamura, H., Yokoya, N.: Image database systems: a survey. Pattern Recogn. **17**(1), 29–43 (1984)
2. Jain, A.K., Vailaya, A.: Image retrieval using color and shape. Pattern Recogn. **29**(8), 1233–1244 (1996)
3. Antani, S., Kasturi, R., Jain, R.: A survey on the use of pattern recognition methods for abstraction, indexing and retrieval of images and video. Pattern Recogn. **35**(4), 945–965 (2002)
4. Islam, M.M., Zhang, D., Lu, G.: Automatic categorization of image regions using dominant color based vector quantization. In: IEEE International Conference on Digital Image Computing: Techniques and Applications (DICTS'2008), pp. 191–198 (2008)
5. Chang, E., Goh, K., Sychay, G., Wu, G.: CBSA: content-based soft annotation for multimodal image retrieval using Bayes point machines. IEEE Trans. Circ. Syst. Video Technol. **13**(1), 26–38 (2003)
6. Cusano, C., Ciocca, G., Schettini, R.: Image annotation using SVM. In: Proceedings of SPIE, SPIE internet imaging, vol. 5304, pp. 330–338 (2004)
7. Carneiro, G., Chan, A.B., Moreno, P.J., Vasconcelos, N.: Supervised learning of semantic classes for image annotation and retrieval. IEEE Trans. Pattern Anal. Mach. Intell. **29**(3), 394–410 (2007)
8. Liu, Y., Zhang, D., Lu, G.: Region-based image retrieval with high-level semantics using decision tree learning. Pattern Recogn. **41**(8), 2554–2570 (2008)
9. Cavus, O., Aksoy, S.: Semantic scene classification for image annotation and retrieval. In: da Vitoria Lobo, N., Kasparis, T., Roli, F., Kwok, J.T., Georgiopoulos, M., Anagnostopoulos, G. C., Loog, M. (eds) Structural, Syntactic, and Statistical Pattern Recognition, pp. 402–410. Springer, Berlin (2008)
10. Rao, C., Kumar, S.S., Mohan, B.C.: Content based image retrieval using exact Legendre moments and support vector machine. Int. J. Multimedia Appl. **2**(2), 69–79 (2010)
11. Index of /imageclef/resources. IAPR TC-12 Benchmark. http://www-i6.informatik.rwth-aachen.de/imageclef/resources/iaprtc12.tgz. Accessed 20 Dec 2014
12. Maree, R., Geurts, P., Piater, J., Wehenkel, L.: Random subwindows for robust image classification. In: IEEE Computer Society Conference on Computer Vision and Pattern Recognition (CVPR'2005), vol. 1, pp. 34–40 (2005)
13. Balasubramanian, G.P., Saber, E., Misic, V., Peskin, E., Shaw, M.: Unsupervised color image segmentation by dynamic color gradient thresholding. In: SPIE 6806, Human Vision and Electronic Imaging XIII, pp. 68061H–68061H-9 (2008)
14. Chan, T.F., Vese, L.A.: Active contours without edges. IEEE Trans. Image Process. **10**(2), 266–277 (2001)
15. Vanhamel, I., Pratikakis, I., Sahli, H.: Multiscale gradient watersheds of color images. IEEE Trans. Image Process. **12**(6), 617–626 (2003)
16. Makrogiannis, S., Vanhamel, I., Fotopoulos, S., Sahli, H., Cornelis, J.: Watershed-based multiscale segmentation method for color images using automated scale selection. J. Electron. Imaging **14**(3), 1–16 (2005)
17. Jung, C.R.: Combining wavelets and watersheds for robust multiscale image segmentation. Image Vis. Comput. **25**(1), 24–33 (2007)

18. Favorskaya, M.N., Petukhov, N.Y.: Comprehensive calculation of the characteristics of landscape images. J. Opt. Technol. **77**(8), 504–509 (2010)
19. Mansouri, A.R., Mitiche, A., Vazquez, C.: Multiregion competition: a level set extension of region competition to multiple region image partitioning. Comput. Vis. Image Underst. **101** (3), 137–150 (2006)
20. Alnihou, J.: An efficient region-based approach for object recognition and retrieval based on mathematical morphology and correlation coefficient. Int. Arab. J. Inf. Technol. **5**(2), 154–161 (2008)
21. Ugarriza, L.G., Saber, E., Vantaram, S.R., Amuso, V., Shaw, M., Bhaskar, R.: Automatic image segmentation by dynamic region growth and multi-resolution merging. IEEE Trans. Image Process. **18**(10), 2275–2288 (2009)
22. Shi, J., Malik, J.: Normalized cuts and image segmentation. IEEE Trans. Pattern Anal. Intell. **22**(8), 88–905 (2000)
23. Carson, C., Belongie, S., Greenspan, H., Malik, J.: Blobworld: image segmentation using expectation-maximization and its application to image querying. IEEE Trans. Pattern Anal. Mach. Intell. **24**(8), 1026–1038 (2002)
24. Tu, Z., Zhu, S.C.: Image segmentation by data-driven Markov chain Monte Carlo. IEEE Trans. Pattern Anal. Mach. Intell. **24**(5), 657–673 (2002)
25. Rui, S., Jin, W., Chua, T.S.: A novel approach to auto image annotation based on pairwise constrained clustering and semi-Naive Bayesian model. In: 11th International Conference on Multimedia Modelling, pp. 322–327 (2005)
26. Alata, O., Ramananjarasoa, C.: Unsupervised textured image segmentation using 2-D quarter-plane autoregressive model with four prediction supports. Pattern Recogn. Lett. **26**(8), 1069–1081 (2005)
27. Cariou, C., Chehdi, K.: Unsupervised texture segmentation/classification using 2-D autoregressive modeling and the stochastic expectation-maximization algorithm. Pattern Recogn. Lett. **29**(7), 905–917 (2008)
28. Mezaris, V., Kompatsiaris, I., Strintzis, M.G.: An ontology approach to object-based image retrieval. In: International Conference on Image Processing (ICIP'2003), vol. 2, pp. 511–514 (2003)
29. Haralick, R.M., Shanmugam, K., Dinstein, I.: Textural features for image classification. IEEE Trans. SMC-3 **6**, 610–621 (1973)
30. Xia, Y., Feng, D., Wang, T., Zhao, R., Zhang, Y.: Image segmentation by clustering of spatial patterns. Pattern Recogn. Lett. **28**(12), 1548–1555 (2007)
31. Hou, Y., Lun, X., Meng, W., Liu, T., Sun, X.: Unsupervised segmentation method for color image based on MRF. In: International Conference on Computational Intelligence and Natural Computing (CINC'2009), vol. 1, pp. 174–177 (2009)
32. Celik, T., Tjahjadi, T.: Unsupervised color image segmentation using dual-tree complex wavelet transform. Comput. Vis. Image Underst. **114**(7), 813–826 (2010)
33. Comaniciu, D., Meer, P.: Mean shift: a robust approach toward feature space analysis. IEEE Trans. Pattern Anal. Mach. Intell. **24**(5), 603–619 (2002)
34. Haykin, S.O.: Neural Networks and Learning Machines, 3edn. McMaster University, Prentice-Hall, Canada (2008)
35. Deng, Y., Manjunath, B.S.: Unsupervised segmentation of color-texture regions in images and video. IEEE Trans. Pattern Anal. Mach. Intell. **23**(8), 800–810 (2001)
36. Komati, K.S., Salles, E.O.T., Filho, M.S.: Unsupervised color image segmentation based on local fractal dimension and J-images. In: IEEE International Conference on Industrial Technology (ICIT'2010), pp. 303–308 (2010)
37. Yang, A.Y., Wright, J., Ma, Y., Sastry, S.S.: Unsupervised segmentation of natural images via lossy data compression. Comput. Vis. Image Underst. **110**(2), 212–225 (2008)
38. Nock, R., Nielsen, F.: Statistical region merging. IEEE Trans. Pattern Anal. Mach. Intell. **26** (11), 1452–1458 (2004)
39. Martınez-Uso, A., Pla, F., Garcıa-Sevilla, P.: Unsupervised color image segmentation by low-level perceptual grouping. Pattern Anal. Appl. **16**(4), 581–594 (2013)

40. Kim, J.S., Hong, K.S.: Color–texture segmentation using unsupervised graph cuts. Pattern Recogn. **42**(5), 735–750 (2009)
41. Yang, Y., Han, S., Wang, T., Tao, W., Tai, X.C.: Multilayer graph cuts based unsupervised color-texture segmentation using multivariate mixed Student's t-distribution and regional credibility merging. Pattern Recogn. **46**(4), 1101–1124 (2013)
42. Vogel, T., Nguyen, D.Q., Dittmann, J.: BlobContours: adapting Blobworld for supervised color- and texture-based image segmentation. In: Multimedia Content Analysis, Management, and Retrieval, SPIE, vol. 6073, pp. 158–169 (2006)
43. Unnikrishnan, R., Hebert, M.: Measures of similarity. In: IEEE Workshop on Application of Computer Vision, vol. 1, pp. 394–400 (2005)
44. Silakari, S., Motwani, M., Maheshwari, M.: Color image clustering using block truncation algorithm. Int. J. Comput. Sci. Issues **4**(2), 31–35 (2009)
45. Kekre, H.B., Mirza, T.: Content based image retrieval using BTC with local average thresholding. In: International Conference on Content Based Image Retrieval (ICCBIR'2008). Niagara Falls, Canada, pp. 5–9 (2008)
46. Chakravarti, R., Meng, X.: A study of color histogram based image retrieval. In: IEEE 6th International Conference on Information Technology: New Generations, pp. 1323–1328 (2009)
47. Rao, P.S., Vamsidhar, E., Raju, G.S.V.P., Satapat, R., Varma, K.V.S.R.P.: An approach for CBIR system through multi layer neural network. Int. J. Eng. Sci. Technol. **2**(4), 559–563 (2010)
48. Long, X., Su, D., Hu, R.: Incremental leaning algorithm for self-organizing fuzzy neural network. In: IEEE 7th International Conference on Computer Science & Education (ICCSE'2012), pp. 71–74 (2012)
49. Tung, W.L., Quek, C.: GenSoFNN: A generic self-organizing fuzzy neural network. IEEE Trans. Neural Netw. **13**(5), 1075–1086 (2002)
50. Malek, H., Ebadzadeh, M.M., Rahmati, M.: Three new fuzzy neural networks learning algorithms based on clustering, training error and genetic algorithm. Appl. Intell. **37**(2), 280–289 (2011)
51. Hao, P., Ding, Y., Fang, Y.: Image retrieval based on fuzzy kernel clustering and invariant moments. In: 2nd International Symposium on Intelligent Information Technology Application (IITA'2008), Shanghai, vol. 1, pp. 447–452 (2008)
52. Wang, H., Mohamad, D., Ismail, N.A.: Semantic Gap in CBIR: automatic objects spatial relationships semantic extraction and representation. Int. J. Image Process. **4**(3), 192–204 (2010)
53. Swain, M.J., Ballard, D.H.: Color indexing. Int. J. Comput. Vis. **7**(1), 11–32 (1991)
54. Dubey, R.S., Choubey, R., Bhattacharjee, J.: Multi feature content based image retrieval. Int. J. Comput. Sci. Eng. **2**(6), 2145–2149 (2010)
55. Pass, G., Zabith, R.: Histogram refinement for content-based image retrieval. In: IEEE Workshop on Applications of Computer Vision, pp. 96–102 (1996)
56. Huang, J., Kuamr, S., Mitra, M., Zhu, W.J., Zabih, R.: Image indexing using color correlogram. In: International Conference on Computer Vision and Pattern Recognition (CVPR'1997), San Juan, Puerto Rico, pp. 762–765 (1997)
57. Manjunath, B.S., Salembier, P., Sikora, T.: Introduction to MPEG-7: Multi-media Content Description Language. John Wiley & Sons Ltd., New York (2002)
58. Tamura, H., Mori, S., Yamawaki, T.: Texture features corresponding to visual perception. IEEE Trans. Syst. Man Cybern. **8**(6), 460–473 (1978)
59. Selvarajah, S., Kodituwakku, S.R.: Analysis and comparison of texture features for content based image retrieval. Int. J. Latest Trends Comput. **2**(1), 108–113 (2011)
60. Haralick, R.M., Shanmugum, K., Dinstein, I.: Textural features for image classification. IEEE Trans. Syst. Man Cybern. **3**(6), 610–621 (1973)
61. Galloway, M.M.: Texture analysis using gray level run lengths. Comput. Graph. Image Process. **4**(2), 172–179 (1975)

62. Favorskaya, M., Damov, M., Zotin, A.: Intelligent texture reconstruction of missing data in video sequences using neural networks. In: Tweedale, J.W., Jain, L.C. (eds.) Advanced Techniques for Knowledge Engineering and Innovative Applications, pp. 163–176. Springer, Berlin (2013)
63. Favorskaya, M., Damov, M., Zotin, A.: Accurate spatio-temporal reconstruction of missing data in dynamic scenes. Pattern Lett. Recogn. **34**(14), 1694–1700 (2013)
64. Favorskaya, M.N., Petukhov, N.Y.: Recognition of natural objects on air photographs using neural networks. Optoelectron. Instrum. Data Process. **47**(3), 233–238 (2011)
65. Mandelbrot, B.B.: The Fractal Geometry of Nature. Freeman, San Francisco (1982)
66. Slabaugh, G., Boyes, R., Yang, X.: Multicore Image Processing with OpenMP. IEEE Sig. Process. Mag. **27**(2), 134–138 (2010)
67. Saleem, S., Lali, I.U., Nawaz, M.S., Nauman, A.B.: Multi-core program optimization: parallel sorting algorithms in Intel Cilk Plus. Int. J. Hybrid Inf. Technol. **7**(2), 151–164 (2014)
68. Furao, S., Ogura, T., Hasegawa, O.: An enhanced self-organizing incremental neural network for online unsupervised learning. Neural Netw. **20**(8), 893–903 (2007)
69. Tangruamsub, S., Tsuboyama, M., Kawewong, A., Hasegawa, O.: Mobile robot vision-based navigation using self-organizing and incremental neural networks. In: International Joint Conference on Neural Networks (IJCNN'2009), pp. 3094–3101 (2009)
70. Najjar, T., Hasegawa, O.: Self-organizing incremental neural network (SOINN) as a mechanism for motor babbling and sensory-motor learning in developmental robotics. In: Rojas, I., Joya, G., Gabestany, J. (eds.) Advances in Computational Intelligence, pp. 321–330. Springer, Berlin (2013)
71. De Paz, J.F., Bajo, J., Rodríguez, S., Corchado, J.M.: Computational intelligence techniques for classification in microarray analysis. In: Bichindaritz, I., Vaidya, S., Jain, A., Jain, L.C. (eds.) Computational Intelligence in Healthcare 4. Studies in Computational Intelligence, pp. 289–312. Springer, Berlin (2010)
72. De Paz, J.F., Rodríguez, S., Bajo, J.: Multiagent systems in expression analysis. In: Demazeau, Y., Pavón, J., Corchado, J.M., Bajo, J. (eds.) 7th International Conference on Practical Applications of Agents and Multi-Agent Systems. Advances in Intelligent and Soft Computing, pp. 217–226. Springer, Berlin (2009)
73. Ester, M., Kriegel, H.P., Sander, J., Xu, X.: Density-based algorithm for discovering clusters in large spatial databases with noise. In: 2nd International Conference on Knowledge Discovery and Data Mining (KDD'1996), pp. 226–231. AAAI Press (1996)
74. Ishioka, T.: An expansion of X-means for automatically determining the optimal number of clusters. In: IASTED International conference on Computational Intelligence (CI'2005), Calgary, Alberta, Canada, pp. 91–96 (2005)

Chapter 5
An Evolutionary Optimization Control System for Remote Sensing Image Processing

Victoria Fox and Mariofanna Milanova

Abstract Remote sensing image analysis has been a topic of ongoing research for many years and has led to paradigm shifts in the areas of resource management and global biophysical monitoring. Due to distortions caused by variations in signal/image capture and environmental changes, there is not a definite model for image processing tasks in remote sensing and such tasks are traditionally approached on a case-by-case basis. Intelligent control, however, can streamline some of the case-by-case scenarios and allow for faster, more accurate image processing to aid in more accurate remote sensing image analysis. This chapter will provide an evolutionary control system via two Darwinian particle swarm optimizations—one a novel application of DPSO—coupled with remote sensing image processing to help in the analysis of image data.

Keywords Darwinian particle swarm optimization · Remote sensing · Intelligent control

5.1 Introduction

In July, 1966, Professor Seymour Papert assigned a summer homework project [1] to a group of graduate students in order to "use our summer workers effectively in the construction of a significant part of a visual system [for computers]" and overcome the issues of "pattern recognition", "figure-ground analysis", "region description", and "object identification." What his students and later researchers

V. Fox (✉)
Department of Mathematics, University of Arkansas at Monticello, 397 University Drive, Monticello, Arkansas 71656, USA
e-mail: fox@uamont.edu

M. Milanova
Computer Science Department, University of Arkansas at Little Rock, 2801 S. University Ave, Little Rock, Arkansas 72204, USA
e-mail: mgmilanova@ualr.edu

© Springer International Publishing Switzerland 2016
R. Kountchev and K. Nakamatsu (eds.), *New Approaches in Intelligent Image Analysis*, Intelligent Systems Reference Library 108,
DOI 10.1007/978-3-319-32192-9_5

discovered is image segmentation, while crucial to higher-level image analysis tasks, is an ill-posed problem subject to perceptual constraints and the overall goal of a specific segmentation. To point, machine vision segmentation is impeded by texture interference, noise, partially occluded objects, blurred edges, and illumination artifacts—all of which are easily processed and discarded by human perceptual vision [2]. Nearly a half-century later, what artificial-intelligence expert Papert thought would be resolved in a few months by graduate students remains a focus and a major sub-discipline of computer science.

In the area of remote sensing—the scanning of an object or area by satellite or aircraft in order to obtain information—image segmentation algorithms must overcome or exploit issues caused by multi-spectral and often multi-scale input data. While there exist remote sensing images with high ground resolutions (e.g. images captures from airborne sensors), a large majority of remote sensing images contain a spatial resolution too low to identify landmarks by shape or spatial detail. Many algorithms and their researchers bypass the spatial domain completely and focus on spectral signatures gathered by the multispectral sensors of the remote sensing platform. However, the use of spectral signatures to distinguish between objects in a remote sensing image is often hindered by the natural variability of a material, granular spectral quantization, and modification of spectral signatures by atmospheric conditions [3]. To add to the complexity of the task, spectral signature data vary from sample to sample, causing researches to compare relative signatures of material within image data rather than search for absolute signatures. As a result, numerous algorithms have been created, compared, and discarded as researchers search for an optimal method.

In 2010, Dey et al. reviewed standard image segmentation techniques that have been applied to remote sensing [4]. The review divided methodologies into three classes: image driven approaches, homogeneity measure approaches, and model driven approaches. Image driven approaches rely upon statistical analysis of the image data and usually incorporate edge information. Approaches that rely upon measures of similarity (or dissimilarity) calculate spectral, textural, spatial, shape, and size measurements and then group objects by some degree of homogeneity. Homogeneity models sometimes incorporate contextual, temporal, and prior knowledge into their algorithms. Model driven segmentation methods include thresholding, Markov random field models, fuzzy set models, neural models, watershed models, and multi-resolution models.

A more recent survey of methods presented at the 10th International Conference on Simulated Evolution and Learning [5], considered the recent advances of evolutionary computation in the realm of image segmentation. While there are a wealth of evolutionary computation techniques for optimization, Liang, Zhang, and Brown found genetic algorithms, genetic programming, differential evolution, and particle swarm optimizations are the most common techniques used in evolutionary computation segmentation methods. Of the four listed techniques, genetic algorithms provided the most simplistic model and particle swarm optimization provided the most accurate optimizations of segmentation algorithms. All of the methods suffered from a low generalization capacity and a relatively high computational complexity. Only two case studies came from the field of remote sensing—one for a multi-objective genetic clustering and another for fuzzy partitioning guided by a genetic algorithm. However, both case studies reported optimal results.

In the area of active learning algorithms for remote sensing segmentation, Tuia et al. [6] surveyed fifty-six active learning algorithms and found that all of the algorithms suffered from a lack of available training sets for researchers, especially concerning the large volume of data to process in the field. Recent adaptations of active learning to the field of remote sensing image segmentation include active selection of unlabeled pixels for classification, spatially adaptive heuristics, and active learning algorithms applied to model adaptation across domains. However, few of the heuristics in the survey use contextual features (e.g. position, texture) and none of the methods are robust to noise. For more discussion concerning the methods listed here, it is suggested to the reader to consult [4–6] and the references listed therein.

From the above discussion, it is obvious remote sensing image segmentation is an area of ongoing research and novel combination of techniques. As such, there is not a definite model for segmentation of remotely sensed images. Intelligent control in image processing, the field of creating adaptive image processing sequences to perform tasks in complex, often varying, and knowledge-intensive applications, can streamline some of the case-by-case scenarios and allow for faster, more accurate image processing. This chapter will provide an intelligent control system coupled with modified image processing techniques to help in the processing of remote sensing image data.

5.2   Background Techniques

5.2.1   Darwinian Particle Swarm Optimization

First used to model social behavior by simulating the movement of individuals in a bird flock or fish school [7], particle swarm optimization quickly made the leap into mathematical optimization of nonlinear functions. The method begins with randomly or heuristically initializing a swarm and then allowing each particle in the swarm to search a given space for a possible solution. Evaluation of the fitness of a particle as well as the neighboring particles occurs on every iteration. As the particles evaluate their fitness and store solutions, they move toward the particle in their neighborhood with the best fitness value. As the swarm continues, it, in theory, approaches the optimal solution and records the solution for the best performing particle in its immediate neighborhood. To model the swarm, each particle n moves in a multidimensional space according to position (x_n) and velocity (v_n) values. Both x_n and v_n are highly dependent on local best (x_b), neighborhood best (n_{nb}) and global best (g_b) information:

$$v_n = Wv_n + p_1 r_1 (g_b - x_n) + p_2 r_2 (x_b - x_n), \quad x_n = x_n + v_n \qquad (5.1)$$

where W represents an inertial weight applied to the velocity, the constants p_1 and p_2 represent "cognitive" and "social" components, and r_1, r_2 are random vectors with each component generally a uniform random number between 0 and 1.

Due to the lack of assumptions in the formulation of the swarm, (i.e. it is a metaheuristic model), particle swarm optimization does not guarantee an optimal solution will be found. However, the formulation does not require the gradient of the cost function, which greatly expands the applications to which it can be applied. Many optimization methods require the optimization function to be differentiable, naturally excluding many applied mathematics functions that are irregular, noisy, or dynamic. As a result, even with its lack of guarantee of an optimal solution, particle swarm optimization is a noteworthy method in areas that cannot use classic optimization methods.

Darwinian particle swarm optimization [8] is an alteration of the classic particle swarm optimization in that it adapts to the fitness landscape and removes areas of stagnation as multiple swarms search for a solution. The assumptions in Darwinian particle swarm optimization are

- In order to assure future generations, each swarm is given a constant, if small, chance of spawning a new swarm. In essence, the longer a swarm lives, the higher the probability it will have offspring.
- The lifetime of a swarm is extended by the swarm finding a more fit solution.
- The lifetime of a swarm is reduced by the swarm failing to find a more fit solution.

By the use of these assumptions modeled from Darwinian natural selection, the method is more likely to arrive at a global optimal solution. The trade-off is, of course, higher computational complexity. Multiple swarms require parallel implementations of particle swarm optimization and the use of natural selection requires the introduction of more parameters in the model.

For traditional particle swarm optimizations, the variables to evaluate the dynamic behavior of a swarm are

v_i	The velocity of the particle.
x_i	The position of the particle.
t	Time.
r_1	A uniform random variable, sampled for each i, t and dimension of the vector x_i.
r_2	A uniform random variable, sampled for each i, t and dimension of the vector x_i.
$x_{i,p}$	The previous position of a particle that resulted in the best fitness (so far).
$x_{i,n}$	The neighborhood position that resulted in the best fitness (so far).
W	An inertia weight applied to the velocity

In Darwinian particle swarm optimization, the method begins by randomly initializing each dimension of each particle over an appropriate range. The velocities are also restricted to an appropriate range and are randomly initialized to encourage exploration of a solution. At each iteration of the algorithm, the decision to evolve or die-off is assessed for each swarm. The evolution of a swarm is determined by evaluating the individual fitness of all the particles in the swarm. Swarms with a new global best fitness are allowed to spawn a new particle. A swarm that fails to find a better fitness solution loses a particle. If a swarm has failed to update to a new best fit after a set number of iterations, the entire swarm is deleted as it has become stagnated (trapped in a local optimum). In practice, as the worst performing particle is deleted when a swarm fails to find a new global best, the population of the swarm decreases. When the population falls beneath a set minimum population level, the swarm is deleted.

The natural selection rules of Darwinian particle swarm optimization allow swarms with good adaptations to continue and swarms with bad adaptations to die-off. One parameter that can negatively affect swarm continuation is the velocity of a particle at initialization. A simple way to control velocity such that a particle is not "exploded" through the search space is to define a maximum velocity and not allow any one velocity to exceed it. Another method to control initial velocity is the introduction of a parameter that simulates friction, causing a particle to experience a "drag" on its velocity measurement. The friction parameter controls the velocity of the particles so that they can better search the solution space for the cost function. The lack of the parameter can result in particles moving large distances in the solution space in one iteration and missing a better fitness value. The selection of which method to use—bounding maximum velocity or introduction of a friction parameter—depends on the fitness landscape of the optimization problem. In general, bounding maximum velocity performs well on complicated landscapes and the friction parameter performs well on more regular landscapes [9]. In optimization problems, Darwinian particle swarm optimization made a significant improvement in the method circumventing local optima and finding the best global optimum.

The first published use of Darwinian particle swarm optimization for image processing occurred in 2012 [10] and was applied to segmentation of remote sensing images. Posing the problem as an optimization of Otsu's Optimal Global Thresholding scheme [11]. Otsu's exhaustive search for $n - 1$ optimal thresholds for n-level image segmentations becomes an evaluation of $n(L - n + 1)^{n-1}$ combinations of thresholds and is computationally expensive for relatively small values of n. In consideration of the size of remote sensing images and the number of potential segmentations in low-resolution remote sensing captures, Otsu's method—formulated as an exhaustive search—is infeasible. However, formulating the task as a multidimensional optimization problem allows the use of Darwinian particle swarm optimization, which greatly lowers the computational expense of the segmentation.

5.2.2   Total Variation for Texture-Structure Separation

A challenging portion of image segmentation is decomposing an image into meaningful features to guide the segmentation process. In traditional region-based active contour segmentation algorithms, for example, the primary feature is the intensity level of the region inside the contour versus the intensity level of the region outside the contour. For the area of remote sensing, separating image information can take the form of removal of irrelevant texture information for the overall segmentation. Consider, as an example, the following images (Fig. 5.1):

Assuming the purpose of the segmentation is to segment the bodies of water, the Fig. 5.1b provides a smoother image surface to apply the active contour and, as a result, achieve the segmentation with a lower probability of becoming "stuck" in local optimums as the contour evolves. Admittedly, the image with the texture removed has areas of new concern, which primarily take the form of artifacts and distortions caused by the smoothing process. From this simple example, however, one can infer the level of attention given to devising algorithms that will remove unwanted texture yet keep structural accuracy of the objects of interest for segmentation tasks depending upon optimization of image energy.

In the structure-texture decomposition problem, an image can be assumed to be composed of a structural part, which represents the objects of interest, and a textural part. The textural part is usually classified by its level of fine scale-details and, in most instances, contains some type of oscillatory nature [12]. A large failing of this intuitive assumption is the definition of texture depends upon the scale of the image. In remote sensing, the scale is often of such a nature that all of the image takes the form of texture (from the above description). A "structure" in a high resolution scale can be regarded as a "texture" in a low resolution scale. As such, when working with low resolution images, care must be taken in setting parameters for the separation of structure and texture.

(a)   **(b)**

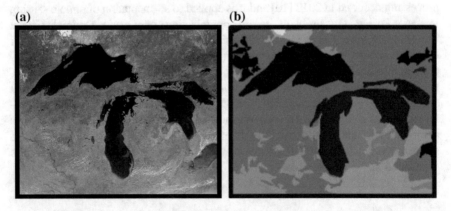

Fig. 5.1 Example of smoothed texture. **a** Original. **b** Image with texture removed

Total variation regularization methods offer a way to separate the structure and textures of an image and provide parameters of which can be tuned to a specific resolution. In image processing, the segmentation task is often formulated as an inverse problem: given the image I, find another image, u, "close" to I such that u represents a simplification of I and can be used in the segmentation task. Generally, u is desired to be an image formed by homogenous regions with sharp boundaries. The image representing the information removed from I can be denoted as v where v is usually a noise or texture. The resulting relation between I, u and v can be expressed as $I = u + v$. Many models disregard v and seek to optimize u, computing optimal piece-wise smooth approximations of I (e.g. active contour methods). If v contains information important to the locating the object of interest, however, these models generally fail. Total variation segmentation methods, extract both u and v in a simple total variational minimization frame [13]. Then, v is represented using two functions (g_1, g_2). This simplification of texture descriptors greatly lowers the computational difficulty in segmenting textured images.

The total variation minimization method for image decomposition begins with mapping a function from $\mathbf{R}^2$ to $\mathbf{R}$. Let $I: \mathbf{R}^2 \rightarrow \mathbf{R}$ be a given image such that I is an element of $L^2(\mathbf{R}^2)$. Allowing that u represents an approximate image of I without additive noise (or texture) v, then the relation between u and I can be expressed as a simple linear model: $I(u, x) = u(x, y) + v(x, y)$. Given that both u and v are unknown, the problem of reconstructing u from I is given as a minimization problem in the space of functions of bounded variation $BV(\mathbf{R}^2)$. This space of functions works well with I, u, and v owing to its property of allowing for edges or discontinuities along curves [14]. Using this space, the model for denoising (or de-texturizing) images while preserving edges is

$$\nabla \inf_{u \in L^2} F(u) = \int |\nabla u| + \lambda \int |I - u|^2 dxdy, \tag{5.2}$$

where λ is a tuning parameter and the second term is a fidelity term. The first term in the formulation is a regularizing term, which keeps important features and edges. For proofs of existence and uniqueness of the above minimization problem, please consult [13]. Allowing $v = I - u$ and then formally minimizing $F(u)$ gives the associated Euler-Lagrange equation:

$$u = I + \frac{1}{2\lambda} div \left(\frac{\nabla u}{|\nabla u|} \right), \tag{5.3}$$

To model v as texture [15], define G as the Banach space consisting of all generalized functions of $v(x, y)$ which can be written as

$$v(x, y) = \partial_x g_1(x, y) + \partial_y g_2(x, y), \quad g_1, g_2 \in L^\infty(\mathbf{R}^2), \tag{5.4}$$

bound of all L^∞ norms of functions

$$\left|\vec{g}\right| = \sqrt{g_1(x,y)^2 + g_2(x,y)^2} \tag{5.5}$$

and the infimum is computed oval all decompositions of v. Showing that, if the v component represents texture, then v is an element of G, Y. Meyer proposed and justified in [15] the following new image restoration model:

$$\inf_u \{E(u) = \int |\nabla u| + \lambda \|v\|_\varphi, I = u + v\} \tag{5.6}$$

With the mathematical underpinnings now stated, we can proceed to the Vese and Osher [13] formulation of a total variation minimization for modeling texture:

$$\inf_{u,g_1,g_2} \left\{ G_p(u, g_1, g_2) = \int |\nabla u| + \lambda \int \left| I - u - \partial_x g_1 - \partial_y g_2 \right|^2 dxdy \right.$$
$$\left. + \mu \left[\int \left(\sqrt{g_1^2 + g_2^2} \right)^p dxdy \right]^{\frac{1}{p}} \right\}, \tag{5.7}$$

where the tuning parameters, λ and μ, are greater than zero and $p \to \infty$. The first term insures u is element of $BV(R^2)$, the second term guarantees that $I \approx u + div$ (g) and the third term operates as a penalty on the norm in G of $v = div(g)$. Minimizing the energy in (5.7) with respect to u, g_1, and g_2 gives the following Euler-Lagrange equations:

$$u = I - \partial_x g_1 - \partial_y g_2 + \frac{1}{2\lambda} div\left(\frac{\nabla u}{|\nabla u|}\right), \tag{5.8}$$

$$\mu\left(\left\|\sqrt{g_1^2 + g_2^2}\right\|\right)^{1-p} \left(\sqrt{g_1^2 + g_2^2}\right)^{p-2} g_1 = 2\lambda\left[\frac{\partial}{dx}(u - I) + \partial_{xx}^2 g_1 + \partial_{xy}^2 g_2\right], \tag{5.9}$$

$$\mu\left(\left\|\sqrt{g_1^2 + g_2^2}\right\|\right)^{1-p} \left(\sqrt{g_1^2 + g_2^2}\right)^{p-2} g_1 = 2\lambda\left[\frac{\partial}{dy}(u - I) + \partial_{xy}^2 g_1 + \partial_{yy}^2 g_2\right], \tag{5.10}$$

Using (5.7)–(5.9), Vese and Osher demonstrate that texture can be effectively modeled and segmented from "well-behaved" texture images. For further discussion on generalizations of structure-texture models based upon the VeseOsher method as well as comments on optimizing the parameter models, the reader is referred to [12] and the references therein.

Recently, Xu et al. [16] presented a relative total variation method for the purpose of extracting structure, u, from a highly textured image, I. In contrast to the previous referenced methods, relative total variation begins without any assumption of specific regularity or symmetry in the texture pattern which allows a high level of randomness in detecting texture, v. The method contains a general pixel-wise local total variation measure, expressed as

$$D_x(p) = \sum_{q \in R(p)} g_{p,q} \cdot \left| (\partial_x S)_q \right|, \quad D_y(p) = \sum_{q \in R(p)} g_{p,q} \cdot \left| (\partial_y S)_q \right|, \qquad (5.11)$$

where q is an element of $R(p)$. the rectangular region centered at pixel p. $D_x(p)$ and $D_y(p)$ are windowed total variations in the x and y directions for pixel p, which counts the absolute spatial difference within the window, $R(p)$. The function $g_{p,q}$ works as a weight defined accordingly to spatial affinity and is proportional to

$$\exp\left(-\frac{(x_p - x_q)^2 + (y_p - y_q)^2}{2\sigma^2} \right), \qquad (5.12)$$

where σ controls the spatial scale of the window. Additionally, the method contains a windowed inherent variation, given as

$$L_x(p) = \left| \sum_{q \in R(p)} g_{p,q} \cdot \partial_x S_q \right|, \quad L_y(p) = \left| \sum_{q \in R(p)} g_{p,q} \cdot \partial_y S_q \right|, \qquad (5.13)$$

in which L captures the global variations. Mathematically, $L(x, y)$ differs from $D(x, y)$ in that $L(x, y)$ does not use the modulus in its formulation. As the result, the sum of ∂S in $L(x. y)$ can change signs depending upon the direction of the gradients. In general, the resulting $L(x. y)$ in a window containing only texture is smaller than $L(x, y)$ $\mathcal{L}(x, y)$. in a window containing texture and edges. Finally, to enhance the contrast between texture and structure, Xu et al. formulated the relation between $D(x, y)$ $\mathcal{D}(x, y)$ and $L(x, y)$ $\mathcal{L}(x, y)$ as the objective function

$$\arg \min_s \sum_p (S_p - I_p)^2 + \lambda \cdot \left(\frac{D_x(p)}{L_x(p) + \varepsilon} + \frac{D_y(p)}{L_y(p) + \varepsilon} \right), \qquad (5.14)$$

in which the term $(S_p - I_p)^2$ prevents the input and result from varying without bound. In the formulation, S is the resulting image structure and I is the input. The term $(S_p - I_p)^2$ is calculated for every iteration to force structural similarity between the output and input. The second term in the above equation is the regulator referred to as relative total variation. The parameter is a weighting parameter that controls the smoothness of the result. Empirically, the larger, the blurrier the result. σ controls the scale of the texture features and is vital in structure-texture separation. The parameters used in [16] are not optimized. Instead, Xu et al. provide suggestions for values to experiment with in order to obtain the desired structure-texture separation.

5.2.3 Multi-phase Chan-Vese Active Contour Without Edges

In general, region based active contours yield more reasonable segmentations than edge-based algorithms when an image has relatively large noise and/or texture values. However, the complexity and computational cost of region-based methods can be large, particularly when considering methods based upon partial differential equations, e.g. active contour models, total variation energy. In order to reduce computational cost, the level set function $\phi(x, y)$ was proposed by Malladi, Sethian, and Vemuri [17] as a formulation to implement active contours. Representing the contour implicitly via two-dimensional Lipschitz-continuous functions defined on the image plane. O a particular level, usually the zero level, the level set function is defined as a contour, such as $C = \{(x, y): \phi(x, y) = 0\}$ or all (x, y) in I. By using the zero level set, the contour can be defined as the border between a positive area and a negative area. As the level set function increases from the initial stage, the corresponding set of contours moves toward the exterior.

In 2001, Tony Chan and Luminita Vese proposed a model for active contours that did not use edge information to segment an image. This method, dubbed the Active Contour without Edges method, has under-gone many alterations, restatements, and simplifications since its initial publication [18]. In general, given a curve $C = \delta\omega$, with ω in Ω an open subset, and two unknown constants c_1 and c_2, with $\Omega_1 = \omega$, $\Omega_2 = \Omega/\omega$, the energy of the curve can be minimized in a level set formulation with respect to c_1, c_2 and $C = \{(x, y)|\phi(x, y) = 0\}$.

$$F(c_1, c_2, \varphi) =$$
$$\int_\Omega (u_0(x,y) - c_1)^2 H(\varphi)\partial x\partial y + \int_\Omega (u_0(x,y) - c_2)^2(1 - H(\varphi))\partial x\partial y + v\int_\Omega |\nabla H(\varphi)|.$$

$$(5.15)$$

$H(\phi)$ is the Heaviside function and c_1, c_2 are resolved, thorough minimization of $F(c_1, c_2, \phi)$ as approximations of intensity within (c_1) and outside (c_2) the level set.

While the above method is groundbreaking in its own right, Chan and Vese further refined [19] their method to include segmentation of color images, four-phase segmentations, and theoretically, eight-phase segmentations. In practice, this refinement, the Multi-Phase Active Contour without Edges, does not exceed four-phases and two level sets. Four a four-phase contour:

$$F_4(c, \varphi) = \int_\Omega (u_0(x,y) - c_{11})^2 H(\varphi_1) H(\varphi_2) \partial x \partial y$$

$$+ \int_\Omega (u_0(x,y) - c_{10})^2 H(\varphi_1)(1 - H(\varphi)) \partial x \partial y$$

$$+ \int (u_0 - c_{01})^2 (1 - H(\varphi_1)) H(\varphi_2) \partial x \partial y$$

$$+ \int (u_0 - c_{00})^2 (1 - H(\varphi_1))(1 - H(\varphi_2) \partial x \partial y + v \int_\Omega |\nabla H(\varphi_1)| + v \int_\Omega |\Delta H(\varphi_2)|,$$

$$(5.16)$$

where $c = \{c_{11}, c_{10}, c_{01}, c_{00}\}$ a constant vector in which each c_{ij} represents the mean of the intensities for each phase and $\phi = \{\phi_1, \phi_2\}$. With these notations, the image function u can be expressed as:

$$u = c_{11} H(\varphi_1) H(\varphi_2) + c_{10} H(\varphi_1)(1 - H(\varphi_2)) + c_{01}(1 - H(\varphi_1)) H(\varphi_2) + c_{00}(1 - H(\varphi_1))(1 - H(\varphi_2)),$$

$$(5.17)$$

5.3 Evolutionary Optimization of Segmentation

5.3.1 Darwinian PSO for Thresholding

With the preliminary discussion of background methods given in Sect. 5.2, we will now proceed to the control system and method used to segment multispectral remote sensing images in this chapter. The primary weakness in Otsu's threshold method is the amount of post processing a user must apply to the thresholded image. Depending upon the complexity of the image in question and the region of interest for the segmentation, a user may have to implement numerous image-processing methods to "clean" the "segmented" thresholded image of unwanted classes. Consider the segmentation performed in Fig. 5.2 with five levels of classification created with a Darwinian particle swarm optimized Otsu thresholding algorithm [10]. From the algorithm, the levels of intensity that will give the maximum between class variance are 58, 99, 145, 199. For an individual image, separating the classes manually utilizing the between class variance and selecting the class of interest is not a difficult task. However multiply the task by a set of images and it quickly becomes time consuming to post-process the images. Additionally, one may notice the texture left in the thresholded image in Fig. 5.2. While the concrete, water, and asphalt all have clear class separations, the surrounding small-scale portions of the image contain a mixture of all five classes. Once the classes are manually separated using the intensity information from the method, a researcher will again use a post-processing procedure to remove the smaller artifacts left over from including the smaller-scale texture portions in the

(a)   (b)

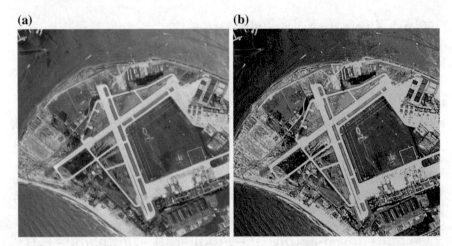

Fig. 5.2 Segmentation via thresholding. **a** Original remote sensed image. **b** Thresholded to five levels

Fig. 5.3 Class 5
segmentation of image (**b**) in
Fig. 5.2

larger regions of interest. Consider segmenting the class associated with the highest intensity band in the classes. Using the information about intensity level from the segmentation, the class 5 segmentation, without post-processing, is given in Fig. 5.3. A far more efficient method would be to transfer the thresholded image to another segmentation method to extract the class of interest. However, the amount of texture remaining in the thresholded image is problematic for further segmentation methods. At this point, it would be advantageous to move to the structure-texture separation algorithms detailed in Sect. 5.2.2. However, the algorithms detailed in Sect. 5.2.2 are not optimized. Using the architecture of Darwinian particle swarm optimization and relative total variation, optimization of the parameters is easily obtained.

5.3.2 Novel Darwinian PSO for Relative Total Variation

As stated in the overview of relative total variation, the parameter involving smoothness is of little importance in determining the structure of the image. As such, lambda will be set to a value of one throughout all experiments. σ, however, is of vital importance. In the formulation, σ appears in the definition of g (p) (Eq. 5.12). In Eq. (5.13), $g(p)$ is used in both the formulations of $D(p)$ and L (p) which are then passed to the relative total variation formulation (5.14). To compute the best σ value for a given image a relative total variation methodology, one must find the σ that represents the size of the background texture in the image. Too large or too small a σ will result in an image without any variation, i.e. a blank image of relatively uniform intensity. Therefore, σ must be bound: $\sigma_{min} < \sigma < \sigma_{max}$. In general, for lower resolution images, $\sigma \rightarrow \sigma_{max}$ and for higher resolution images, $\sigma \rightarrow \sigma_{min}$. Using Eq. (5.12) as the fitness function, the following Darwinian PSO is applied to a given remote sensing image (Table 5.1).

Table 5.1 Parameters for optimization of σ in Eq. 5.12 with the Darwinian PSO

Parameter	DPSO
Num of iterations	100
Population	30
ρ_1	0.8
ρ_2	0.8
W	1.2
V_{max}	1.5
V_{min}	−1.5
X_{max}	255
X_{min}	0
σ_{max}	20
σ_{min}	1
Max population	50
Min population	10
Initial num of swarms	4
Max swarms	6
Min swarms	2
Stagnancy	10

(a) (b)

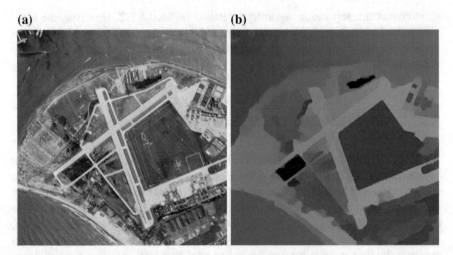

Fig. 5.4 The effect of optimal relative total variation. **a** Original image. **b** Result from relative total variation with optimal

where W, ρ_1, ρ_2 are coefficients that assign weights to the inertial influence (friction), the global best and the local best when determining the new velocity, respectively. X represents the (x, y) position within the image. To demonstrate the effectiveness of the optimized relative total variation, consider Fig. 5.4 in which the optimal value for the image in Fig. 5.1a was computed with the above parameters and found to be 4.3017. The cartoon result, u of the original image, I, will be much easier to segment with traditional segmentation methods. However, applying the optimized total variation to the original image, before classification, removes areas of interest. Therefore, for the procedure detailed in Sect. 5.3.4, the Darwinian PSO total variation algorithm is applied after processing the original image in the Darwinian PSO threshold method. Since both methods obtain optimal thresholding or σ automatically, the flow of the control system is entirely automatic.

5.3.3 Multi-phase Active Contour Without Edges with Optimized Initial Level Mask

Due to the low importance placed upon the initial contour in the original formulation [18, 19], it has become the practice of traditional Active Contours without Edges (ACWE) to begin the level set evolution with an arbitrarily defined ϕ—usually in the form of a circle or rectangle. Multiphase ACWE are traditionally initiated with multiple arbitrarily defined ϕ_i that completely cover the image plane. The lack on emphasis placed upon the initial contour is partially caused by the sampling of intensities across the entire image domain. In original experiments,

Chan and Vese set v, the parameter that weights the curve length, at 0, evolving their experimental equation to

$$F(c_1, c_2, \varphi) =$$
$$\int_\Omega (u_0(x,y) - c_1)^2 H(\varphi)\partial x \partial y + \int_\Omega (u_0(x,y) - c_2)^2 (1 - H(\varphi))\partial x \partial y$$

$$(5.18)$$

By removing a requirement for where to begin the initial contour, Chan and Vese distanced their method from the traditional active contour method which depends upon initial contour information to evolve. However, recent research has underscored optimization benefits that can be gained from using edge maps to initiate the contour in an ACWE [20].

Our method for segmentation begins with applying the Darwinian PSO threshold segmentation routine to an image and then passing the thresholded image through the Darwinian PSO relative total variation scheme to separate structure and texture. As an intermediate step, the image has clear delineations between the optimized thresholded regions in its intensity profile. The Darwinian PSO threshold segmentation routine outputs a three-layer image with the weights of each categorization embedded in each layer. If there are only three threshold levels, then each layer of the output image represents one class. By using the maximum variance between class information provided by the optimization, a mask with sharp edges can be extracted from the layers of the thresholded image.

As an example, consider an image that has been thresholded with the Darwinian PSO threshold optimization scheme into three classes. The maximum variance between each class is the global best fitness value for each class and can be used to choose which layer to use for an initial contour. Figure 5.5 shows the output of such a segmentation given by the optimized thresholding scheme.

The intensity levels that reflect the maximum variance between classes for each of the layers in Fig. 5.5 are given in Table 5.2.

As mentioned in Sect. 5.2.3, Chan and Vese gave a theoretical formulation for an eight-phase ACWE with three level sets. To the author's knowledge, the eight-phase ACWE has not been implemented in the traditional Multiphase ACWE formulation. Therefore, reflecting the common practice of using two level sets—resulting in four phases—for multiphase ACWE, only two of the class segmentations can be used as initial masks for the optimized ACWE algorithm. By inspection, it is noted the general results that produce the best segmentation in the three level threshold segmentations with the Darwinian PSO are the classes with the largest variance. For the example in Fig. 5.5, then, Class 2 and Class 3 would be selected as the masks to feed into the optimized Multiphase ACWE. The selection of the classes to use as masks becomes a simple matter of initiating an *if-then* statement after evaluating the difference in the variances.

Furthermore, since the optimized Multiphase ACWE is operating on a roughly segmented image, it is advantageous to use the edge map of the initial masks to help

Fig. 5.5 Three-level segmentation results of Darwinian PSO thresholding

Table 5.2 Intensity levels for maximum variance between classes for Fig. 5.5

Class 1	Class 2	Class 3
110	59	43
183	147	150

regulate the energy of the method. Recording the binary edge maps of the initial masks as $m = \{m_1, m_2\}$, then the contour regulation term of the contour can be expressed as

$$v \int_\Omega \left(H(\varphi) - m\right)^2 dxdy. \tag{5.19}$$

where $\phi = \{\phi_1, \phi_2\}$ and H is the Heaviside function, given by

$$H_\varepsilon(u) = \frac{1}{2}\left(1 + \frac{2}{\pi}\arctan\left(\frac{u}{\varepsilon}\right)\right). \tag{5.20}$$

Furthermore, by expressing $c = \{c_{11}, c_{10}, c_{01}, c_{00}\}$ as the mean of each phase, we can write the generalized formulation for the optimized Multiphase ACWE with a mask regularizing term as

$$F_4(c, \varphi) =$$

$$\int_\Omega (u_0(x,y) - c_{11})^2 H(\varphi_1) H(\varphi_2) \partial x \partial y + \int_\Omega (u_0(x,y) - c_{10})^2 H(\varphi_1)(1 - H(\varphi_2)) \partial x \partial y$$

$$+ \int_\Omega (u_0(x,y) - c_{01})^2 H(\varphi_2)(1 - H(\varphi_1)) \partial x \partial y + \int_\Omega (u_0(x,y) - c_{10})^2 H(\varphi_1)(1 - H(\varphi_2)) \partial x \partial y$$

$$+ \int_\Omega (u_0(x,y) - c_{00})^2 (1 - H(\varphi_2))(1 - H(\varphi_1)) \partial x \partial y + v \int_\Omega (H(\varphi_1) - m_1)^2 + v \int_\Omega (H(\varphi_2) - m_2)^2$$

$$(5.21)$$

As an example of the effectiveness of this algorithm, Fig. 5.6 demonstrates the resulting segmentation of the original image in Fig. 5.5 by the optimized Multiphase ACWE and setting Class 2, Class 3 segmentations as initial masks. As the results demonstrate, the method can quickly segment the simple image.

5.3.4 Workflow of Proposed System

Due to the evolutionary particle swarm algorithm used in two of the intermediate steps—one a novel application of Darwinian particle swarm optimization—the entire system is automated. Figure 5.7 shows workflow of the proposed system. The optimization algorithms provide computationally low-cost solutions for

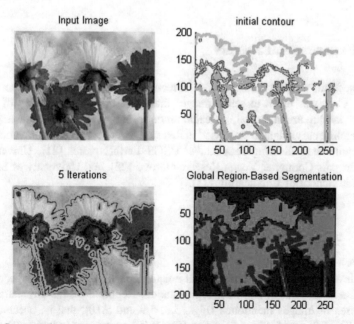

Fig. 5.6 Segmentation results of a simple image using the optimized Multiphase ACWE

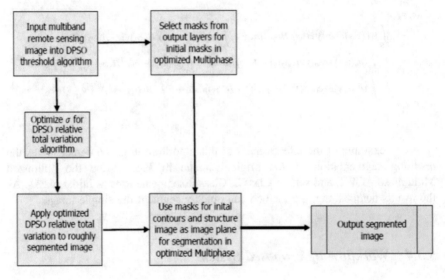

Fig. 5.7 Workflow of proposed system

traditionally, high-computational cost methods. The information provided by the best fitness values of intensity thresholding and smoothing window directly influence the information given to the segmentation process.

5.4   Experimental Results

In this section, we provide experimental results applied to various remote sensing images. Care was taken to select images that represent a large variety of remote sensing tasks: urban, rural, oceanic monitoring, vegetation indexing, and aerial/satellite surveillance. Data used in this study were collected from various sites and include the following data banks: USGS EarthExplorer [21], University of Massachusetts Computer Vision Research Group [22], and University of Southern California School of Engineering [23].

5.4.1   Results

The following tables reflect the original image, roughly segmented image, selected masks, structure image, and final segmentation. Reported results are restrained to the classes of images mentioned (Figs. 5.8, 5.9 and 5.10), that is, oceanic monitoring (Fig. 5.8), rural (Fig. 5.9), urban (Fig. 5.10), vegetation indexing (Fig. 5.11), and aerial/satellite images (Figs. 5.12, 5.13 and 5.14). The images represent aerial

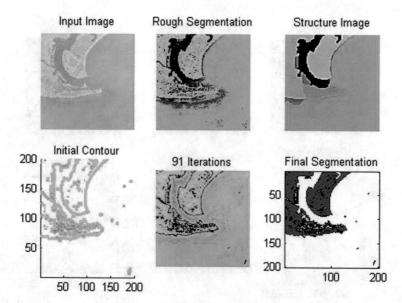

Fig. 5.8 Coastline segmentation

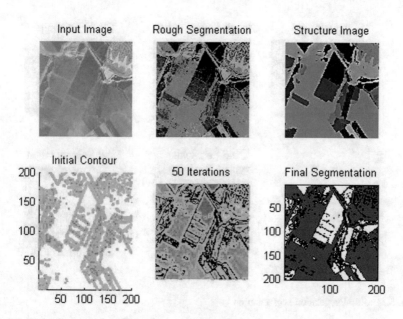

Fig. 5.9 Rural segmentation. Here, *dark* fields are the regions of interest

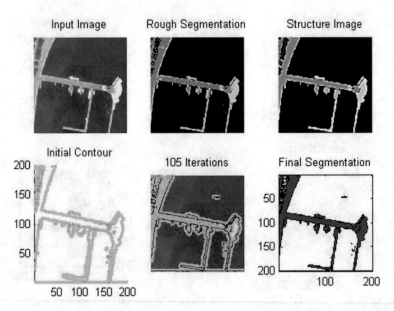

Fig. 5.10 Urban architecture segmentation

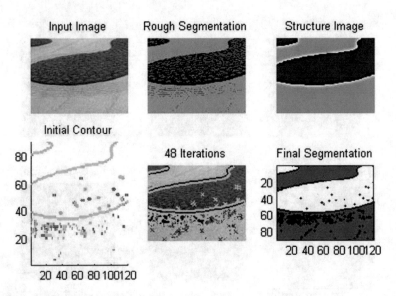

Fig. 5.11 Rural/vegetation segmentation

remote sensing resolution, low-resolution satellite images, and high-resolution satellite images. The parameters used in the system are as defined in Sect. 5.3. As a final note, the images were not preprocessed from their obtained state. Haze, blur, and other artifacts were left intact in the original image.

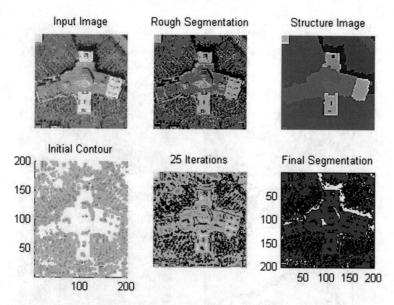

Fig. 5.12 Surveillance segmentation

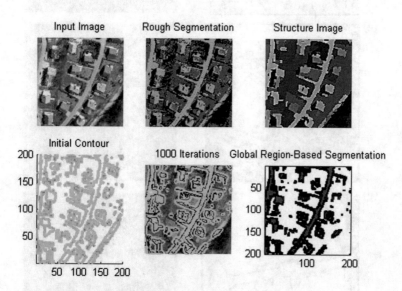

Fig. 5.13 Urban/Suburban architecture segmentation

By using the information provided by the evolutionary algorithms, i.e. maximum between class variance and optimum size for the soothing window, the workflow combines to preprocess images for the segmentation algorithm. The threshold segmentation gives a rough segmentation based upon intensity levels and the smoothing

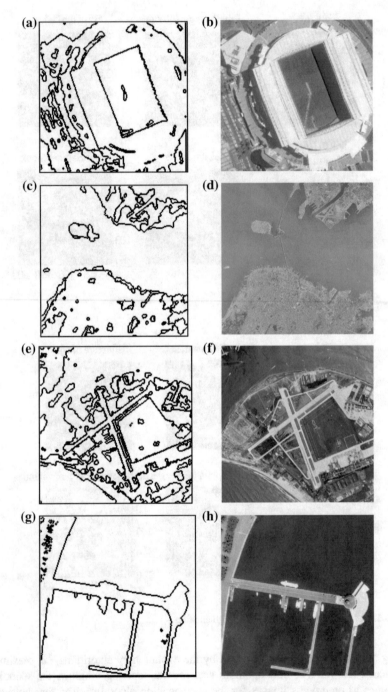

Fig. 5.14 Segmentation final contours and the original images for the segmentations

window operator removes texture from the image that could influence the active contour. The active contour method integrates the optimal mask from the threshold-optimizing module and the structure-texture separation from the smoothing module into a multiphase active contour with the edge map of the mask penalizing the energy to keep the contour in the neighborhood of the region of interest.

5.4.2 Discussion

An immediate notable observation is the lack of sophistication for the heuristic used in the region analysis portion of the active contour. To provide a method that focuses on the advantages of evolutionary schemes, a conscious decision was made by the researchers to ignore task specific heuristics to determine c_{11}, c_{10}, c_{01}, c_{00}. It is a recommendation, especially in the field of remote sensing image processing, to use all of the spectral and radiation information made available from the satellite sensors to improve and refine the segmentation of remote sensing images. Many of the images in the test set would have benefitted from the use of known heuristics to differentiate classes of objects of interest.

Still, the contour images showed a clear segmentation of complicated images. The active contour method, using intensity values in the phases, was attracted to objects with intensity levels on the white end of the color spectrum. Fabricated structures were more easily segmented, partially due to their uniformity in color and their inherent smooth texture. However, vegetation dense images were still partially segmented even with the extreme noise in the images. It is the opinion of the researchers that the use of known heuristics for vegetation would greatly help the segmentation of such images with this method. It is relatively easy to incorporate such heuristics into active contour energy formulations and the method benefits from the wealth of research relating to modifications to Active Contours without edges. Readers are directed to [21, 22], and the references in them to learn more about known heuristics that have been shown to aid in classification/segmentation of remote sensing images. For information on incorporating heuristics into active contour models, [23] and its references provide a good beginning for understanding the topic. However, the authors wish to stress the use of more specific heuristic to influence the active contour energy would not result in a change in the evolutionary algorithms that drive the system. If a researcher chooses to use heuristics that are more sophisticated for optimizing the energy of the contour, the system will still operate with the same workflow.

5.4.3 Conclusion and Future Research

The control system for processing a remote sensing image with the use of two Darwinian particle swarm optimizations of state-of-art methods and the passing of the optimization to a traditionally computationally expensive method executed well. Without the intelligent optimization method, the task would have been too computationally expensive to compute without switching to parallel processing. As such, the experiments were conducted on an IvyBridge5 processor machine with 64 bit architecture and 8 gigabits of SDRAM and each experiment was completed in under twenty seconds. The use of Darwinian particle swarm optimization implemented in this chapter to determine the optimum parameter value for smoothing is a new application of evolutionary programming. Additionally, the use of penalizing the contour energy with the edge map of the mask for each level set helped optimize the minimization of the contour energy. For future experimentation, the researchers plan to incorporate heuristics other than intensity values into the active contour formulation as well as extend the method into classification of objects in remote sensing images. The authors' novel use of Darwinian particle swarm optimization to determine the optimal smoothing window for the structure-texture separation will be explored further for other parameter optimization applications. The fact remains that the new combination of evolutionary algorithms' best-fit values provided by the Darwinian particle swarm optimization for thresholding and the novel use of the DPSO for determining optimal smoothing window size provided an automatic, computationally inexpensive segmentation of cluttered, distorted, and poorly sampled remote sensing images.

References

1. Papert, S.: The Summer Vision Project. Artificial Intelligence Group, Cambridge, MA (1966)
2. Wagemans, J., Elder, J.H., Kubovy, M., Palmer, E.S., Peterson, M.A., Singh, M., von der Heydt, R.: A century of Gestalt psychology in visual perception: I. Perceptual grouping and figure-ground organization. Psychol. Bull. **138**(6), 1172 (2012)
3. Schowengerdt, R.A.: Remote Sensing: Models and Methods for Image Processing, 3rd edn, pp. 13–15. Elsevier, Burlington, MA (2007)
4. Dey, V., Zhang, Y., Zhong, M.: A review on image segmentation techniques with remote sensing perspectives. In: Wagner, W., Szekely, B. (eds.) International Society for Photogrammmetry and Remote Sensing, XXXVIII (2010)
5. Yuyu, L., Zhang, M., Browne, W.N.: Image segmentation: a survey of methods based on evolutionary computation. In: Simulated Evolution and Learning, pp. 847–859. Springer International Publishing (2014)
6. Tuia, D., Volpi, M., Copa, L., Kanevski, M.F., Munoz-Mari, J.: A survey of active learning algorithms for supervised remote sensing image classification. IEEE J. Sel. Top. Sign. Proces. **5**(3), 606–617 (2011)
7. Kennedy, J., Eberhart, R.: Particle swarm optimization. In: Paper presented at the IEEE International Conference on Neural Networks, Perth, WA (1995)
8. Tillet, J., Rao, T., Sahin, F., Rao, R.: Darwinian particle swarm optimization. In: Proceedings of the 2nd Indian International Conference on Artificial Intelligence (2005). http://scholarworks.rit.edu/other/574, Accessed 3 Mar 2014
9. Panda, S., Padhy, N.P.: Comparison of particle swarm optimization and genetic algorithm for FACTS-based controller design. Appl. Soft Comput. **8**(4), 1418–1427 (2008)

10. Bhamisi, P., Couceiro, M.S, Ferreira, N.M., Kumar, L.: Use of Darwinian particle swarm optimization technique for the segmentation of remote sensing images. In: 2012 IEEE International Geoscience and Remote Sensing Symposium (IGARSS), pp. 4295–4298 (2012)
11. Otsu, N.: A threshold selection method from gray-level histograms. IEEE Trans. Syst. Man Cybern. B. Cybern. 9(1), 62–66 (1979)
12. Aujol, J.F., Gilboa, G., Chan, T., Osher, S.: Structure-texture image decomposition—modeling, algorithms, and parameter selection. Int. J. Comput. Vision 67(1), 111–136 (2006)
13. Vese, L.A., Osher, S.J.: Modeling textures with total variation minimization and oscillating patterns in image processing. J. Sci. Comput. 19(1−3), 553–572 (2003)
14. Rudin, L.I., Osher, S., Fatemi, E.: Nonlinear total variation based noise removal algorithms. Physica D 60(1), 259–268 (1992)
15. Meyer, Y.: Oscillating patterns in image processing and nonlinear evolution equations. In: The Fifteenth Dean Jacqueline B. Lewis Memorial Lectures, vol. 22, no.1. American Mathematical Society (2001)
16. Xu, L., Yan, Q., Yang, X., Jia, J.: Structure extraction from texture via relative total variation. ACM Trans. Graph. 31(6), 139 (2012)
17. Malladi, R.J.A., Sethian, J.A., Vemuri. B.C.: Shape modeling with front propagation: a level set approach. IEEE Trans. Pattern Anal. Mach. Intell. 17(2), 158–175 (1995)
18. Chan, T.F., Vese, L.A.: Active contours without edges. Image Proc. IEEE Trans. 10(2), 266–277 (2001)
19. Vese, L.A., Chan, T.F.: A multiphase level set framework for image segmentation using the Mumford and Shah model. Int. J. Comput. Vision 50(3), 271–293 (2002)
20. Huiyan, J., Tan, H., Yang, B.: A priori knowledge and probability density based segmentation method for medical CT image sequences. BioMed Res. Int. (2014)
21. USGS, USGS Earth Explorer. http://earthexplorer.usgs.gov/ (2014). Accessed 3 Oct 2014
22. UMass Computer Vision Research Group. http://vis-www.cs.umass.edu/~vislib/Aerial/directory.html (2014). Accessed 23 Sept 2014
23. USCViterbi, University of Southern California. http://sipi.usc.edu/database/ (2014). Accessed 15 Sept 2014

Chapter 6
Tissue Segmentation Methods Using 2D Histogram Matching in a Sequence of MR Brain Images

Vladimir Kanchev and Roumen Kountchev

Abstract MR brain image sequences are characterized by a specific structure and intra- and inter-image correlation but most of the existing histogram segmentation methods do not consider them. We address this issue by proposing a method for tissue segmentation using 2D histogram matching (TS-2DHM). Our 2D histogram is produced from a sum of co-occurrence matrices of each MR image. Two types of model 2D histograms are constructed: an intra-tissue 2D histogram for separate tissue regions and an inter-tissue edge 2D histogram. Firstly, we divide a MR image sequence into a few subsequences using wave hedges distance between 2D histograms of consecutive MR images. Then we save and clear out inter-tissue edge entries in each test 2D histogram, match the test 2D histogram segments in a percentile interval and extract the most representative entries for each tissue, which are used for kNN classification after distance learning. We apply the matching using LUT and two ways of distance metric learning: LMNN and NCA. Finally, segmentation of the test MR image is performed using back projection with majority vote between the probability maps of each tissue region, where the inter-tissue edge entries are added with equal weights to the corresponding main tissues. The proposed algorithm has been evaluated with IBSR 18 and 20, and BrainWeb data sets and showed results comparable with state-of-the-art segmentation algorithms, although it does not consider specific shape and ridges of brain tissues. Its benefits are modest execution time, robustness to outliers and adaptation to different 2D histogram distributions.

Keywords MR image segmentation · Transductive learning · 2D histogram matching · LUT · KNN classification · Back projection

V. Kanchev (✉) · R. Kountchev
Department of Radio Communications and Video Technologies,
Technical University of Sofia, 8 Kl. Ohridski Blvd., 1000 Sofia, Bulgaria
e-mail: v_kanchev@tu-sofia.bg

R. Kountchev
e-mail: rkountch@tu-sofia.bg

© Springer International Publishing Switzerland 2016   183
R. Kountchev and K. Nakamatsu (eds.), *New Approaches in Intelligent
Image Analysis*, Intelligent Systems Reference Library 108,
DOI 10.1007/978-3-319-32192-9_6

6.1   Introduction

Magnetic resonance imaging (MRI) possesses excellent soft-tissue contrast and high spatial resolution, which makes it a widely used method for anatomical imaging. One of its main tasks is medical image segmentation—to partition an image into different regions, corresponding to different tissues and organs. Currently, the segmentation task, in the case of MR images, is far from being solved because of non-consistent parameters, intra- and inter-subject intensity variability, low contrast and the presence of different types of noise, caused by time and equipment limitations. Moreover, the increased amount of MRI data requires the segmentation algorithms to be more robust to different intensity tissue ranges, resolution, and number of MR images in a sequence. A review of the current problems, methods, and future trends in MR brain image segmentation is given in [3, 11], while the interested reader is referred to [6] for more information about the necessary mathematics for biomedical imaging.

The image histogram is a well-known feature in medical imaging and it is used for segmentation, recognition, and image retrieval. One of its main properties is that there is no unique mapping between a histogram and a given image. Classic histogram segmentation approaches rely on the intensity distribution of individual pixels. They are based on the assumption that the intensity levels of each tissue stay within a separate interval and each pixel represents a single tissue type. In fact, this assumption is seldom satisfied: intensity ranges of separate tissues often overlap, pixels from a single tissue belong to different ranges due to intensity inhomogeneity (INH) artifact, and a single pixel includes a signal from more than one tissue due to the partial volume effect (PVE) artifact [44]. Therefore, there exists significant overlapping between separate histogram segments, which correspond to different tissue regions in the MR image. The histogram also changes its shape in the presence of INH and PVE artifacts. Another problem is the non-normal properties of MRI data, which leads to the use of more complex segmentation models. Hence, it is problematic to use the histogram to extract accurate intensity thresholds or to fit distributions based on its form. Another problem is that it does not contain spatial information from a MR image or a straightforward way to include prior information.

In order to use spatial correlation, existing between neighboring pixels in MR images, we construct a 2D histogram as a sum of co-occurrence matrices; thus, after normalization, the 2D histogram describes the probability distribution of pixel pairs at a given offset in the MR image. Until now, the co-occurrence matrices were used mostly for texture classification by extracting texture features [4] and medical image retrieval [58]: in general, they described structures in images. We use the 2D histogram as a global feature, do not quantize it which allows at the end segmentation using back projection into a local window due to its direct correspondence with pixel pairs from the test MR image. The global nature of the 2D histogram leads to lower computational load, as well.

Histogram matching [16] (or also called histogram specification) is a method used to generate a processed image that has a specified histogram. It is a classical preprocessing technique, used mostly for image enhancement in the case of degraded images. It aims to reshape the histogram of an image to stress certain intensity levels, characteristic of certain objects, and to suppress other intensity levels, characteristic of noise or image background. In our work, we have proposed a new application of the matching operation—we match a test 2D histogram and use it to construct a training set of each 2D histogram segment. Then we classify the test 2D histogram using a k-Nearest Neighbor (kNN) classifier after distance metric learning and finally we perform MR image segmentation using back projection.

Another common problem is integration of prior knowledge into the segmentation algorithm: in our method we integrate it using 2D histogram matching—the prior knowledge, presented by region MR images, corresponds to automatically selected MR images from the MR image sequence. For this purpose, we divide the MR image sequence into subsequences using a similarity distance coefficient between test 2D histograms of consecutive MR images. We construct model 2D histograms for each subsequence using their first and end region MR images. Thus, better accuracy is achieved, since tissue parameters vary smoothly along a MR image sequence.

So our main contribution consists of developing a tissue segmentation method using 2D histogram matching, applied to MR subsequences; the segmentation method consists of the following stages:

- divide a MR image sequence into few subsequences by using a similarity distance coefficient between 2D histograms.
- match a test 2D histogram in a 1D domain using model region 2D histograms.
- classify the test 2D histogram with a kNN classifier after distance metric learning.
- segment the test MR image by back projection using the classified 2D histogram.

In the next section we will provide more information about the existing research, related to our developed algorithm and a few state-of-the-art approaches in medical brain segmentation such as Markov random fields (MRF)/hidden Markov models (HMM), variational and level set methods, mixture model methods, and fuzzy methods, related to fuzzy c-means (FCM) clustering.

6.2   Related Works

The common histogram methods for medical segmentation, based on the intensity values of individual pixels, use information, extracted directly from the histogram: for example, some of the methods perform thresholding [50], while others analyze a histogram outline or model its peaks [32], relative positions [42], or fit probability distributions to a histogram [25]. Zagorodnov and Ciptadi [51] consider tissue intensity probabilities as a blind source separation problem: unknown distributions are treated as sources and the histograms of subvolumes as mixtures. Another current approach is the extraction of a feature vector from the image histogram and then training another classifier as SOM [34] in order to perform classification and finally image segmentation.

There are two main groups of methods, which use spatial correlation (spatial smoothness) within a given region in the MR image: methods, related to MRF and variational (level set) methods. In the first group, the discrete energy model embraces contextual information and spatial continuity. The energy function is optimized by another method like simulated annealing (SA) or iterated conditional modes (ICM). For example, spatial information is incorporated into a segmentation model by contextual constraints on neighboring pixels through hidden Markov random field (HMRF) [57]. The current MRF methods such as [31] set weights to different classes during maximum a posteriori (MAP) estimation or introduce specific clonal selection algorithm (CSA) and Markov chain Monte Carlo (MCMC) to perform HMRF model estimation [55].

In the second group of the variational methods, the MR image is modeled using piecewise smooth function, while in the level set methods contours or surfaces are represented as the zero level set of a higher dimensional function; the second group of methods achieves sub-pixel accuracy, closed contour of segmentation and automatically extracts complex boundaries. It also allows incorporation of different prior information as shape prior constraints [29], intensity distribution [8], or specific information for INH artifact [27], existing in the MR image. Another method for region segmentation in medical volume images using deformable model and level set theory is given in [26].

The Bayesian model provides another mathematical approach to introduce prior information in the segmentation process: it consists of a prior model and a generative observation model. Another approach introduces an anatomical atlas as prior information, an average estimate from one or few subjects of similar age; the anatomical atlas can be also binary or gray-scale and can be used as ground truth, as well. For example, in [30] an adaptive kernel bandwidth is calculated using a Bayesian approach; a review of atlas segmentation methods is given in [7], while the authors in [24] suggest a method for the integration of the atlas in a MRF model. Although the atlas segmentation methods provide decent segmentation results, they require a registration of each test MR image with the given atlas and thus a registration error is introduced; some problems also arise with missing tissues, presence of noise, and artifacts.

Another suggested histogram is a 2D histogram for the segmentation of renal-biopsy images [52], where general object-background segmentation is performed using Otsu thresholding of a 2D histogram. The coordinates of the 2D histogram entries are intensity value of a central pixel and an averaged intensity value of other pixels in a neighboring window. The method is extended by a 2-phase 2D thresholding algorithm [9] and a curvilinear thresholding method [53]. The authors take advantage of the discriminative position of the background, object, and the image edges in the 2D histogram. Although they perform back projection segmentation, they neither exploit all existing information from the 2D histogram, nor construct more complex models, nor train them.

As we already mentioned in the introduction, histogram matching [16] is a method used to generate a processed image that has a specified histogram. It is implemented usually as a single-valued mapping between source and target histograms using a look-up table (LUT) or dynamic programming [39]. In medical imaging it is applied to various operations: histogram standardization [35], an issue, arising due to inter-scan intensity variation, MR image normalization [13], registration [41, 42], and image enhancement [38, 48]. Here, histogram matching is usually performed before segmentation: it is not an intrinsic part of the segmentation process.

MR image segmentation also implies pixel classification and depending on the used prior knowledge, it can be divided into three groups: unsupervised, supervised, and semi-supervised methods, which are set into inductive and transductive learning framework. The supervised methods require complete labeled training data to construct a model and therefore achieve better results but make rigid assumptions of MRI data, which are often not met. The unsupervised methods group similar pixels without preliminary information and have higher reproducibility and adaptability: they also make fewer assumptions to MRI data. The semi-supervised methods are situated between both groups and use limited labeled training data, incorporated in different approaches: ensemble framework [2], Bayesian transductive learning [25], and spectral clustering [56]. The transductive methods are slightly different from the semi-supervised methods: they do not construct a predictive function and use all test data during classification; they also employ intrinsic test data structure but are computationally intensive.

A mixture model is a type of unsupervised learning: it considers data as a set of observations coming from different probability distributions. It might also be parametric or non-parametric. The parametric models use a known function of parametric distribution, for example, Gaussian mixture models (GMM) [12] or Rician mixture models (RMM) [37] and their parameters are often determined statistically by maximum-likelihood (ML) or MAP approaches. Local interaction between pixels is incorporated using a regularization term into the parametric model [12]. On the other hand, non-parametric models, which are based, for example, on the intensity distribution of pixels, are more accurate but slower, compared with the parametric models. Mean-shift [30] and kNN classification [46] are popular non-parametric methods in medical segmentation.

The fuzzy segmentation methods suggest another approach to solve the problem of overlapping intensity ranges of tissue regions in MR images. They assume that each pixel belongs to more than one tissue/class, which is consistent with the PVE artifact. Currently, this is a very popular approach in medical segmentation. The recent trend is to introduce local and global spatial information in the fuzzy segmentation models—local pixel neighborhood into GMM [20, 22] or into FCM algorithm [21], weights of image patches into fuzzy clustering [19]. The drawbacks are slow speed of convergence, sensitivity to the weights of different features, and lack of a general way to incorporate prior knowledge.

Our work aims to develop a tissue segmentation method within a transductive learning framework using 2D histogram matching. We aim to show that our segmentation results are at least comparable with the most popular current state-of-the-art methods, presented in this section.

6.3   Overview of the Developed Segmentation Algorithm

In our work, we propose a MR image segmentation method using a 2D histogram matching. The method has the following main contributions: (1) introduction of a 2D histogram for tissue segmentation, (2) algorithm for 2D histogram matching and classification, (3) algorithm for MR image segmentation using back projection of the classified 2D histogram.

We construct our 2D histogram using a sum of eight co-occurrence matrices of a MR image; it incorporates correlation between adjacent pixel pairs, and correlation with other MR images from the sequence by matching. From probabilistic point of view, it provides probability occurrence of the pixel pairs in a MR image.

Our algorithm consists of the following steps: (1) we divide the whole sequence into a few subsequences, as we calculate a similarity coefficient based on wave hedges distance metric between 2D histograms of consecutive MR images; we make the subsequence division by keeping the similarity coefficient into a predefined interval. Then (2) we use region MR images, corresponding to a start and end MR images of the subsequence, to build two 2D histogram models for each subsequence: model intra-tissue 2D histograms for separate tissue regions and inter-tissue edge 2D histograms. Then (3) we match segments of the test 2D histogram, corresponding to the different tissue, in a 1D domain using a LUT with model intra-tissue 2D histograms. Next, (4) we subtract the initial from the matched vector of each segment of the test 2D histogram, quantize their difference using k-means clustering and preserve the largest values, which correspond to the most common pixel pairs between test and different region MR images. Thus, we achieve stable discrimination between the 2D histogram segments. After that, (5) we apply one of two distance metric learning methods: Large Margin Nearest Neighbor (LMNN) or Neighbourhood Component Analysis (NCA) before the k-Nearest Neighbor (kNN) classification of the test 2D histogram. Then, (6) we construct a probability map for each tissue from the classified test 2D histogram through back projection into a local window, while we also add inter-tissue edge pixel pairs with equal weights to the main corresponding tissues. Finally, (7) segmentation is performed with a majority vote between the probability maps.

We apply our developed algorithm to IBSR 18, IBSR 20, and BrainWeb data sets. The results are comparable with benchmark and state-of-the-art MR segmentation methods due to the employed correlation; on the other hand, the specific shape of brain tissues is not considered and many edge pixel pairs of the ridges are misclassified. Other problems, such as sensitivity to different artifacts, e.g. PVE and INH, remain as future tasks for the algorithm improvement.

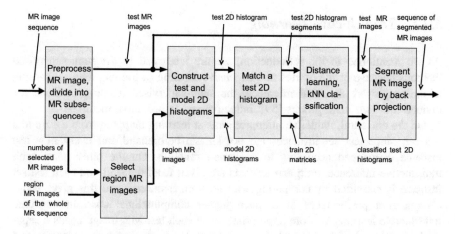

Fig. 6.1 Algorithm for MR image segmentation using 2D histogram matching

The present work shows in the next sections our segmentation method using 2D histogram matching, which consists of the following stages (Fig. 6.1): MR image preprocessing and division into MR image subsequences, calculation of model and test 2D histograms, matching of the test 2D histogram, then distance metric learning and classification of the test 2D histogram with kNN classifier, and finally segmentation of the test MR image using back projection. In the results section we provide description of parameters of our algorithm, the test results after its application to IBSR 18, IBSR 20, and BrainWeb data sets and we also highlight the differences between them. Finally, the results are discussed and main conclusions are made in the last two sections, and a few ideas for the algorithm improvement are given, as well.

6.4 Preprocessing and Construction of a Model and Test 2D Histograms

In our paper, we introduce a 2D histogram model to segment three main neuroanatomical tissues of a healthy brain in MR images: white matter (WM), gray matter (GM), and cerebrospinal fluid (CSF). The other structures such as the scalp, skin, etc. are of no interest to us. We also use three supplementary edge classes for adjacent pixel pairs in the current work: gray matter-white matter (GM-WM), cerebrospinal fluid-gray matter (CSF-GM), and cerebrospinal fluid-white matter (CSF-WM).

Our segmentation model is applied locally to MR image subsequences; for this purpose each MR image sequence is divided into few MR subsequences and additional preprocessing operations are applied to each MR image, test and model 2D histograms, as well. Since our segmentation model is included into a transductive learning framework, we also describe semi-supervised and transductive learning methods.

6.4.1 Transductive Learning

As we mentioned in the introduction, we are interested in the semi-supervised methods because of the limited and expensive labeling of training data. According to the employed statistical inference, the semi-supervised learning methods are divided into two main groups [59]: inductive and transductive methods.

On the one hand, inductive inference aims at learning mapping from a train to a test set, since test set instances are not known beforehand and every new test instance is mapped according to a learned function. On the other hand, with transductive inference, train and test sets are given from the start and each test set instance is classified by comparing with all train instances, and this gives better classification performance at a much higher computational cost. In our case transductive learning is more appropriate since each test MR image needs a separately developed ad hoc model that best fits it due to its structure variability and possible presence of artifacts using a more complicated single general model for all MR images from the MR image sequence.

More formally, we have labeled train data $L = \{x_l, y_l\}_{l=1}^{L}$ and unlabeled test data $U = \{x_{l'}\}_{l'=L+1}^{L+U}$ (usually $U \gg L$, where x and y are data and label vectors, respectively. In the case of inductive inference the main aim is to learn such a function f_1, which predicts labels of the future test data:

$$f_1 : X \rightarrow Y, \tag{6.1}$$

so f_1 is a predictor function on future data, beyond $\{x_{l'}\}_{l'=L+1}^{L+U}$.

In the case of transductive inference we do not learn an explicit function f_1, but only calculate labels for test data U:

$$f_1 : X^{L+U} \rightarrow Y^{L+U}, \tag{6.2}$$

such that f_1 is a predictor function on the unlabeled data $\{x_{l'}\}_{l'=L+1}^{L+U}$.

6.4.2 MRI Data Preprocessing

The aim of MRI data preprocessing is to transform MR image sequences so that subsequent operations produce more accurate results and to place the MR image sequences from different test data sets under equal conditions for validation. It consists of the following three operations:

- removing unnecessary tissues.
- transforming separate MR images into coronal plane.
- optional non-linear mapping of intensity values of each MR image (gamma correction).

We remove redundant tissues from the data sets in a different way: while real MR brain image sequences from IBSR 18 and 20 data sets are provided with their pre-processed versions, with suppressed noise, homogenized background and removed undesired structures, in the case of BrainWeb data sets, additional undesired structures are particularly labeled as different classes, which facilitates their removal.

We also transform separate MR images from the original to the coronal plane: for instance, MR images from the real and simulated MR image sequences from IBSR 18 and BrainWeb data sets are transformed from the sagittal to the coronal plane. The main idea is all 2D histograms to be constructed from pixel pairs in the coronal plane, where tissues have more compact representation.

It is optional to apply gamma correction independently to each MR image from the real MR image sequences (IBSR 18 and 20) in order to achieve greater compactness of lower intensity levels and slightly worse distinction of higher intensity levels. Otherwise the CSF segment in a test 2D histogram would have many scattered entries and larger distance to GM segment.

Gamma correction can be described using the following power-law function:

$$I_{out} = c_1 \cdot I_{in}^{\gamma}, \tag{6.3}$$

where I_{out} and I_{in} are intensity ranges of output (corrected) and input (uncorrected) MR images, respectively; $c_1 = \left(\frac{1}{\max(I_{in})}\right)^{-(1-\gamma)}$ is a normalization coefficient that aims to preserve the initial intensity range after non-linear mapping, γ is the gamma correction coefficient, when $\gamma > 1$, we have gamma compression and $\gamma < 1$— gamma expansion.

6.4.3 Construction of a 2D Histogram

We use a 2D histogram of a MR image as a basic structure in our segmentation model. We construct the 2D histogram of a MR image (Fig. 6.2a) as a sum of K co-occurrence matrices of adjacent pixels (Fig. 6.2b), corresponding to the K-neighbor connectivity of a pixel. Our 2D histogram (Fig. 6.2c) is informative since adjacent pixel pairs of uniform regions, edges, and noise are distinctive; averaging their intensity values will concentrate them too much around the diagonal, as shown (Fig. 6.2d) [52], and this would reduce the information it contains.

Let us have an MR image, represented by a 2D intensity function $f(x, y)$, which gives a gray level of pixel (x, y), ranging from 0 to $B - 1$; $x = 1, \ldots, N_x$, $y = 1, \ldots, N_y$, B—maximum number of intensity levels in the entire intensity range, and N_x, N_y—dimensions of the MR image. Let $g_k(x, y)$ be the function of gray levels of neighboring pixels of (x, y) (Fig. 6.2b), where $k = 0, \ldots, K - 1$, K defines pixel connectivity and k—connectivity direction. Let C_{ij}^k represent the number of intensity transition between a pixel (x, y) and its neighboring pixels with intensity

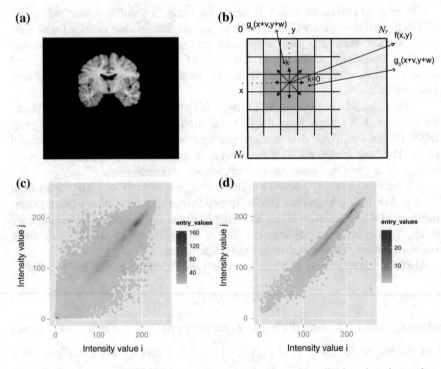

Fig. 6.2 Construction of a 2D histogram (preprocessed and non-normalized, projected on a plane, inverted) of a MR image (IBSR 20): **a** Input MR image, **b** Construction of our 2D histogram by summing up K co-occurrence matrices, **c** Our 2D histogram, **d** 2D histogram from [52]

levels $f(x, y) = i$ and $g_k(x+v, y+w) = j$, respectively, where $0 \leq C_{ij}^k \leq N_x \cdot N_y \cdot K$, $(v, w) \in \{-1, 0, 1\}^2 \backslash \{(0,0)\}$, and $i, j = 0, \ldots, B-1$.

To describe the probability distribution of pixel pairs in a given connectivity direction k, we introduce the following joint probability mass function p_k:

$$p_k(i,j) = \frac{C_{ij}^k}{N_x \cdot N_y \cdot K}, \qquad (6.4)$$

where $\sum_{k=0}^{K-1} \sum_{i=0}^{B-1} \sum_{j=0}^{B-1} p_k(i,j) = 1$. We get $P_{ij} = \sum_{k=0}^{K-1} p_k(i,j)$ and obtain the final result—a normalized 2D histogram.

6.4.4 Separation into MR Image Subsequences

Since tissues change their properties smoothly along the MR image sequence and artifacts and noise do not appear in isolation, it is not suitable to apply a single segmentation model to the whole sequence. In order to use correlation between MR images, we split the MR image sequence into a few subsequences by measuring the current similarity coefficient based on wave hedges distance D_c between the normalized 2D histograms E and F of the current and reference MR images— (Fig. 6.3a) respectively [18]:

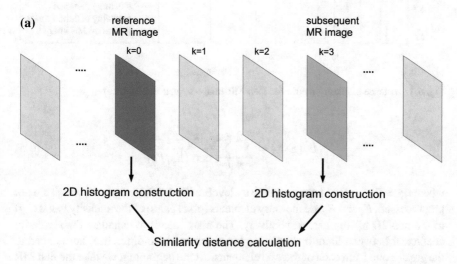

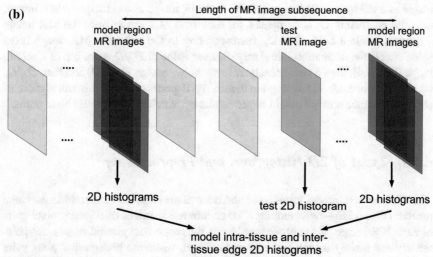

Fig. 6.3 MR image sequence: **a** subsequence division using similarity coefficient between consecutive 2D histograms, **b** test, model intra-tissue and inter-tissue edge 2D histograms construction in a given MR image subsequence

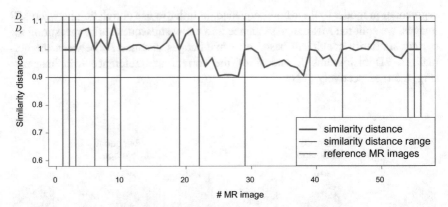

Fig. 6.4 Distance similarity metric for 19th MR image sequence (IBSR 20)

$$D_c(E, F) = \sum_{i=0}^{B-1} \sum_{j=0}^{B-1} \frac{|E_{ij} - F_{ij}|}{\max\left(E_{ij}, F_{ij}\right)}, \qquad (6.5)$$

where $i, j = 0, \ldots, B - 1$ are the intensity levels of both 2D histograms, which are not preprocessed; F_{ij} and E_{ij} are the entry elements (pixel pairs) of the intensity value (i, j) of the two 2D histograms, respectively. The main idea is to calculate the similarity coefficient between them by summing their element-wise difference, normalized to the greater one from each of the two elements. At the beginning, we take the first MR image as a reference MR image and the next MR image as a subsequent MR image, then the coefficient D_c is computed for each next MR image from the MR image sequence, while a coefficient D_r, corresponding to the reference MR image takes value one: if the normalized similarity distance value of D_c/D_r goes out of a certain range (Fig. 6.4), then a new subsequence starts, we assign $\boldsymbol{D_c}$ to $\boldsymbol{D_r}$ and then D_c/D_r takes value one. A 2D histogram matching is performed by the calculation of absolute distance and we obtain better results by matching similar 2D histograms.

6.4.5 Types of 2D Histograms and Preprocessing

We construct test, model intra-tissue and inter-tissue edge 2D histogram in the same manner (Fig. 6.3b)—by summing K co-occurrence matrices of adjacent pixel pairs of each MR image of the sequence. Hence the tissue-background edge pixel pairs are situated mainly on the first row and column, while the background pixel pairs stay mostly at the entry $(1, 1)$; we apply different types of preprocessing to remove them, considering their function in the segmentation algorithm.

The test 2D histogram is made using the calculation of a 2D histogram of a test MR image. It has a larger area and a few clusters with more peaks on the diagonal. The clusters are comparatively close and there is a considerable overlapping between them. The test 2D histogram is not compactly distributed due to its larger intensity range; additionally the presence of inter-tissue edge pixel pairs and sometimes artifacts deteriorates its compactness.

The model intra-tissue 2D histograms (Fig. 6.5a–c) are made by summing 2D histograms of corresponding model region MR images. Their entries are situated on the diagonal, forming a cluster with a few peaks within their intensity range, which slightly overlap; there are outliers, as well.

We introduce a model inter-tissue edge 2D histogram to hold pixel pairs of inter-tissue edges (Fig. 6.5d), which occur in model MR images for a given sub-sequence. We construct it in the following way: firstly, we construct a 2D histogram of a model MR image, by combining all its model region MR images, then we subtract from it 2D histograms of the separate model region MR images to obtain an inter-tissue edge 2D histogram, whose entries belong only to CSF-GM, GM-WM, and CSF-WM edge classes and finally we sum the two inter-tissue edge 2D histograms, corresponding to the two model MR images; the edge pairs are scattered

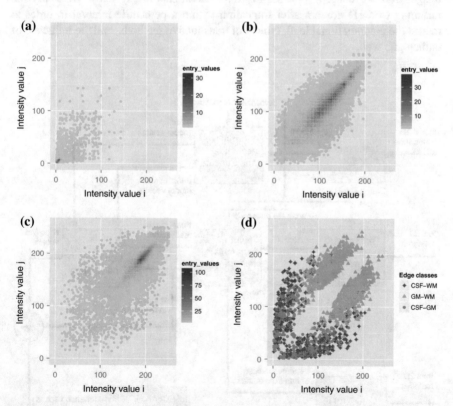

Fig. 6.5 Model 2D histograms (IBSR 20) (preprocessed and non-normalized, projected on a plane, inverted): **a**, **b**, **c** of CSF, GM, and WM tissues, respectively, **d** of CSF-GM, GM-WM, and CSF-WM edges

on both sides of the diagonal (Fig. 6.5d) and their position reflects the fact that some of them contain a PVE artifact.

During preprocessing all kinds of 2D histograms, we firstly set to zero the first row and column entries. Then the inter-tissue edge 2D histogram avoids any further preprocessing. In the case of a test 2D histogram we keep for back projection all inter-tissue edge entries using indices of non-zero entries of the model inter-tissue edge 2D histogram and after that we set them to zero. Thus, only intra-tissue pixel pairs remain in the test 2D histogram. This operation preserves the outline and its initial form and does not change the positions of peaks and valleys.

6.5   Matching and Classification of a 2D Histogram

Using 2D histogram matching we localize and extract the most representative 2D histogram entries of each segment, corresponding to each tissue region in a test MR image; thus we make a train set for kNN classification (Fig. 6.6). We apply the matching to a 1D domain after truncation within a percentile interval in order to reduce the computational load, typical of transductive methods, and the influence of outliers, as well.

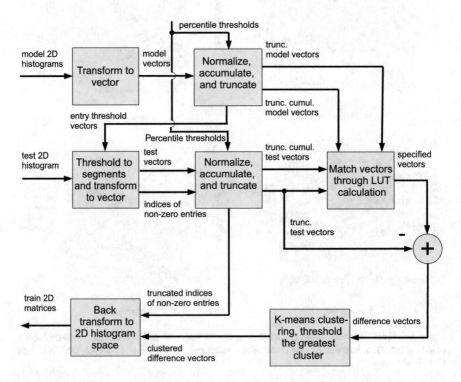

Fig. 6.6 Construction of train 2D histogram segments using 2D histogram matching

6.5.1   Construct Train 2D Histogram Segments Using 2D Histogram Matching

The 2D histogram matching using LUT can be described in two stages: in the first stage we process a model 2D histogram of each region MR image, while in the second stage we match each test 2D histogram segment for each model 2D histogram for a given MR image subsequence.

1. First we threshold the model 2D histogram to remove outliers. Since a 2D histogram is not suitable for direct matching due to its sparse nature and large memory requirements, we extract all non-zero entries by zig-zag reordering (Fig. 6.7) of a 2D histogram M_r of the rth region MR image, similar to the one used in JPEG encoding [47], into a model vector $\boldsymbol{m_r}$, as shown for the tth element:

$$m_{rt(i,j)} = \begin{cases} M_{rij}, & \text{if } M_{rij} > 0, \\ \emptyset, & \text{if } M_{rij} = 0, \end{cases} \tag{6.6}$$

where $r = 1, \ldots, R$, $t(i,j) = 1, \ldots, N_r$, $i, j = 0, \ldots, B-1$, R is the number of regions, while N_r—the number of non-zero entries, and B—the maximum number of intensity levels in the entire intensity range of M_r. We apply the zig-zag reordering to exploit the existing correlation and symmetry in M_r.

2. Calculate the tth element of a normalized vector $\hat{\boldsymbol{m}}_r$ of $\boldsymbol{m_r}$ and its cumulative model vector $\hat{\boldsymbol{c}}_r$:

$$\hat{m}_{rt} = \frac{m_{rt}}{n_r}, \tag{6.7}$$

$$\hat{c}_{rt} = T(\hat{m}_{rt}) = \sum_{e=1}^{t} \hat{m}_{re}, \tag{6.8}$$

where $t = 1, \ldots, N_r$ and n_r is the sum of all $\boldsymbol{m_r}$ elements.

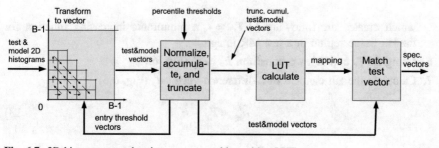

Fig. 6.7 2D histogram match using vector matching with a LUT

3. Truncate $\hat{c}_r$ to $\hat{c}'_r$ ($N'_r < N_r$ elements) by the lower and upper thresholds η_{rl} and η_{ru} (percentile values) of a given percentile interval, calculate their corresponding entry threshold vectors in F_r : θ_{rl} and θ_{ru}. Truncate $\hat{m}_r$ to $\hat{m}'_r$ alike.

Next, we threshold each test 2D histogram using θ_{rl} and θ_{ru} and match with truncation using $\hat{c}'_r$ for each model region, assuming that the number of model regions and tissue regions in each test MR image in a given subsequence is equal.

1. Threshold a test 2D histogram H (Q elements) by θ_{rl} and θ_{ru} to obtain the rth segment H_r, where $r = 1, \ldots, R$.
2. Extract by zig-zag reordering (Fig. 6.7) non-zero entries of the H_r into a vector h_r (Q_r elements). After that, indices of the extracted non-zero entries are preserved.
3. Calculate the zth element of a normalized vector $\hat{h}_r$ of h_r and its cumulative vector $\hat{v}_r$:

$$\hat{h}_{rz} = \frac{h_{rz}}{q_r}, \qquad (6.9)$$

$$\hat{v}_{rz} = G(\hat{h}_{rz}) = \sum_{e=1}^{z} \hat{h}_{re}, \qquad (6.10)$$

where $z = 1, \ldots, Q_r$ and q_r is the sum of all h_r elements.
4. Truncate $\hat{v}_r$ to $\hat{v}'_r$ ($Q'_r < Q_r$ elements) by η_{rl} and η_{ru}. Truncate $\hat{h}_r$ to $\hat{h}'_r$ and the indices of the non-zero entries alike.
5. Calculate a specified vector $\hat{s}'_r$:

$$\hat{s}'_r = G^{-1}(\hat{v}'_r) = G^{-1}(\hat{c}'_r) = G^{-1}[T(\hat{m}'_r)]. \qquad (6.11)$$

For that purpose we build a LUT by relating the two mappings for cumulative vectors $\hat{v}'_r$ and $\hat{c}'_r$. For each zth index of $\hat{v}'_r$, we find out the corresponding tth index of $\hat{c}'_r$, where $z = 1, \ldots, Q'_r$ and $t = 1, \ldots, N'_r$, so that the best match between $\hat{v}'_{rz}$ and $\hat{c}'_{rt}$ regarding L1 min distance exists:

$$\left| \hat{v}'_{rz} - \hat{c}'_{rt} \right| = \min_{z'} \left| \hat{v}'_{rz'} - \hat{c}'_{rt} \right|, \qquad (6.12)$$

which creates the following $LUT_{rz} = t$, a monotonic increasing function for the rth tissue region in a test MR image.
6. Using LUT_r and $\hat{m}'_r$, we transform $\hat{h}'_r$ to $\hat{s}'_r$ (Fig. 6.7).
7. Calculate the zth element of a difference vector $\hat{d}'_r$ (Fig. 6.6):

$$\hat{d}'_{rz} = \hat{s}'_{rz} - \hat{h}'_{rz}. \qquad (6.13)$$

8. Partition $\hat{d}'_r$ into O $(O \leq N'_m)$ sets $S = \{S_1, S_o, \ldots, S_O\}$ by k-means clustering (Fig. 6.6) with a $L1$ distance objective:

$$\arg \min_S \sum_{o=1}^{O} \sum_{\hat{d}'_{rz} \in S_o} |\hat{d}'_{rz} - \mu_o|, \tag{6.14}$$

where μ_o is the mean value of cluster S_o, $o = 1, \ldots, O$, o—a cluster index. We set all elements of $\hat{d}'_r$, which are not members of the largest S_O cluster, to zero.

9. Back transform $\hat{d}'_r$ to a matrix $\hat{H}'_r$ (Q'_{O_r} non-zero elements) using the truncated indices of non-zero entries of the test 2D histogram segment H_r (Fig. 6.6).

6.5.2 2D Histogram Classification After Distance Metric Learning

To classify the test 2D histogram H (Fig. 6.8a), we first transform a multi-class problem into a group of binary classification problems using the all-by-all method [17] and after kNN classification and distance metric learning we obtain a classified test 2D histogram (Fig. 6.8b). Our test set consists of all entries of all test 2D histogram segments H_r (Fig. 6.8c–e), while the train set consists of the all calculated entries of the train 2D histogram segments $\hat{H}'_r$ (Fig. 6.8f–h). Feature space is constructed only from the x and y coordinates of the train and test set entries, to decrease the computational burden and to avoid the impact of segment size. We selected two types of distance metric learning methods: NCA (Neighbourhood Component Analysis) [15] and LMNN (Large Margin Nearest Neighbour) [49] to improve the kNN classification results. In the first case it learns a global metric, which is applied equally over the entire feature space, while in the second case it learns also a global metric that satisfies only the local constraints: both methods consider statistics of input space and the importance of separate features instead of relying on fixed distances in feature space.

6.5.2.1 Pairwise (All-Versus-All) Classification

The classification of the test 2D histogram is reduced to a choice among all existing pairs of R tissue classes: each binary classifier discriminates each pair of classes and discards the rest of them [17]:

$$f(x) = \arg \max(\sum_{r'} f_{rr'}(x)), \tag{6.15}$$

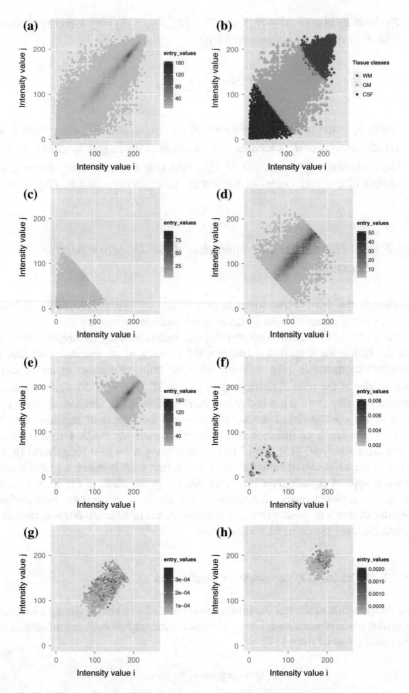

Fig. 6.8 Classification of a test 2D histogram of an MR image (IBSR 20) (preprocessed, projected on a plane, inverted): **a** input test 2D histogram, **b** classified test D histogram—CSF, GM, and WM tissues, **c**, **d**, **e** test segments of CSF, GM, and WM tissues, respectively, **f**, **g**, **h** train segments of CSF, GM, and WM tissues (after matching)

where $r, r' = 1, \ldots, R(r \neq r')$, $f_{rr'}$ represents a decision function between tissue classes r and r', $f_{rr'} = -f_{r'r}$ and the whole operation requires $\frac{R \cdot (R-1)}{2}$ binary classifiers.

6.5.2.2 Distance Metric Learning

The main idea of distance metric learning is to adapt a distance metric to a given data. For example, we show fixed distance L_k, which is a generalization of L_1 (Manhattan distance) and L_2 (Euclidean distance):

$$L_k = (\sum_{j'=1}^{d} |a_{mj'} - a_{nj'}|^k)^{\frac{1}{k}}, \qquad (6.16)$$

where a_m, a_n are data vectors and $a_m, a_n \in \mathbb{R}^d$, $i = 1, \ldots, n'$ and n'–number of vectors, d—number of features.

Most distance learning methods are based on the learning of Mahalanobis distance D_Q:

$$D_Q(a_m, a_n) = (a_m - a_n)^T Q^{-1}(a_m - a_n), \qquad (6.17)$$

where $Q \in \mathbb{R}^{d \times d}$ is a covariant, positive definite or positive semi-definite, matrix. If Q is constrained to be a positive matrix, it allows convex optimization but it has a computationally prohibitive learning. Hence, most methods prefer to regard it as a positive semi-definite matrix and thus refusing to use the favorable properties of the convex optimization. In the case of LMNN distance metric learning, the matrix Q is denoted usually as A.

In our case, we have a training set $\{a_1, a_{i'}, \ldots, a_{n'}\}$ with known class labels $\{f_1, f_{i'}, \ldots, f_{n'}\}$, where $a_{i'j'} \in \{0, \ldots, B-1\}$, test set $\{u_1, u_{k'}, \ldots, u_{m'}\}$, where $u_{i'j'} \in \{0, \ldots, B-1\}$, $f_{i'} \in \{1, \ldots, R\}$, $i' = 1, \ldots, n'$, $k' = 1, \ldots, m'$, $j' = 1, \ldots, d$, $n' = \sum_{r=1}^{R} Q'_{lr}$, $m' = Q' - n'$, Q' is the number of all test elements from the test matrix H, and d—number of the features of the 2D histogram (d = 2).

6.5.2.3 Large Margin Nearest Neighbor

Here, we have a supervised global metric, which enforces local constraints: it learns a Mahalanobis distance measure for kNN classification. The main idea is to enforce k-nearest neighbor train set elements around a given test element to make it belong to the same class, while the train elements from different classes be separated by different margin; this is an optimization problem, solved by semi-definite programming approach.

Let us describe how LMNN distance metric learning is applied in our case: we have vectors a_m, a_n, and a_q from the training set. The set S_a consists of all pairs

(a_m, a_n), where a_q is one of the k-nearest neighbors in the same class (target) as a_m. Similarly, the set R_a consists of all triples (a_m, a_n, a_q), where a_m and a_n are target members of the same class, while a_q is an element with a different label (impostor). We optimize the distance metric D_A over the space of positive semi-definite matrices A, so we minimize the following pseudo distance:

$$D_A(a_m, a_n) = (a_m - a_n)^T A(a_m - a_n), \qquad (6.18)$$

where A should be a positive semi-definite matrix in order D_A to be a well-defined metric.

Hence, we optimize the distance metric D_A in Eq. (6.18) over the space of positive semi-definite matrices A:

$$\min_{A \succeq 0} \sum_{(m,n) \in S_a} D_A(a_m, a_n) + c \sum_{(m,n,q) \in R_a} \left[1 + D_A(a_m, a_n) - D_A(a_m, a_q) \right]_+, \qquad (6.19)$$

where in the second term $[z']_+ = \max(z', 0)$ denotes the standard hinge loss and the tradeoff constant $c > 0$, defined by cross-validation, weighs the two terms: the lower value of c shrinks the distance to target neighbors, ignoring the number of included impostors, and vice versa. We aim to find out the distance metric D_A—a linear transform of train set entries, which minimizes distances to correctly classified entries (first term) and the number of incorrectly classified entries (second term).

6.5.2.4   Neighbourhood Component Analysis

This method aims to find out such a linear transform (projection matrix) A of the feature space of the test 2D histogram segments, after which the kNN classifier would perform better, scaling up useful directions for discrimination. The NCA method does not imply any specific class distribution and shape of separating surfaces.

Here the covariance matrix Q is a symmetric, positive semi-definite real matrix, which can be decomposed by Cholesky decomposition as $Q = A^T A$ (where $A \in \mathbb{R}^{d \times d}$) and the Mahalanobis distance between vectors a_m, a_n is given as follows:

$$d_A(a_m, a_n) = (a_m - a_n)A^T A(a_m - a_n) = (Aa_m - Aa_n)^T \times (Aa_m - Aa_n) =$$
$$= |Aa_m - Aa_n|_2^2. \qquad (6.20)$$

In general, a pseudo-metric D_A can be represented as a squared Euclidean distance after applying a linear transformation.

Each train vector a_m selects stochastically another vector a_n as its neighbor with a probability value p_{mn}, based on the softmax of the distance $D_A(a_m, a_n)$:

$$p_{mn} = \frac{e^{[-D_A(a_m,a_n)]}}{\sum\limits_{q \neq m} e^{[-D_A(a_m,a_q)]}} . \tag{6.21}$$

We calculate probability value p_m that the given vector a_m will be correctly classified:

$$p_m = \frac{1}{N} \sum_{n \in C_m} p_{mn}, \tag{6.22}$$

where $C_m = \{n/z_n = z_m\}$ gives a set of vectors of the same class of m and $p_{mm} = 0$.

Finally, the projection matrix A is calculated by maximizing an expected number of correctly classified points $f_2(A)$:

$$f_2(A) = \sum_m p_m = \sum_m \sum_{n \in C_m} p_{mn}, \tag{6.23}$$

where $p_{mm} = 0$.

Briefly, the NCA algorithm consists of the following steps:

1. Initialize a linear projection matrix A with identity matrix.
2. Optimize $f(A)$ using conjugate gradient method in the following steps:

   a. Project the training vector a_m by projection matrix A to yield $A^T a_m$, where $m = 1, 2, \ldots, n'$ and $n' = \sum_{r=1}^{R} Q'_{Or}$.
   b. Calculate the square Euclidean distance between the training vectors a_m and a_n in the transformed space $|A^T x_m - A^T x_n|^2$ for $m, n = 1, 2, \ldots, n'(m \neq n)$ (Eq. 6.20).
   c. Compute probability values p_{mn} (Eq. 6.21) and p_n (Eq. 6.22).
   d. Calculate gradient $\frac{df_2}{dA}$ of $f_2(A)$ (Eq. 6.23) and update the projection matrix A by a conjugate gradient optimizer. Repeat all four steps of the current point, predefined number of times, until reaching convergence of $f_2(A)$.

6.5.2.5   K-Nearest Neighbour Classification

From the distance metric learning methods, we obtain a global metric—linear matrix A (matrix $A \equiv Q$ in the case of LMNN method) to project train and test data vectors a and u, respectively. We perform the projection for each pair of classes.

So we project the i'th training vector $a_{i'}$:

$$\tilde{a}_{i'} = A a_{i'}, \tag{6.24}$$

where $i' = 1, \ldots, n'$.

Then we project the k'th test vector $\boldsymbol{u}_{k'}$:

$$\tilde{\boldsymbol{u}}_{k'} = A\boldsymbol{u}_{k'}, \tag{6.25}$$

where $k' = 1, \ldots, m'$ and $m' = Q' - n'$.

The goal is to predict the class label of each k'th test vector $\tilde{\boldsymbol{u}}_{k'}$, using training set vectors $\tilde{\boldsymbol{a}}_{i'}$. We apply a kNN classifier to estimate the posterior probability $\hat{P}(\omega_r/\tilde{\boldsymbol{u}}_{k'})$ of belonging of $\tilde{\boldsymbol{u}}_{k'}$ to a given segment—class ω_r, within a modified (after distance metric learning) neighbourhood $N_0(\tilde{\boldsymbol{u}}_{k'})$ of $\tilde{\boldsymbol{u}}_{k'}$:

$$\hat{P}(\omega_r/\tilde{\boldsymbol{u}}_{k'}) = \frac{\sum_{i'=1}^{n'} \theta(\tilde{\boldsymbol{a}}_{i'} \in N_0(\tilde{\boldsymbol{u}}_{k'}))(\theta(f_{i'} = \omega_r))}{\sum_{i'=1}^{n'} \theta(\tilde{\boldsymbol{a}}_{i'} \in N_0(\tilde{\boldsymbol{u}}_{k'}))}, \tag{6.26}$$

where $\theta(.)$ is an indicator function that returns one if the input argument is true and zero otherwise. The kNN classifier is considered to be a non-parametric and semi-supervised, since it does not assume specific distribution of input data; it is a simple and accurate classifier with enough training data, it can also produce non-linear decision surface. Finally, we get a classified test 2D histogram (Fig. 6.8b) after combining all binary classifiers results.

6.6 Segmentation Through Back Projection

Segmentation through back projection is implemented as we calculate for each pixel b_{00} in a test MR image (Fig. 6.10a) the K adjacent pixel pairs in a local window (Fig. 6.9), find out their labels from the classified test 2D histogram

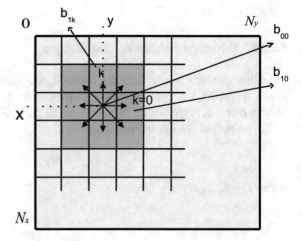

Fig. 6.9 Construction of a sequence of K adjacent pixel pairs in the local window around element b_{00}

(Fig. 6.10c). We show the original test 2D histogram (Fig. 6.10b). We assign equal weights to the pixel pairs of CSF-GM, GM-WM, and CSF-WM edge classes (Fig. 6.10d) and add them to the main tissue classes. The algorithm consists of the following steps:

1. Make a sequence of K adjacent pixel pairs in the local window (Fig. 6.9):
   $\{(b_{00}, b_{10}), (b_{00}, b_{1k}), \ldots, (b_{00}, b_{1(K-1)})\}$, where $k = 0, \ldots, K - 1$.

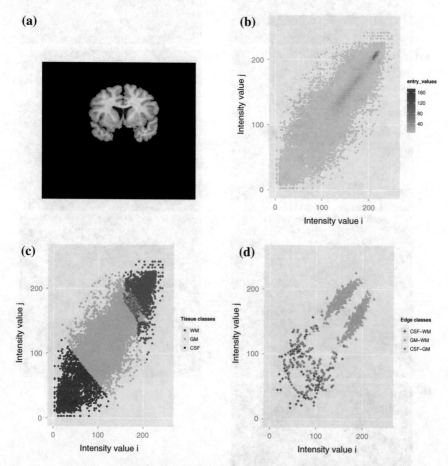

Fig. 6.10 Segmentation by back projection of an MR image (IBSR 20): **a** input MR image, **b** its (preprocessed and non-normalized, projected on a plane, inverted) 2D histogram, **c** classified test 2D histogram—CSF, GM, and WM tissues, **d** classified CSF-GM, GM-WM, and CSF-WM edges, **e**, **f**, **g** grayscale probability maps (inverted, quantized to 8 levels) of CSF, GM, and WM tissues, respectively, **h**, **i**, **j** grayscale probability maps (inverted, quantized to 3 levels) of CSF-GM, CSF-WM, and GM-WM edges, respectively, **k** segmented CSF, GM, and WM tissues—final result

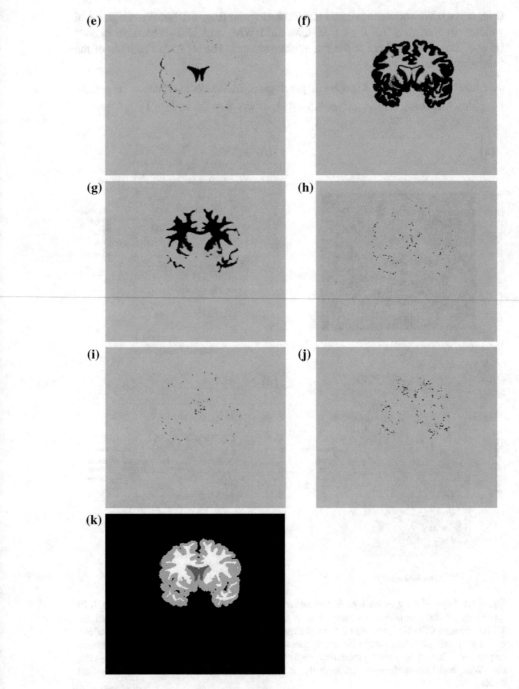

Fig. 6.10 (continued)

2. Let's calculate a weight map $V_{r,k}(x,y)$ for the r-th tissue.

$$V_{r,k}(x,y) = \begin{cases} 1 & \text{if } (b_{00}, b_{1k}) \text{ picks class } \omega_r, \\ 0.5 & \text{if } (b_{00}, b_{1k}) \text{ picks } \omega_{rr'}, \\ 0 & \text{otherwise}, \end{cases} \quad (6.27)$$

where $r, r' = 1, \ldots, R(r \neq r')$, (b_{00}, b_{1k}) is an entry in the classified 2D histogram, (x,y) are coordinates x and y of the pixel b_{00} in the MR image, ω_r—the rth tissue class (segment), and $\omega_{rr'}$—inter-tissue edge class between the rth and the r'th tissue.

3. Calculate a probability map $V_r(x,y)$ for the rth tissue:

$$V_r(x,y) = \frac{\sum_{k=0}^{K-1} V_{r,k}(x,y)}{K}. \quad (6.28)$$

Separate back projected grayscale probability maps of the main tissue classes (before the addition of edges) (Fig. 6.10e, g) and the supposed inter-tissue edge classes (Fig. 6.10h, j) are shown.

4. Select the final label $l_1(x,y)$ after majority vote:

$$l_1(x,y) = \arg \max_{r=1,\ldots,R} \{r : V_r(x,y)\}. \quad (6.29)$$

5. Perform post-processing on binary masks of the segmented tissues: morphological opening by area and morphological filling of CSF and WM tissues [16]. After we unite binary masks, we get a segmented test MR image (Fig. 6.10k).

6.7 Experimental Results

Here we present all parameter values and results of our developed segmentation algorithm with baseline comparison.

6.7.1 Test Data Sets and Parameters of the Developed Algorithm

Algorithms were applied to three data sets of MR images: IBSR 18 [36], IBSR 20, and BrainWeb [10]. The first two data sets consist of real scans from human subjects and the third one—of simulated anatomical models, all of them with corresponding expert-labeled masks (Table 6.1). As it can be seen from Table 6.1, IBSR 20 or IBSR v1.0 consists of 20 MR low resolution normal brain images,

Table 6.1 Specification of available test data sets

Data sets	IBSR 18	IBSR 20	BrainWeb
No. of data	18	20	20
Total no. of images	2304	1250	1938
Slice resolution (mm^2)	from 0.8371 × 0.8371 to 1 × 1	1 × 3.1	1 × 1
Interslice resolution (mm)	1.5	3.1	1–9
Orig. data size	256 × 128 × 256	256 × 65 × 256	256 × 256 × 181
Bit depth (bits)	16(8)	8	16(8)
No. of male cases	14	10	na
No. of female cases	4	10	na
Patient age (years)	7–71	20–38	na
Image format	Analyze	Analyze	Raw

while IBSR 18 or IBSR v.2.0 consists of 18 high resolution 1.5 mm T1-weighted MR scans, where MRI data sets were also filtered and artifacts suppressed or removed. MR images from the first two data sets were also spatially normalized into Talairach orientation, of 8-bit depth, with already removed scalp/skull. The twenty normal MR brain data sets and their manual segmentations were provided by the Center for Morphometric Analysis at Massachusetts General Hospital and are available at http://www.cma.mgh.harvard.edu/ibsr/. The third BrainWeb data sets consists of 20 MR images of T1-weighted simulated data with the following specific parameters: SFLASH (spoiled FLASH) sequence with TR = 22 ms, TE = 9.2 ms, flip angle = 30°, and 1 mm isotropic voxel size.

We set several parameters during the different stages of the developed segmentation algorithm as MR image preprocessing and sequence division, 2D histogram preprocessing, matching and classification, and finally MR image segmentation by back projection.

The whole sequence was divided into several subsequences by an empirically selected threshold interval $[0.9, 1.1]^T$, for a similarity coefficient based on a wave hedges distance (Eq. (6.5)) between normalized 2D histograms of consecutive MR images. This led to the division of subsequences of MR images by comparatively similar 2D histograms; longer in the middle and shorter at both ends. We also introduced an upper threshold for subsequence length of 10 MR images—segmentation accuracy dropped considerably in the case of longer subsequences. The chosen range of the similarity coefficient required labeled data for 20–30 % for all MR images in a sequence. A greater value of the range would make subsequences longer and vice versa. We performed also transform from the sagittal to the coronal plane by basic Matlab operations as 3D matrix indexing and image rotation in the case of IBSR18 and Brainweb data sets.

To make a denser 2D histogram, we summed eight co-occurrence matrices ($K = 8$) and tested our algorithm on MR images in a coronal plane of 256 intensity levels ($L = 256$) of three tissue classes ($R = 3$) with the following dimensions:

($M = N = 256$) for IBSR 18 and 20 data sets and ($M = 181$, $N = 256$) for BrainWeb data sets and labeled mask sets (after interpolation). We used 8-bit version of IBSR 18 data sets and after MRI data preprocessing we had the following length of MR image sequences: 128 (IBSR18), 65 (IBSR 20), and 256 (BrainWeb) MR images. Back projection was implemented with eight pairs of adjacent pixels ($K = 8$) in the local window of size 3 × 3 pixels.

The model intra-tissue 2D histograms M_r were normalized after division by two and all entries smaller than a threshold two were removed before transform to a 1D vector. In order to obtain wider test segments H_r, we calculated thresholds θ_{rl} and θ_{ru} without the last thresholding of M_r. We set a minimum number of 2D histogram entries—20—to perform 2D histogram matching and classification and 20 pixels in a model region MR image—to build a segmentation model and as a requirement to perform segmentation of the test MR image, as well.

We applied 2D histogram matching in a 1D domain—the direction of zig-zag ordering had no impact on the final results. Since we wanted to localize better segments of a test 2D histogram and to overcome outliers, we matched after truncation within the following percentile intervals: [5, 95] (IBSR 18 and 20) and [2.5, 97.5] (BrainWeb). Thus the remaining unclassified areas, at the beginning and the end of the test 2D histogram, were assigned after the classification mostly to the CSF and WM segments, respectively. The matching with a truncation was decisive to achieve accurate results of classification.

We also set default parameter values for the LMNN and NCA distance metric learning algorithm; and we set for kNN classification 1 neighbor, L2 distance parameter values. We employed k-means clustering using the following parameters: L1 distance, 2 clusters for CSF and WM tissues, and for GM tissue we set 2 clusters (BrainWeb) or 3 clusters (IBSR 18 and 20), uniform selection of entries, 3 times replication. The number of clusters was decisive to obtain enough entries for successful 2D histogram classification and to consider train data imbalance in real MRI data. Probabilistic selection of start points increased robustness to the variability in MRI data and led to slightly different results at each application of the algorithm on the same MR image data.

6.7.2 Segmentation Results

The algorithms were implemented on Matlab 2013a and evaluated on the following hardware configuration: 64 bits, Intel Core i5-440/3.1 GHz with RAM 8 GB DDR3L at 1600 MHz. Efficient external libraries were also used: reading input medical images [33], mLMNN2.4 implementation of LMNN algorithm [49], and NCA implementation [15].

To estimate the segmentation results, we used the following similarity coefficients with segmented S_1 and a corresponding ground truth S_2 regions: Jaccard and Dice coefficients [43], precision and recall measures.

The Jaccard similarity coefficient (JSC) measures the similarity of the two corresponding sets as a ratio of the size of their intersection to the size of their union:

$$D_{JSC}(S_1, S_2) = \frac{|S_1 \cap S_2|}{|S_1 \cup S_2|} \cdot 100 \ \%, \tag{6.30}$$

where $|.|$ represents region area (in pixels).

The second metric is the Dice similarity coefficient (DSC), which measures the similarity of the two corresponding sets as a ratio of the size of their intersection to their sum:

$$D_{DSC}(S_1, S_2) = \frac{2|S_1 \cap S_2|}{|S_1| + |S_2|} \cdot 100 \ \%. \tag{6.31}$$

Both metrics range from zero for sets with no common elements to one for fully identical sets; although they give slightly different values, they are interrelated and can be derived from one another.

The other common approach for estimation of classification as receiver operating characteristics (ROC) is not appropriate for our case, since areas of the separate tissue regions differ a lot. Therefore, we will use precision and recall measure metrics to evaluate the results of our segmentation algorithms to separate MR image sequences from IBSR 18 data sets.

So the third measure metric precision $Pr(S_1, S_2)$ gives a ratio of the size of correctly segmented pixels (true positive) to the size of all segmented pixels from S_1:

$$Pr(S_1, S_2) = \frac{|S_1 \cap S_2|}{|S_1|} \cdot 100 \ \%. \tag{6.32}$$

The fourth metric measure recall $Rec(S_1, S_2)$ provides a ratio of the size of correctly segmented pixels (true positive) to the size of all ground truth pixels from S_2:

$$Rec(S_1, S_2) = \frac{|S_1 \cap S_2|}{|S_2|} \cdot 100 \ \%. \tag{6.33}$$

The segmentation results of our algorithm differ slightly depending on the type of test data set: whether they are artificially generated MRI data (BrainWeb), real non-filtered MRI data (IBSR 20), or real filtered MRI data (IBSR 18). Firstly, our algorithm showed slightly inferior segmentation results on simulated MR images of BrainWeb data sets compared with the results of software packages SPM8-Seg, FSL, and Brainsuite [23] and with the results of published segmentation methods (Table 6.2) (the bold type indicates the best result): APRS [28] and RiCE [36]. We selected the following software packages because they implemented the main segmentation approaches to MRI data in the literature, as follows:

Table 6.2 The mean and standard deviation of DSC of benchmark algorithms for BrainWeb data sets

Segmentation algorithm	WM (%)	GM (%)	CSF (%)
SPM8-Seg	92 ± 4	90 ± 3	62 ± 5
FSL	93 ± 2	92 ± 2	74 ± 6
Brainsuite	91 ± 6	80 ± 16	46 ± 16
APRS	96	**96**	**96**
RiCE	**97**	95	96
Our method	91 ± 0.5	90 ± 1	87.5 ± 1.6

1. SP8-Seg—a Matlab software package which implements improved unified segmentation [1] and models image intensities as a mixture of Gaussian distributions and tissue probability maps as prior information for Bayesian estimation.
2. FSL—a FSL-FAST segmentation tool, based on a HMRF model, optimized using the EM algorithm [57], which is oriented to MRI brain data. Here the image histogram is presented as a mixture of Gaussian distributions with mean and variance values for each tissue class.
3. BrainSuite—a suite of image analysis tools, incorporates bias field corrector and partial volume classifier (PVC) [40] for tissue segmentation with three or six tissue classes.

Some authors claim that the BrainWeb data sets do not incorporate correctly PVE, have some histogram artifacts and they are not suitable for evaluation of segmentation models, which rely on intensity distribution [5]. For this purpose, we show 2D histograms from BrainWeb, IBSR 18 and 20 data sets (Fig. 6.11). In the case of BrainWeb, the 2D histogram (Fig. 6.11a) is predominantly concentrated on the main diagonal with three high peaks, corresponding to three tissue classes, and this hinders successful 2D histogram matching. It lacks diversity of 2D histograms of real MRI data (Fig. 6.11b, c) with presence of artifacts of different strength and various properties of MRI data distribution.

Some specific characteristics should be considered during work with artificially generated MR images. For example, we did not apply gamma correction to the artificially generated MR images. Since the resolution of discrete model data sets (labeled masks) after preprocessing (362, 362, 434) was higher than the simulated data sets (181, 256, 256) (also referred to as multiple anatomical model) and we did not have enough information about the original resampling method, we interpolated using the nearest neighbor method simulated data sets to two times their original size in three dimensions, centered them around their corresponding labeled masks and we removed the redundant elements. Then we reduced the enlarged simulated data sets and labeled masks to the original size of the simulated data sets.

So segmentation of the real MR images (IBSR 20) represent a more challenging task: as it can be seen from Table 6.3 and Fig. 6.13, our algorithm performed better than the benchmark algorithm results, reported on the IBSR site: adaptive MAP (amap), biased MAP (bmap), MAP, fuzzy c-means (fuzzy), tree-structure k-means

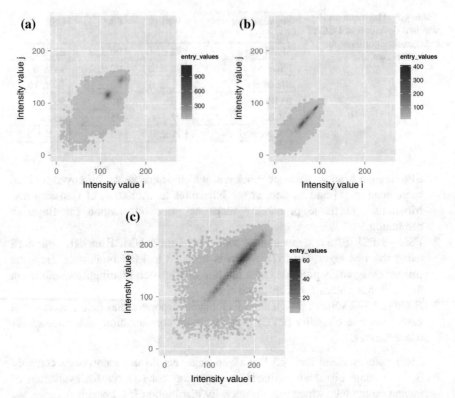

Fig. 6.11 Types of 2D histograms of different data sets: **a** BrainWeb, **b** IBSR 18, **c** IBSR 20—stronger INH artifact

Table 6.3 The mean and standard deviation values of JSC of benchmark algorithms for IBSR 20 data sets

Segmentation algorithm	WM (%)	GM (%)	CSF (%)
Manual	83	88	na
aMap	57 ± 18	56 ± 13	7 ± 3
bMap	56 ± 21	56 ± 17	7 ± 3
FCM	57 ± 20	47 ± 12	5 ± 2
MAP	55 ± 21	55 ± 16	7 ± 4
tskmeans	57 ± 20	48 ± 12	5 ± 2
ML	55 ± 21	54 ± 16	6 ± 3
APRS	**74 ± 3**	**83 ± 3**	**71 ± 6**
MLHMM	56	70	na
FLGMM	**74**	77	na
Our method	70 ± 7	78 ± 7	50 ± 19

(tskmeans), and maximum likelihood (MLC). Since we also wanted to make a comparison with state-of-the-art algorithms from the main approaches to MR image segmentation: fuzzy methods, mixture model methods, and MRF/HMM, we

Table 6.4 The mean and standard deviation values of DSC of benchmark algorithms for IBSR 18 data sets

Segmentation algorithm	WM (%)	GM (%)	CSF (%)
MLHMM	77	86	na
GMM-CSA w/o atlas	87	81	17
RiCE	87 ± 2	**94 ± 1**	na
Our method	**88 ± 1.3**	87 ± 1.3	**24 ± 9.6**

included published results from the following methods (Tables 6.3, 6.4): FLGMM [20], GMM-CSA [54], MLHMM [14], RiCE, and APRS. It can be seen that our method showed comparable results. The use of a separate segmentation model for each subsequence provided us with more accurate results, while the presence of stronger INH and PVE artifacts changed the distribution of a 2D histogram and deteriorated segmentation results. Our algorithm failed to segment correctly many of the ridges of WM in MR images (Fig. 6.12c, f) but the addition of inter-tissue edges improved segmentation accuracy, in general. It can be seen that our algorithm showed considerably better results for GM tissues (Figs. 6.13a), while it was favorably evaluated for WM tissue (Fig. 6.13b) with other benchmark methods for IBSR 20 data sets. It should also be noted that our method did not segment tissue regions with area and 2D histogram entries below the already mentioned thresholds. It can also be seen that certain MR image sequences from IBSR 20 data sets—4, 7, 11, etc.—have lower segmentation results for the benchmark and our developed methods, since they possess stronger artifacts.

The MR image sequences from IBSR 18 data sets were pre-filtered and PVE and INH artifacts were reduced or removed, which led to considerably better results than

IBSR 20 data sets (Table 6.4), since they possess a more stable shape of their 2D histograms. From (Fig. 6.14a, b) it can be seen that real MRI data sets (IBSR 18) are characterized either by higher precision and lower recall values of WM tissue or vice versa; the same is valid for GM tissue—this is caused by over- or under-segmentation of ridge areas of WM tissue. Another important fact is the lower precision values for CSF tissues particularly for IBSR 18 (Fig. 6.14a) data sets.

It can be seen from the test results (Tables 6.2, 6.3, 6.4) that the segmentation result for the CSF tissue is considerably lower, compared with the other GM and WM tissues. In [45] an interesting investigation about segmentation approaches to Sulcal cerebrospinal fluid (SCSF) tissue is given. The authors claim that the segmentation results of benchmark algorithms of CSF and WM tissue vary notably depending on whether SCSF voxels are included into CSF tissue. In the case of IBSR 18 and 20 data sets SCSF tissue is considered as a part of the GM tissue, while it is labeled as a CSF tissue in BrainWeb artificially generated data sets. In our study we regarded SCSF pixels as part of GM or CSF tissues, in accordance with ground truth labeled masks of the given test data sets. In the final stage of our algorithm, during back projection segmentation in the case of IBSR 18 and 20 data sets, we applied morphological binary opening on a CSF tissue mask and thus removed all small groups of pixels, including some segmented SCSF pixels, since

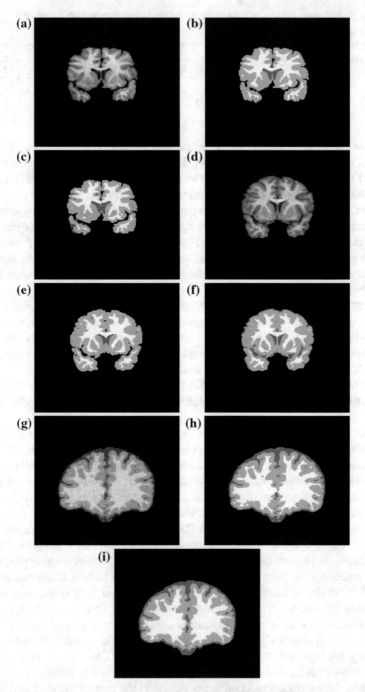

Fig. 6.12 Segmentation results of a real MR image (IBSR 20), (IBSR 18), an artificially generated MR image (BrainWeb), as follows: **a, d, g** input MR image (preprocessed), **b, e, h** ground truth segmented CSF, GM, and WM tissues, respectively, **c, f, i** segmented CSF, GM, and WM tissues (our algorithm)

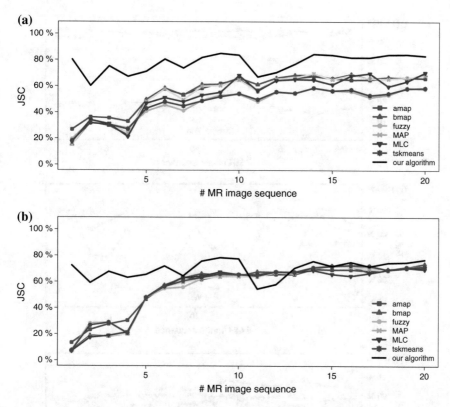

Fig. 6.13 Comparison with benchmark algorithms (IBSR 20) for the following tissues: **a** GM, **b** WM

we considered them to be noise; in the case of BrainWeb data sets we did not apply morphological opening in order to preserve segmented SCSF tissue as CSF tissue. We think this contributed to the slightly better results of our method with BrainWeb (Table 6.2) data sets, compared with real MRI data (Tables 6.3, 6.4).

The application of distance metric learning led to slightly better results, compared to the case with straightforward application of kNN classification. Although the three types of the classification of the test 2D histogram look quite similar at first glance (Fig. 6.15), they differ in a few aspects: LMNN provides the most steady borders between different tissue classes (Fig. 6.15a) and it is about five times faster than the NCA method (Fig. 6.15b), which is prohibitively slow; although kNN classification without distance metric learning (Fig. 6.15c) is the fastest classification method (20–30 % faster than LMNN method), it provides the most inaccurate results—curved borders between separate tissue segments and unreliable classification of the 2D histograms of some test MR images.

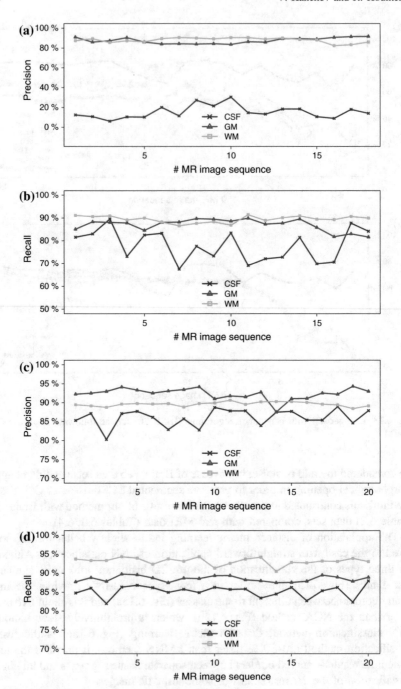

Fig. 6.14 Evaluation of our algorithm of CSF, GM, and WM tissues: **a** Precision (IBSR 18), **b** Recall (IBSR 18), **c** Precision (BrainWeb), **d** Recall (BrainWeb)

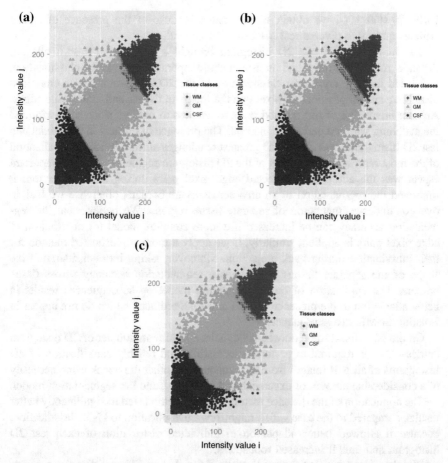

Fig. 6.15 A classified 2D histogram (CSF, GM, and WM tissues) of an MR image (IBSR 20) after the following distance metric learning methods: **a** LMNN, **b** NCA, **c** direct kNN classification without distance metric learning

6.8 Discussion

Our results suggest that the tissue segmentation method with a 2D histogram model can be applied successfully to a MR image sequence, giving comparable results with the traditional histogram, standard benchmark, and state-of-the-art segmentation methods in terms of accuracy. Overall, it overcomes the most common drawbacks of the conventional histogram of individual pixels of a MR image: it introduces spatial information in itself, has increased discrimination from overlapping segments, corresponding to different tissue regions in a MR image. When a 2D histogram has a consistent distribution, it can be reliably classified and hence gives better results in the case of filtered real MRI data (IBSR 18); in the case of

IBSR 20 data sets, we obtain worse results because of the presence of strong artifacts and more diverse real MRI data, in general.

The larger area of our 2D histogram brought a few advantages: it allowed distance metric learning and non-parametric operations as kNN classification, which led to more consistent classification of the 2D histogram, it also embraced distinctive spatial positions of pixel pairs of the main tissues and inter-tissue edges. Another advantage of our method is its robustness to outliers since we performed the matching within a percentile interval. The presence of outliers in the model and test 2D histograms was one of the greatest challenges since it hindered localization of the most representative entries of the 2D histogram segments. Another important aspect was the addition of detected edge pixel pairs during back projection: it improved the segmentation of an area across tissue borders, otherwise we had an over- or under-segmentation of separate tissue regions. We believe that the segmentation accuracy can be increased if a more complex model for distribution of edge pixel pairs is applied; similarly, if we apply a more sophisticated method for their introduction during back projection. Moreover, taking into consideration the shape of the separate tissues will increase segmentation accuracy across tissue borders. The application of our segmentation model on subsequences results in better adaptation to the presence of INH and PVE artifacts, which do not appear in isolation in MR image sequences.

On the other hand some drawbacks remain: a minimum number of 2D histogram entries—20—is required to perform successfully and reliably classification of 2D histograms of all MR images from the sequence. Another drawback is the necessity of a considerable amount of expert-labeled masks to build the segmentation model.

The application of the distance metric learning methods led to significantly better results, compared to the case with straightforward application to kNN classification, because it allowed better adaptation to individual distribution of each test 2D histogram and thus it increased robustness.

We also found out that the application of weights to different directions during 2D histogram construction and back projection did not improve the segmentation model and the results. The reason for that is a lack of dominant direction of distributions of pixel pairs of tissues and edges in MR brain images.

The results are comparable with recent brain segmentation algorithms due to the richer information of intra- and inter-image correlation it uses. We believe that further refinement of the segmentation model with respect to the used correlation can lead to better results. Another aspect we can improve in our algorithm is to select a more accurate similarity distance metric between 2D histograms of the MR images from the sequence.

We also think that our method can be applied to other types of medical images, characterized by adjacent pixels with close intensity values and existing strong correlation between consecutive images but not natural or man-made images due to their less compact 2D histograms.

6.9 Conclusion

In the present work, we proposed a method with 2D histogram matching for MR image segmentation. Our model is non-parametric and adjusted to the local characteristics of MR images. Other segmentation methods with 2D histograms do not build models to perform segmentation. Considering the limits of use of a traditional image histogram for image segmentation, we introduced in our 2D histogram model inter-image correlation and intra-image spatial information of MR images in the MR image sequence. Prior knowledge was introduced by the matching operation and its application was possible due to the larger area of the 2D histogram. Another important merit of our 2D histogram model is that it allows also application of non-parametric classifiers as kNN and modern machine learning techniques as distance learning. A further advantage comes from its application to MR subsequences: it adapts naturally to the presence of INH and PVE artifacts since they do not appear in isolation in MR image sequences. It can be seen that our segmentation results are comparable with more complex state-of-the-art segmentation methods in terms of accuracy, although there is still room for improvement of our method.

In future work, we plan to extend the segmentation model with multi-scale information of a 2D histogram, to include co-occurrence matrices with larger offsets and to consider the shape of the different tissues in MR image using a more complicated decision rule in the local window during back projection.

References

1. Ashburner, J., Friston, K.J.: Unified segmentation. Neuroimage **26**(3), 839–851 (2005)
2. Azmi, R., Pishgoo, B., Norozi, N., Yeganeh, S.: Ensemble semi-supervised frame-work for brain magnetic resonance imaging tissue segmentation. J. Med. Signals. Sens. **3**(2), 94–106 (2013)
3. Balafar, M.A., Ramli, A.R., Saripan, M.I., Mashohor, S.: Review of brain MRI image segmentation methods. Artif. Intell. Rev. **33**(3), 261–274 (2010)
4. Barley, A., Town, C.: Combinations of feature descriptors for texture image classification. J. Data Anal. Inf. Process. **2**, 67–76 (2014)
5. Bromiley, P.: Problems with the Brainweb MRI simulator in the evaluation of medical image segmentation algorithms, and an alternative methodology. Technical report, ISBE, Manch Univ Med Sch, tina Memo No. 2008-002, Internal Memo (2008)
6. Burger, M., Modersitzki, J., Tenbrinck, D.: Mathematical methods in biomedical imaging. GAMM-Mitt. **37**(2), 154–183 (2014)
7. Cabezas, M., Oliver, A., Lladó, X., Freixenet, J., Bach Cuadra, M.: A review of atlas-based segmentation for magnetic resonance brain images. Comput. Meth. Prog. Bio. **104**(3), 158–177 (2011)
8. Chen, S., Radke, R.J.: Level set segmentation with both shape and intensity priors. In: 2009 IEEE 12th International Conference Computer Vision, pp. 763–770. IEEE (2009)
9. Chen, W., Cao, L., Qian, J., Huang, S.: A 2-phase 2-D thresholding algorithm. Digit. Signal Proc. **20**(6), 1637–1644 (2010)
10. Cocosco, C.A., Kollokian, V., Kwan, R.S., Evans, A.C.: BrainWeb: online interface to a 3D MRI simulated brain database. NeuroImage **5**(4), 425 (1997)

11. Despotović, I., Goossens, B., Philips, W.: MRI segmentation of the human brain: challenges, methods, and applications. Comput. Math. Methods Med. 2015 (2014)
12. Dong, F., Peng, J.: Brain MR image segmentation based on local Gaussian mixture model and nonlocal spatial regularization. J. Vis. Commun. Image Represent. **25**(5), 827–839 (2014)
13. Ekin, A.: Pathology-robustmr intensity normalizationwith global and local constraints. In: Proceedings of IEEE International Symposium on Biomedical Imaging (ISBI), pp. 333–336. IEEE (2011)
14. Foruzan, A.H., Kalantari Khandani, I., Baradaran Shokouhi, S.: Segmentation of brain tissues using a 3-D multi-layer Hidden Markov Model. Comput. Biol. Med. **43**(2), 121–130 (2013)
15. Goldberger, J., Roweis, S., Hinton, G., Salakhutdinov, R.: Neighbourhood components analysis. In: Saul, L.K., Weiss, Y., Bottou, l. (eds.) Advances in Neural Information Processing Systems 17, vol. 17, pp. 513–520. MIT Press, Cambridge, MA (2005)
16. Gonzalez, R.C., Woods, R.E.: Digital image processing. 3rd edn (2007)
17. Hastie, T., Tibshirani, R.: Classification by pairwise coupling. Ann. Statist. **26**(2), 451–471 (1998)
18. Hedges, T.: Technical note. an empirical modification to linear wave theory. In: ICE Proceedings, Thomas Telford, vol. 61, pp. 575–579 (1976)
19. Ji, Z., Xia, Y., Chen, Q., Sun, Q., Xia, D., Feng, D.D.: Fuzzy c-means clustering with weighted image patch for image segmentation. Appl. Soft Comput. **12**(6), 1659–1667 (2012)
20. Ji, Z., Xia, Y., Sun, Q., Chen, Q., Xia, D., Feng, D.D.: Fuzzy local Gaussian mixture model for brain MR image segmentation. IEEE Trans. Inf. Technol. Biomed. **16**(3), 339–347 (2012)
21. Ji, Z., Liu, J., Cao, G., Sun, Q., Chen, Q.: Robust spatially constrained fuzzy c-means algorithm for brain MR image segmentation. Pattern Recogn. **47**(7), 2454–2466 (2014)
22. Ji, Z., Xia, Y., Sun, Q., Chen, Q., Feng, D.: Adaptive scale fuzzy local Gaussian mixture model for brain MR image segmentation. Neurocomputing **134**, 60–69 (2014)
23. Kazemi, K., Noorizadeh, N.: Quantitative comparison of SPM, FSL, and brainsuite for Brain MR image segmentation. J. Biomed. Phys. Eng. **4**(1), 13–26 (2014)
24. Ledig, C., Wolz, R., Aljabar, P., Lotjonen, J., Heckemann, R.A., Hammers, A., Rueckert, D.: Multi-class brain segmentation using atlas propagation and EM-based refinement. In: 2012 9th IEEE International Symposium on Biomed Imaging (ISBI), pp. 896–899. IEEE (2012)
25. Lee, J.D., Su, H.R., Cheng, P., Liou, M., Aston, J., Tsai, A., Chen, C.Y.: MR image segmentation using a power transformation approach. IEEE Trans. Med. Imaging **28**(6), 894–905 (2009)
26. Lee, M., Cho, W., Kim, S., Park, S., Kim, J.H.: Segmentation of interest region in medical volume images using geometric deformable model. Comput. Biol. Med. **42**(5), 523–537 (2012)
27. Li, C., Huang, R., Ding, Z., Gatenby, J., Metaxas, D.N., Gore, J.C.: A level set method for image segmentation in the presence of intensity inhomogeneities with application to MRI. IEEE Trans. Image Process. **20**(7), 2007–2016 (2011)
28. Lin, L., Garcia-Lorenzo, D., Li, C., Jiang, T., Barillot, C.: Adaptive pixon represented segmentation (APRS) for 3D MR brain images based on mean shift and Markov random fields. Pattern Recogn. Lett. **32**(7), 1036–1043 (2011)
29. Liu, W., Shang, Y., Yang, X., Deklerck, R., Cornelis, J.: A shape prior constraint for implicit active contours. Pattern Recogn. Lett. **32**(15), 1937–1947 (2011)
30. Mahmood, Q., Chodorowski, A., Mehnert, A., Persson, M.: A novel Bayesian approach to adaptive mean shift segmentation of brain images. In: 2012 25th IEEE International Symposium on Computer-Based Medical Systems (CBMS), pp 1–6. IEEE (2012)
31. Monaco, J.P., Madabhushi, A.: Class-specific weighting for Markov random field estimation: Application to medical image segmentation. Med. Image Anal. **16**(8), 1477–1489 (2012)
32. Morin, J.P., Desrosiers, C., Duong, L.: Image segmentation using random-walks on the histogram. In: Proceedings of SPIE, vol. 8314, p. 83140 (2012)
33. Oostenveld, R., Fries, P., Maris, E., Schoffelen, J.M.: FieldTrip: open source software for advanced analysis of MEG, EEG, and Invasive Electrophysiological Data. Intell. Neurosci. **2011**(1), 1–9 (2011)

34. Ortiz, A., Górriz, J., Ramrez, J., Salas-Gonzalez, D., Llamas-Elvira, J.M.: Two fully-unsupervised methods for MR brain image segmentation using SOM-based strategies. Appl. Soft Comput. **13**(5), 2668–2682 (2013)
35. Robitaille, N., Mouiha, A., Crépeault, B., Valdivia, F., Duchesne, S.: Tissue-based MRI intensity standardization: application to multicentric datasets. Int. J. Biomed. Imaging **2012** (347), 120 (2012)
36. Rohlfing, T.: Image similarity and tissue overlaps as surrogates for image registration accuracy: widely used but unreliable. IEEE Trans. Med. Imaging **31**(2), 153–163 (2012)
37. Roy, S., Carass, A., Bazin, P.L., Resnick, S., Prince, J.L.: Consistent segmentation using a Rician classifier. Med. Image Anal. **16**(2), 524–535 (2012)
38. Rubin, S., Kountchev, R., Todorov, V., Kountcheva, R.: Contrast enhancement with histogram-adaptive image segmentation. In: Proceedings of IEEE International Conference on Information Reuse and Integration, pp. 602–607. IEEE (2006)
39. Shapira, D., Avidan, S., Hel-Or, Y.: Multiple histogram matching. In: Proceedingsof 20th International Conference on Image Processing (ICIP), pp. 2269–2273. IEEE (2013)
40. Shattuck, D.W., Sandor-Leahy, S.R., Schaper, K.A., Rottenberg, D.A., Leahy, R.M.: Magnetic resonance image tissue classification using a partial volume model. NeuroImage **13**(5), 856–876 (2001)
41. Shen, D.: Image registration by local histogram matching. Pattern Recognit. **40**(4), 1161–1172 (2007)
42. Solanas, E., Duay, V., Cuisenaire, O., Thiran, J.: Relative anatomical location for statistical non-parametric brain tissue classification in MR images. In: Proceedings of 7th International Conference on Image Processing (ICIP), vol. 2, pp. 885–888. IEEE (2001)
43. Theodoridis, S., Koutroumbas, K.: Pattern Recognition, 4th edn. Academic Press (2008)
44. Tohka, J.: Partial volume effect modeling for segmentation and tissue classification of brain magnetic resonance images: a review. World J. Radiol. **6**(11), 855–864 (2014)
45. Valverde, S., Oliver, A., Cabezas, M., Roura, E., Lladó, X.: Comparison of 10 brain tissue segmentation methods using revisited IBSR annotations. J. Magn. Reson. Im **41**(1), 93–101 (2015)
46. Vrooman, H., der Lijn, F.V., Niessen, W.: Auto-kNN: Brain Tissue Segmentation using Automatically Trained k-Nearest-Neighbor Classification (2013)
47. Wallace, G.K.: The JPEG Still Picture Compression Standard. Commun. ACM **34**(4), 30–44 (1991)
48. Wang, Q., Chen, L., Shen, D.: Fast histogram equalization for medical image enhancement. In: Engineering in Medicine and Biology Society, 2008. EMBS 2008. 30th Annual International Conference of the IEEE, pp 2217–2220. IEEE (2008)
49. Weinberger, K.Q., Saul, L.K.: Distance metric learning for large margin nearest neighbor classification. J. Mach. Learn. Res. **10**, 207–244 (2009)
50. Wu, J., Pian, Z., Guo, L., Wang, K., Gao, L.: Medical image thresholding algorithm based on fuzzy sets theory. In: 2nd IEEE Conference on Industrial Electronics and Applications, 2007. ICIEA 2007, pp. 919–924. IEEE (2007)
51. Zagorodnov, V., Ciptadi, A.: Component analysis approach to estimation of tissue intensity distributions of 3D images. IEEE Trans. Med. Imaging **30**(3), 838–848 (2011)
52. Zhang, J., Hu, J.: Image segmentation based on 2D otsu method with histogram analysis. In: Proceedings of 2008 International Conference on Computer Science and Software Engineering, CSSE '08, vol. 6, pp. 105–108. IEEE (2008)
53. Zhang, J., Hu, J.: Curvilinear thresholding method for noisy images based on 2D histogram. In: Proceedings of 2008 IEEE International Conference on Robotics and Biomimetics, ROBIO 2008, pp. 1014–1019. IEEE (2008)
54. Zhang, T., Xia, Y., Feng, D.D.: A clonal selection based approach to statistical brain voxel classification in magnetic resonance images. Neurocomput **134**, 122–131 (2014)
55. Zhang, T., Xia, Y., Feng, D.D.: Hidden Markov random field model based brain MR image segmentation using clonal selection algorithm and Markov chain Monte Carlo method. Biomed. Signal Proces. **12**, 10–18 (2014)

56. Zhang, X., Dong, F., Clapworthy, G., Zhao, Y., Jiao, L.: Semi-supervised tissue segmentation of 3D brain MR images. In: 2010 14th International Conference on Information Visualisation (IV), pp 623–628. IEEE (2010)
57. Zhang, Y., Brady, M., Smith, S.: Segmentation of brain MR images through a hidden Markov random field model and the expectation-maximization algorithm. IEEE Trans. Med. Imaging **20**(1), 45–57 (2001)
58. Zhou, Y., Huang, Y., Ling, H., Peng, J.: Medical image retrieval based on texture and shape feature co-occurrence. In: SPIE Medical Imaging, International Society for Optics and Photonics, vol. 8315, p. 83151 (2012)
59. Zhu, X., Goldberg, A.B.: Introduction to semi-supervised learning. Synth. Lect. Artif. Intell. Mach. Learn. **3**(1), 1–130 (2009)

Chapter 7
Multistage Approach for Simple Kidney Cysts Segmentation in CT Images

Veska Georgieva and Ivo Draganov

Abstract In the chapter is presented a multistage approach for segmentation of medical objects in Computed Tomography (CT) images. Noise reduction with consecutive applied median filter and wavelet shrinkage packet decomposition, and contrast enhancement based on Contrast limited adaptive histogram equalization (CLAHE) are applied in preprocessing stage. As a next step is used a combination of 2 basic methods for image segmentation such as split and merge algorithm, following by color based K-mean clustering. For refining the boundaries of the detected objects additional texture analysis is introduced based on limited Haralick's feature set and morphological filters. Due to the diminished number of components for the feature vectors the speed of the segmentation stage is higher in comparison with the full feature set. Some experimental results are presented, obtained by computer simulation in the MATLAB environment. The experimental results give detailed information about detected simple renal cysts and their boundaries in axial plane of CT images which are presented in native, arterial and venous phases. The proposed approach can be used in real time for precise diagnosis or in monitoring the disease progression.

Keywords Bio-medical informatics · CT images · Kidney cysts detection · Noise reduction · Wavelet transformations · Image segmentation · Texture analysis

V. Georgieva (✉) · I. Draganov
Department of Radio Communications and Video Technologies,
Technical University of Sofia, 8 Kliment Ohridski Blvd, 1756 Sofia, Bulgaria
e-mail: vesg@tu-sofia.bg

I. Draganov
e-mail: idraganov@tu-sofia.bg

© Springer International Publishing Switzerland 2016
R. Kountchev and K. Nakamatsu (eds.), *New Approaches in Intelligent Image Analysis*, Intelligent Systems Reference Library 108,
DOI 10.1007/978-3-319-32192-9_7

7.1 Introduction

7.1.1 Medical Aspect of the Problem for Kidney Cyst Detection

A kidney cyst is a round or oval fluid-filled pouch with a well-defined outline. They are the most common space-occupying lesions of the kidney [1]. Kidney cysts typically grow on the surface of a kidney, but some may develop inside kidney. They can be associated with serious disorders that may impair kidney function. The diagnosis of many of the "cystic kidneys" requires clinical, genetic, radiological, and pathological information. A precise diagnosis is important for prognosis and treatment. Simple cysts are discrete lesions within the kidney that are typically cortical, extending outside the parenchyma and distorting the renal contour. One typical example of simple cyst is presented in Fig. 7.1.

Their importance stems from their increased detection in aging populations with widespread use of abdominal ultrasonography and computed tomography (CT) or magnetic resonance imaging (MR), with the aim of evaluating renal cyst morphology and volume and estimating the amount of residual renal parenchyma [1].

In our investigations we use CT images. CT has the advantages of widespread availability, more rapid examination time in comparison with MR imaging, and lower cost than MR imaging. The diagnostic challenges they present is their differentiation from the atypical features of the far less common complex cysts associated with malignancy. A distinct characteristic of simple cysts is their

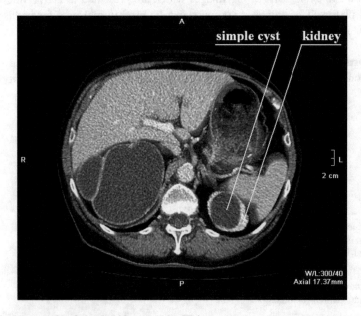

Fig. 7.1 An example of simple kidney cyst in CT image

increased occurrence with aging. The cyst wall is characteristically smooth, transparent, avascular, yellowish or bluish white in color, and formed by a thin layer of fibrous tissue lined by a single layer of flattened or cuboidal epithelia [2].

7.1.2 Review of Segmentation Methods

Image segmentation is typically used to locate objects and their boundaries (lines, curves, etc.) in medical images. Hence, region-growing, or edge detection algorithms are unable to effectively cope with this data. Although a large variety of segmentation methods have been developed for medical image processing. There are some publications based only on basic methods of segmentation for detection of kidney cysts [3, 4]. Some proposed methods start with an initial rough segmentation of a cyst in only one 2D slice from the full stack of the volume scan. This may be quickly done by a radiologist. But the quality of the initialization does not greatly affect the quality of the final segmentation, hence granting good repeatability properties of the measurement [5]. In [6] is presented a system for processing CT images and MRI kidney, which is based on the split and merges method for segmentation step, and application of SVM (Support Vector Machine) for detecting abnormalities in these images.

7.1.2.1 Texture Based Segmentation

As a major process at the segmentation stage of a particular abnormality texture analysis is widely used in medical image processing [7–13]. Most of these methods have grounds which lay way back within the pioneering works of Haralick [14, 15] and further in expanded form by others [16–18] specifically for medical applications. Generally there are 8 distinctive measures which are mostly incorporated in forming the feature vectors for a particular texture discrimination based on the early studies [14]—energy, entropy, correlation, difference moment, inertia (contrast), cluster shade, cluster prominence, and Haralick's correlation. In his later study [15] the author touches the direct physical relation between the final pictorial representation of texture and its natural forming stages. There the optical processing methods are placed as an important factor for the textural edginess and the spatial gray tone dependence from which features are formed. It is also shown that a large set of primitives and some more general texture measures become obsolete when getting in a set the listed above.

Depending on the organ being investigated with a specific probable disease only a portion of the measures are used, e.g. the contrast is being analyzed in relation of its degree of isotropic distribution introducing directional estimation for it along with Fourier domain energy sampling [17] when the object of interest are the lungs. Some more recent studies taking an advantage of obtaining higher dimensionality of the image data retrieved in the form of 3D CT scans actually extend the number of features up from the original Haralick's set [19, 20]. There Tesar et al. introduce co-occurrence measures for the voxels in a small cubic region which allows them to define a multidimensional Gaussian mixture model with parameters ready to be found by the EM algorithm. As an additional refining step for higher accuracy of the segmentation inference phase is used along with other probabilistic methods merging. Degree of correspondence among the targeted tissues is reported as high as a little over 0.8.

Some authors relying only on the two-dimensional data from a single CT slice place variety of additional parameters for improving the results of the segmentation. In [11] Castellano et al. describe histogram, absolute gradient and a run-length matrix to supplement Haralick's features for more accurate segmentation. All of them are combined together in order to run an auto-regressive model with the use of wavelets in discovering obstructive lung diseases and others.

In large variety of cases CT imaging is the preferred mean for segmenting bone and bone tumors [7]. This is done by thresholding and region growing operations as basic steps which further are being combined within Markov random fields, deformable models and fuzzy region growing to carry out the training stage. This approach proves to be efficient also in reconstruction of bronchial trees while deformable models alone are thought to be promising for segmentation of the abdominal area and the heart.

Similar set of features as the above are selected in other approaches where the training stage is performed with the use of artificial neural networks with additional processes such as relaxation [16]. As Pal et al. reveal such combination copes with the noises typical for the images being segmented better in many cases.

7.1.2.2 Incorporating Texture Segmentation in M-CBIR

When considering a complete medical content based image retrieval system (M-CBIR) an important phase of selecting the features used to match images or portions from them based on textural analysis is to discriminate them to local and global. Glatard et al. [12] render an account of these two groups where each of them has a particular impact over diminishing the effect of present noises and the lower resolutions of medical images in general. While local features are more appropriate for training in lots of cases for getting the patterns of single tissue with or without any abnormalities in it considering the noises after the consecutive segmentation phase global features may very successfully be used for measurements of sizes including volume. An extensive study on global features performance is made by Gueld et al. in [21]. The image corpus consist of nearly 6000 images most of which are from radiography and arranged in more than 80 categories. The features used include Tamura's, Castelli's and Ngo's texture descriptors taken from downscaled image representations. Another novelty is the application of Zhou and Huang image structure—a physical model of filling with water to a binarized gradient image. Then k-NN classifier is used where k = {1, 5} and also classifier combinations of parallel, serial and hierarchical type. When testing the single classifier within 5 neighbors the accuracy varies between 40 and 80 % when using different distance measure—Euclidian and Mahalanobis. When taking 10 neighbors the accuracy increases considerably from 89 to 97 %. Classifier combination raises the accuracy around 92 % at 5 neighbors and up to 95 % for 10.

Further development of M-CBIR systems introduces metric data structures [22] by Chuctaya et al. They expand the feature set with 5 more of the Haralick's descriptors increasing them up to 13 parameters combining them with Gabor filter output and gray level histograms. Thus a classification of pixels to border and interior related for a particular area becomes possible. Using the slim-tree indexing technique a similarity measure based on distance provides the final decision for selection the most related images from the database. The number of feature components rises up to 435 which assures precision of up to 0.85 at 0.18 recall when the test set contains over 28,000 images with more than 1000 query images run 70 times over again.

Some other approaches are based only on local texture analysis as the algorithm for automatic detection of abnormalities in chest radiographs developed by van Ginneken et al. [9]. There the exact nature of the abnormality is not preliminary defined but rather is found as discerned area to its neighborhood. Subdivision of the lungs is done into overlapping regions of various sizes by applying texture analysis preceded by more rough area division by active shape models. Multiscale filter banks are used monitoring their response and finding the respective moment. Difference features are calculated by subtracting feature vectors from corresponding regions in the left and right lung fields. Then selection is performed by k-NN classifier followed by weighted multiplier for combining the results of each region. The final result is abnormality score per image which in the case of database consisting of nearly 400 images a classification sensitivity of 0.86 is achieved given specificity of 0.50. A test with a second database having 200 images half of which contain abnormalities leads to sensitivity 0.97 with specificity 0.90. In [23] is presented a hybrid technique for the classification of the magnetic resonance images (MRI). The medical decision making system designed by the wavelet transform, the principal component analysis, and the supervised learning methods (FP-ANN and k-NN) give very promising results in classifying the healthy and brain patient.

7.1.2.3 Combining Haralick Features and Morphology

A separate group of methods for medical image segmentation combine Haralick's features with morphological parameters [18, 24]. Chaddad et al. [24] try to decrease the number of feature components while preserving the accuracy of image segmentation. Starting from the original snake approach they use progressive division reducing the dimensions of the image to achieve smaller execution time. Nine morphological parameters were added to the Haralick's set and as a learning structure a probabilistic neural network. Speed up of 50 % is reported where only three of the morphological parameters are thought to be most effective—area, xor convex and contrast. From the second group parameters—orrelation, entropy and contrast tend to be most promising for discrimination. In addition to incorporating the morphological set of features Malpica et al. [18] use splitting of the initial image into *n* channels by generating *n* texture features from around each pixel's neighborhood. Most discriminative features are applied based on previous studies. Then a gradient image of all the channels is constructed and minima selection is done from where a watershed process begins leading to watershed segmentation where touching flooded regions are merged in iterative fashion. The way of finding the initial minima is dynamics with its advantages over the interactive approach or the one based on a priori knowledge of the image content as well as the waterfall algorithm. The results from testing when using the mean approach show around 90 % correctly classified pixels at (8–18) % of dynamic range (dynamics). When using the weighted mean and the Hotelling test the results are even higher.

7.1.2.4 Pre- and Post-processing Strategies

In [8] Freixenet et al. put a special attention to the post-processing integration stage when developing a complete medical images segmentation system. They view this stage as a closing part of the whole process after applying region-based and boundary-based algorithms. The goal is to have the region and the edge information put back as accurate as possible. To achieve these three paths could be followed. The first one is over-segmentation—the segmentation stage is carried out in such a way that by fixing a set of parameters preliminary an over-segmented image is always the result. Then another segmentation algorithm is applied leading to different result in terms of number of segmented areas and this information is used to get more accurate boundaries for them from the first stage. The second path is to have boundary refinement—at first again a pre-segmented image is obtained with roughly found boundaries and then using edge detection techniques help in refining them. The third path is selection-evaluation—it is similar to the previous one with the difference that edge detection is implemented over several pre-segmented versions of the input image and the more accurate result is generated from the most accurate refinement. Freixenet et al. report that multiresolution analysis and the "snake" approach could be also embedded to have the boundary refinement stage accurately done.

A detailed comparison of texture models for automatic liver segmentation [13] is presented by Pham et al. Beginning with window level CT images and passing through texture feature extraction and then pixel-level classification a set of liver probability images is received. After that seed sets detection is preformed followed by adaptive region growing resulting in liver segmented images. From experimental testing is found that co-occurrence texture model performs better in comparison to Gabor filters and it is considered perspective for wide variety of other organs' tissue other than the liver.

In [10] Felippe et al. test a set of texture features for tissue identification from medical images. In a 4-step approach the authors manage to compare the use of a set of Haralick features over images containing different types of tissue. In step 1 the co-occurrence matrices of each image is calculated, then in step 2—the values of selected descriptors are found followed by image signatures generation in step 3 and finally in step 4—comparison of the images through their signatures. Haralick's features under investigation here are energy, entropy, variance, homogeneity, 3rd order moment, and inverse variance. The combined set of features leads to the most precise segmentation in of almost 90 % of the total tissue surface which needs to be selected. Then the gradient and homogeneity features used alone give around 72 to 74 % accurately segmented tissue followed by the others with less precision. As for the precision vs. recall function about different types of tissue when using the combined feature muscle is the most accurately segmented, then heart, breast, etc. Liver and spine are at the bottom of the list for precisely segmented tissues.

Another autonomous system for particular liver diagnosis from CT images is presented in [25] by Chen et al. Here normalized fractional Brownian motion model feature curves are employed. The stages through which the boundary of the liver is found particularly called "detect-before-extract" include initial detection and contour modification using a deformable active model. At the stage of classification of liver tumors spatial gray-level co-occurrence matrices are used with probabilistic neural networks.

In [26] another extensive comparison through experimentation is made for the most popular texture features in this case over wider range of images. Here the co-occurrence feature along with the one from Gabor filter tends to produce almost identical results.

7.1.3 Proposed Approach

In accordance with the main properties of the examined algorithms above we propose to use a multistage segmentation approach, based on combination of split and merge algorithm and color based K-mean clustering, with the goal to obtain more information and better defined boundaries of the cysts. As first is proposed a stage of preprocessing in order to reduce the noise and enhance the image. For discrimination of one object from another in terms of more accurate boundaries a texture analysis is introduced based on reduced Haralick's feature set.

The chapter is arranged as follows: In Sect. 7.2 is given the stapes in preprocessing stage; in Sect. 7.3 are presented the steps of multistage segmentation approach; Sect. 7.4—some experimental results, obtained by computer simulation; then their interpretation in Sect. 7.5—Discussion and finally in Sect. 7.6—a Conclusion is made.

7.2 Preprocessing Stage of CT Images

The quality of CT images varies depending on penetrating X-rays in a different anatomically structures. Noise in CT arises from the fundamentally statistical nature of photon production and can appear as thin bright and dark streak artifact in the reconstructed image preferentially in the direction of greatest attenuation. With increased noise, low contrast soft tissue boundaries may be obscured [6]. This quantum noise is dominant and comes from the quantization of energy into photons. It is Poisson distributed and independent of measurement noise and has the characteristic of multiplicative noise [27]. The measurement noise is additive Gaussian noise and usually negligible relative to the quantum noise.

To obtain a better quality of investigated medical object in CT images is proposed to reduce the noise and enhance the contrast.

Our investigations in area of noise reduction in CT images of abdominal structures show that some organs (for example the liver), may have density variations within them that have the appearance of random noise. For this reason we propose to be applied a combination of median filter and wavelet denoising on the base of wavelet packet shrinkage decomposition and adaptive threshold.

The general algorithm of preprocessing consists of three consecutive basic stages, used to improve the image quality:

- Noise reduction with median filter for elimination of distortion or blurring by impulse noise;
- Noise reduction based on wavelet packet decomposition and adaptive threshold;
- Contrast limited adaptive histogram equalization (CLAHE) for contrast enhancement.

7.2.1 Noise Reduction with Median Filter

Median filter is an example of non-linear filters. In median filter, the ranking of the neighboring pixels is done according to the intensity or brightness level and value of the pixel under evaluation is replaced by the median value of surrounding pixel values. Median filter can therefore effectively denoise medical images. The CT images distorted or blurred by shot or impulse noise can be denoised using this filter. Median filters have many advantages over smoothening filters [28]:

- In median filter the output values consist of only those present in the neighborhood (median value) so there is no reduction in contrast across the steps;
- The boundaries are also not shifted when median filter is used;
- The edges are minimum degraded and hence median filter can be repeatedly applied.

7.2.2   *Noise Reduction Based on Wavelet Packet Decomposition and Adaptive Threshold*

The algorithm for noise reduction based on wavelet packet transform contains the following basic stages [29]:

(1) Decomposition of the CT image

The wavelet packet methods for noise reduction give a richer presentation of the image, based on functions with wavelet forms, which consist of 3 parameters: position, scale and frequency of the fluctuations around a given position. They propose numerous decompositions of the image, that allows estimate the noise reduction of different levels of its decomposition. Based on the organization of the wavelet packet library, the decomposition can be determined from a given orthogonal wavelets. Commonly used wavelet functions are daubechies, coiflet, and symmlet. The wavelets are chosen based on their shape and their ability to analyse the signal in a particular application. Various wavelet shrinkage algorithms denoise image by reduce wavelet coefficient. An optimal decomposition is used with respect to a conventional criterion. In case of denoising the 2D joint entropy of the wavelet co-occurrence matrix is used as the cost function to determine the optimal threshold. In this case 2D Discrete Wavelet Transform (DWT) is used to compose the noisy image into wavelet coefficients [30].

We use another adaptive approach. The criterion is a minimum of three different entropy criteria: the energy of the transformed in wavelet domain image, Shannon entropy and the logarithm of energy [29, 31].

(2) Determination of the threshold and thresholding of detail coefficients

By determination of the global threshold it is used the strategy of Birge-Massart [32]. It uses spatial-adapted threshold, which allows to determinate the thresholds in three directions: horizontal, vertical and diagonally. The threshold can be hard or soft. The soft-thresholding method is chosen over hard-thresholding, because it yields more visually pleasant images over hard-thresholding. To become more precisely determination of the threshold for noise reduction in the image we can penalize adaptively the sparsity parameter α. Choosing the threshold too high may lead to visible loss of image structures, but if the threshold is too low the effect of noise reduction may be insufficient.

(3) Restoration of the image

The restoration of the image is on the base on 2D Inverse Wavelet Packet Transform. The reconstructions level of the denoised image is dependent on the level of its best shrinkage decomposition.

(4) Estimation of filtration.

The procedure for noise reduction can be determined on the base of the calculated estimation parameters. All adaptive procedures in the proposed algorithm are made automatically, based on calculated estimation parameters. PSNR and SNR values are higher for better denoised CT image where the value of NRR is lower.

7.2.3 Contrast Limited Adaptive Histogram Equalization (CLAHE)

Contrast limited adaptive histogram is a generalization of ordinary histogram equalization and adaptive histogram equalization. CLAHE does not operate on the whole image works like ordinary Histogram Equalization (HE), but it works on small areas in images, named tiles. Each tile's contrast is enhanced, so that the histogram of the output area roughly matches the histogram determined by the 'Distribution' parameter. This parameter can be selected depending on the type of the input image. The adjacent tiles are then combined using bilinear interpolation to eliminate artificially induced boundaries. The contrast, particularly in homogeneous regions, can be limited to avoid amplifying any unwanted information like noise which could be existed in images. The algorithm CLAHE limits the slope associated with the gray level assignment scheme to prevent saturation. This process is accomplished by allowing only a maximum number of pixels in each of the bins associated with the local histograms. For limiting the maximum slope is to use a clip limit β to clip all histograms. This is a contrast factor that prevents over-saturation of the image specifically in homogeneous areas. These areas are characterized by a high peak in the histogram of a particular image tile due to many pixels falling inside the same gray level range.

Finally, cumulative distribution functions (CDF) of the resultant contrast limited histograms are determined for grayscale mapping. The result mapping at any pixel is interpolated from the sample mappings at the four surrounding sample grid pixels. Pixels in the borders of the image outside of the sample pixels need to be processed specially. The neighboring tiles were combined using bilinear interpolation and the gray scale values were altered according to the modified histograms [33].

7.3 Segmentation Stage

7.3.1 Segmentation Based on Split and Merge Algorithm

The split and merge algorithm attempts to divide an image into uniform regions. The basic representational structure is pyramidal, i.e. a square region of size m by m at one level of a pyramid has 4 sub-regions of size $m/2$ by $m/2$ below it in the

(a)   (b)

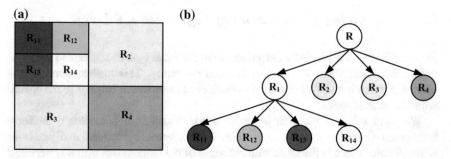

Fig. 7.2 Illustration of split algorithm: **a** a splited image; **b** the corresponded quad tree

pyramid. Usually the algorithm starts from the initial assumption that the entire image is a single region, and then computes the homogeneity criterion to see if it is TRUE. If FALSE, then the square region is *split* into the four smaller regions. This process is then repeated on each of the sub-regions until no further splitting is necessary. The algorithm may be summarized by the following steps [34]:

(1) Split any region Ri into four almost equal regions, where P(Ri) = FALSE.
(2) Merge any adjacent regions Ri and Rj for which P(Ri ∪ Rj) = TRUE.
(3) Stop when no further merging or splitting is possible.

Otherwise repeat steps (1) and (2). P(Ri) is a logical predicate over the set of pixels in the set of pixels in Ri and ∅ is the empty set.

For gray-level images, this condition can be that the variance of the gray-levels within a region is smaller than a given threshold value T. When this condition for a region is not met, this region is further split up. Figure 7.2 illustrates an example for split algorithm.

There R indicates the entire image. Each node corresponds to a (sub) region, whereby in this example only region R_1 was further divided up. If the image is divided up only into regions, adjoining regions will be similar in the final division.

These small square regions are then *merged* if they are similar to give larger irregular regions. The problem (at least from a programming point of view) is that any two regions may be merged if adjacent and if the larger region satisfies the homogeneity criteria, but regions which are adjacent in image space may have different parents or be at different levels (i.e. different in size) in the pyramidal structure. The process terminates when no further merges are possible. The predicate of homogeneity for a region R is based on two criteria:

• The average gray level of the region R is lower than a threshold.
• The variance of the gray levels in the region R is greater than a threshold.

This approach tends to be computationally intensive. This stage is applied to selected ROI of CT images by interactive procedure.

7.3.2   Clustering Classification of Segmented CT Image

In order to obtain a precisely detection of simple renal cyst in segmented image is used a color classification based on k-mean clustering. This method aims to partition n observations into k clusters in which each observation belongs to the cluster with the nearest mean.

We have selected the L*a*b* color space which is a perceptually uniform orthogonal Cartesian coordinate system. The differences between two pixels in L*a* b* color space is the same with the sense of the human eyes visual system and this color space enables doctors to quantify these visual differences. Color-Based Segmentation using K-mean clustering segments colors in an automated fashion using the L*a*b* color space and K-means clustering method. K-means clustering treats each object as having a location in space. It finds partitions such that objects within each cluster are as close to each other as possible. K-means requires that the number of clusters to be partitioned should be specified and also a distance metric to quantify how close two objects are to each other [35, 36]. Clustering is the process for grouping data points with similar feature vectors into a single cluster and for grouping data points with dissimilar feature vectors into different clusters. Let the feature vectors derived from l clustered data be X = {x_i, i = 1,2…., l}. The generalized algorithm initiates k cluster centroids C = {c_j, j = 1,2,….k} by randomly selecting k feature vectors are grouped into k clusters using a selected distance measure such as Euclidean distance so that

$$d = \|x_i - c_j\|. \tag{7.1}$$

The next step is to precompute the cluster centroids based on their group members and then regroup the feature vector according to the new cluster centroids. The clustering procedure stops only when all cluster centroids tend to converge.

7.3.3   Segmentation Based on Texture Analysis

Given the input CT image in its grayscale representation $g(i,j)$ with its dimensions $i = \overline{1,N}$ and $j = \overline{1,M}$ along the rows and columns respectively first the set of the following Haralick features [14] could be calculated:

$$f_1 = \sum_{i=-4}^{4} \sum_{j=-4}^{4} g^2(i,j), \tag{7.2}$$

where f_1 is the energy parameter found for every pixel inside a 9 × 9 neighborhood. Next the entropy feature denoted by f_2 is selected following the expression in the same neighborhood around each current image pixel:

$$f_2 = \begin{cases} -\sum_{i=-4}^{4} \sum_{j=-4}^{4} g(i,j) \log_2 g(i,j) & \text{if } g(i, j) \neq 0 \\ 0 & \text{if } g(i, j) = 0 \end{cases} \tag{7.3}$$

Then the correlation parameter f_3 is presented as follows:

$$f_3 = \sum_{i=-4}^{4} \sum_{j=-4}^{4} \frac{(i-\mu)(j-\mu)g(i,j)}{\sigma^2}, \tag{7.4}$$

where μ is the weighted pixel average and assuming the probable symmetry of the image along vertical and horizontal directions without any a priori knowledge it is thought to be:

$$\mu = \sum_{i=-4}^{4} \sum_{j=-4}^{4} ig(i,j) = \sum_{i=-1}^{1} \sum_{j=-1}^{1} jg(i,j). \tag{7.5}$$

In (7.4) σ is the weighted pixel variance and from the same symmetry assumptions it could be represented as:

$$\sigma = \sum_{i=-4}^{4} \sum_{j=-4}^{4} (i-\mu)^2 g(i,j) = \sum_{i=-4}^{4} \sum_{j=-4}^{4} (j-\mu)^2 g(i,j). \tag{7.6}$$

Defined this way μ and σ are actually the mean and standard deviation of the row and respectively column sums.

After that the difference moment f_4 could be obtained according to:

$$f_4 = \sum_{i=-4}^{4} \sum_{j=-4}^{4} \frac{1}{1+(i-j)^2} g(i,j). \tag{7.7}$$

The contrast also presented as inertia in some sources is given by:

$$f_5 = \sum_{i=-4}^{4} \sum_{j=-4}^{4} (i-j)^2 g(i,j). \tag{7.8}$$

The cluster shade is then included following the expression:

$$f_6 = \sum_{i=-4}^{4} \sum_{j=-4}^{4} ((i-\mu)+(j-\mu))^3 g(i,j). \tag{7.9}$$

Cluster prominence is the next parameter chosen in correspondence with:

$$f_7 = \sum_{i=-4}^{4} \sum_{j=-4}^{4} ((i-\mu)+(j-\mu))^4 g(i,j). \tag{7.10}$$

Finally Haralick's correlation is taken from the processing neighborhood:

$$f_8 = \frac{\sum\limits_{i=-4}^{4} \sum\limits_{j=-4}^{4} (i.j)g(i,j) - \mu^2}{\sigma^2}. \tag{7.11}$$

Taking the entropy feature f_2 from the input image $g(i, j)$ a new array with the same dimensions is generated $e(i, j)$. Then normalization is done for it to fit the dynamic range of the original image:

$$e_n(i,j) = \frac{e(i,j)}{\max\{e(i,j)\}} . \max\{g(i,j)\}. \tag{7.12}$$

Typically double precision is used for preserving accuracy.

In the next stage a histogram $h(e_n)$ of the resulting image is calculated and multi threshold segmentation is performed in order to get rough masks for each candidate texture area. Since some of the separate areas may take dominant part of the whole image and there may be no expressed maximum in the histogram so maximum entropy segmentation is selected for the purpose [37]. For a particular threshold θ_k separating two modes from the histogram a maximum in the entropy is desired for most accurate splitting to mode A and mode B:

$$\left|\begin{array}{l} h_A(\theta_k) = - \sum\limits_{i=\theta_{k-1}}^{\theta_k} h(e_{ni}) \log_2(e_{ni}) \\ h_B(\theta_k) = - \sum\limits_{i=\theta_k}^{\theta_{k+1}} h(e_{ni}) \log_2(e_{ni}) \end{array}\right. . \tag{7.13}$$

The a priori probabilities for both regions then are p_A and p_B found by:

$$p_A = \sum_{i=\theta_{k-1}}^{\theta_k} h(e_{ni}), \tag{7.14}$$

$$p_B = (1/n) - p_A = (1/n) - \sum_{i=\theta_{k-1}}^{\theta_k} h(e_{ni}), \tag{7.15}$$

where n is the number of regions. The normalized entropy then is:

$$H_A^0(\theta_k) = -\sum_{i=\theta_{k-1}}^{\theta_k} \frac{h(e_{ni})}{p_A} \log_2\left(\frac{h(e_{ni})}{p_A}\right), \tag{7.16}$$

$$H_B^0(\theta_k) = -\sum_{i=\theta_k}^{\theta_{k+1}} \frac{h(e_{ni})}{p_B} \log_2\left(\frac{h(e_{ni})}{p_B}\right) =$$

$$= -\sum_{i=\theta_k}^{\theta_{k+1}} \frac{h(e_{ni})}{(1/n) - p_A} \log_2\left(\frac{h(e_{ni})}{(1/n) - p_A}\right). \tag{7.17}$$

The total entropy for the group is given by:

$$H_{AB} = -\sum_{i=\theta_{k-1}}^{\theta_{k+1}} h(e_{ni}) \log_2 h(e_{ni}). \tag{7.18}$$

The target function then for a particular threshold is:

$$f(\theta_k) = H_A^0(\theta_k) + H_B^0(\theta_k) =$$

$$= \log_2 p_A (1 - p_A) + \frac{H_A(\theta_k)}{p_A(\theta_k)} + \frac{H_{AB} - H_A(\theta_k)}{1 - p_A(\theta_k)}. \tag{7.19}$$

From the maximum of f the current optimal threshold θ_{kopt} is found:

$$\theta_{kopt} = \arg\{\max\{f(\theta)\}\}. \tag{7.20}$$

If no apparent maximum is found an average over the investigated interval is taken.

Then a new segmented (indexed) image from $e_n(i, j)$ is generated based on the found thresholds:

$$s(i,j) = \begin{cases} 0, & \text{for } e_n(i,j) \in [0, \theta_1] \\ \quad \cdots \\ k, & \text{for } e_n(i,j) \in [\theta_k, \theta_{k+1}] \\ \quad \cdots \\ n-1, & \text{for } e_n(i,j) \in [\theta_{n-1}, \theta_n] \end{cases}. \tag{7.21}$$

Now from the indexed image each region can be represented by a binary mask $b_k(i, j)$:

$$b_k(i,j) = \begin{cases} 0, & \text{if } s(i,j) \neq k \\ 1, & \text{otherwise} \end{cases}. \tag{7.22}$$

For each $b_k(i, j)$ a binary representation is taken from $e_n(i, j)$ using the Otsu algorithm to have $e_{bk}(i, j)$. Then a series of morphological operations are performed over each e_{bk}. It starts with an opening according to the equation:

$$e_{bk} \circ S_1 = (e_{bk} \ominus S_1) \oplus S_1, \tag{7.23}$$

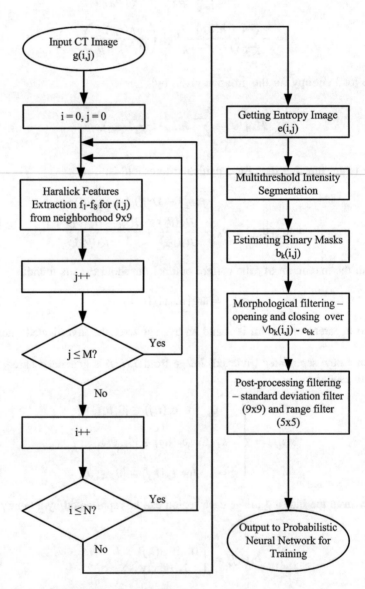

Fig. 7.3 Proposed approach for texture segmentation

where the structural element S_1 assures removal any object smaller than 50 pixels in total. Afterwards a morphological closing is performed:

$$e_{bk} \bullet S_1 = (e_{bk} \oplus S_2)\ominus S_2, \tag{7.24}$$

where now the structural element S_2 being a square with dimensions 9x9 coinciding with the initial size of the neighborhood from which the entropy image had been generated actually helps to refine any sharp corners inside the texture profile. It also removes any too small openings in the forms of dots, typically caused by non-suppressed noise.

Two additional processing steps are then applied for further cleaning of the image from known types of distortions. The first one is standard deviation filtering with a neighborhood of 9×9 again:

$$g_s(i,j) = \frac{g(i,j)}{2\pi\sigma^2} e^{-\frac{(i^2+j^2)}{2\sigma^2}}. \tag{7.25}$$

Finally a range filtration is done within a neighborhood of 5x5 in a progressive fashion over the whole resulting image.

So far a set of binary masks define the rough boundaries of totally n regions found. Now for each region the remaining 7 Haralick features are calculated separately—f_1, f_3–f_8. Then comparing the newly formed 7-components feature vectors among all the regions proper merging in type could be implemented by using the well-known probabilistic neural network approach [24]. The whole process is presented in generalized form in Fig. 7.3. The benefit of using preliminary the entropy feature is the considerable reduction of areas to be compared and eventually selected as separate texture fields rather than going directly to the general approach by forming the full-scale feature vectors consisting typically of 13 components.

7.4 Experimental Results

The formulated stages of processing are presented by computer simulation in MATLAB, version 7.14 environment with using the IMAGE PROCESSING and WAVELET TOOLBOXES. In analysis are used real 20 grayscale abdominal CT images in axial plane with size 667×557 pixels in native, arterial and venous phase of kidney. The original images are obtained in DICOM, but have been archived in jpeg file format. By pre-processing they are converted in bmp format. In Fig. 7.4 are presented original images in axial plane for the native, arterial and venous phase of kidney. The images in these three phases can be processed in parallel.

The 3×3 template is used to achieve median filtering. The obtained average results for NRR are around 0.5 and shows that the noise is two times reduced.

The best results by noise reduction of Poisson noise are obtaining by coiflet wavelet packet functions, adaptive shrinkage decomposition (best tree) on the base

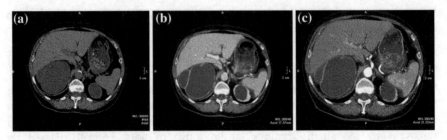

Fig. 7.4 Original CT presented in: **a** native phase; **b** arterial phase; **c** venous phase

Table 7.1 Simulation results for pre-processing stages of CT images

Stage of processing	PSNR (dB)	SNR$_Y$ (dB)	SNR$_F$ (dB)	E$_{FF}$ (dB)
1. Median filter	28.052	20.548	22.329	1.781
2. Noise reduction on WPT	33.814	22.329	24.522	2.193
3. CLAHE	35.035	–	–	–

of the third level and minimum of the Shannon entropy criteria, by using of hard penalized threshold. By using of the log energy and energy criteria the effectiveness of the filtration is smaller. In order to quantify how much noise is suppressed by the proposed noise reduction approach, the noise reduction rate is computed. The obtained average results for NRR are around 0.3 and shows that the noise is three times reduced. The values of PSNR and Effectiveness of filtration (E$_{FF}$) are sufficient.

The best results by contrast enhancement using CLAHE are obtained by bell-shaped form of histogram (Rayleigh distribution) and clip limit 0.04. Higher clip limit values will clip fewer values and thus they will be spread out more, hence more contrast.

The obtained averaging results by preprocessing of CT images in native phase are shown in Table 7.1. It presents the values of the objective quantitative estimation parameters such as PSNR, Signal to noise ratio in the noised image (SNR$_Y$), Signal to noise ratio in the filtered image (SNR$_F$), Effectiveness of filtration (E$_{FF}$) in the different stages of the algorithm. The variances are about ±0.005 by the particular images. The obtained averaging results by preprocessing of CT images in the other phases are similar and show insignificantly differences in the values.

A visual presentation of original CT image in native phase and its modifications as a result of pre-processing stags can be seen in Fig. 7.5. The CT images and their modifications in the arterial and venous phases are presented respectively in Figs. 7.6 and 7.7.

In Fig. 7.8 are shown the original CT ROI image with size 87 × 88 pixels in native phase with simple cyst in the left kidney and corresponding results obtained by split and merge segmentation and clustering of the segmented image.

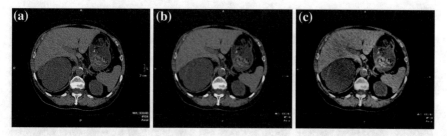

Fig. 7.5 The original CT image in native phase and its modifications as a result of pre-processing: **a** original; **b** after noise reduction stage; **c** after CLAHE

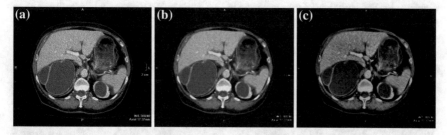

Fig. 7.6 The original CT image in arterial phase and its modifications as a result of pre-processing: **a** original; **b** after noise reduction stage; **c** after CLAHE

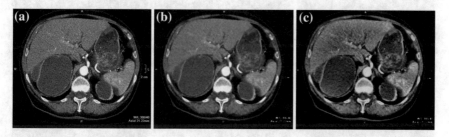

Fig. 7.7 The original CT image in venous phase and its modifications as a result of pre-processing: **a** original; **b** after noise reduction stage; **c** after CLAHE

Figure 7.9 presents the processed CT ROI image with size 87 × 88 pixels in native phase with simple cyst in the left kidney and corresponding results obtained by split and merge segmentation and clustering of the segmented image.

In Figs. 7.10 and 7.11 are given respectively original CT ROI image with size 87 × 88 pixels in arterial phase with simple cyst in the left kidney and corresponding results obtained by split and merge segmentation and clustering of the segmented image.

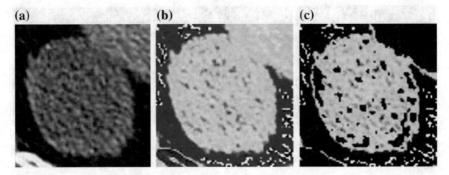

Fig. 7.8 The original CT ROI image in native phase and corresponding results of split and merge segmentation and clustering of the segmented image: **a** original ROI image; **b** after split and merge segmentation; **c** cluster 3 of the segmented ROI image

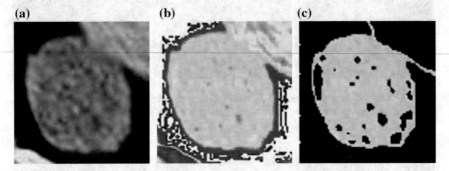

Fig. 7.9 The processed CT ROI image in native phase and corresponding results of split and merge segmentation and clustering of the segmented image: **a** processed ROI image; **b** after split and merge segmentation; **c** cluster 3 of the segmented ROI image

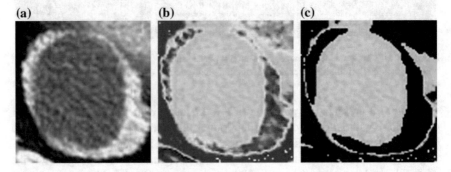

Fig. 7.10 The original CT ROI image in arterial phase and corresponding results of split and merge segmentation and clustering of the segmented image: **a** original ROI image; **b** after split and merge segmentation; **c** cluster 3 of the segmented ROI image

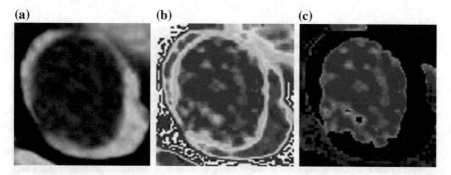

Fig. 7.11 The processed CT ROI image in arterial phase and corresponding results of split and merge segmentation and clustering of the segmented image: **a** processed ROI image; **b** after split and merge segmentation; **c** cluster 3 of the segmented ROI image

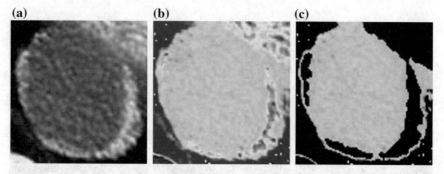

Fig. 7.12 The original CT ROI image in venous phase and corresponding results of split and merge segmentation and clustering of the segmented image: **a** original ROI image; **b** after split and merge segmentation; **c** cluster 3 of the segmented ROI image

In Figs. 7.12 and 7.13 are shown respectively original CT ROI image with size 97 × 98 pixels in venous phase with simple cyst in the left kidney and corresponding results obtained by split and merge segmentation and clustering of the segmented image.

The experiments showed differences by the clusters from original and processed CT ROI images. They are illustrated in Fig. 7.14.

After getting the output images for the three phases of the ROI containing the single cyst the texture segmentation based on Haralick's features along with the series of morphological operators are applied. These steps are done once over the non-filtered images and then identically over the preprocessed ones for all the phases. The aim is to have means for visual and quantitate comparison based on located surface for the cyst itself.

In Fig. 7.15 are given the input image from the arterial phase without being preprocessed and then the entropy map followed by the local standard deviation map and the final resulting image after range filtration which returns the most

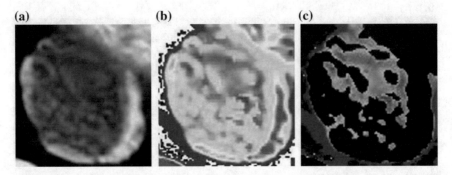

Fig. 7.13 The processed CT ROI image in venous phase and corresponding results of split and merge segmentation and clustering of the segmented image: **a** processed ROI image; **b** after split and merge segmentation; **c** cluster 3 of the segmented ROI image

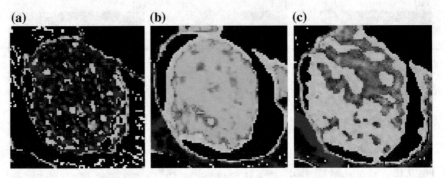

Fig. 7.14 Illustration of differences between the original and segmented clusters, respectively in: **a** native phase; **b** arterial phase; **c** venous phase

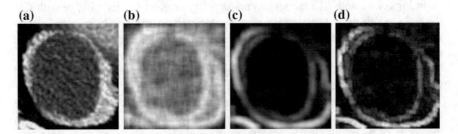

Fig. 7.15 Main stages of the segmentation algorithm for refining the boundaries of the simple cyst in the arterial phase without preprocessing: **a** input image; **b** entropy map; **c** local standard deviation map; **d** output image after final range filtration

(a)   **(b)**   **(c)**   **(d)**

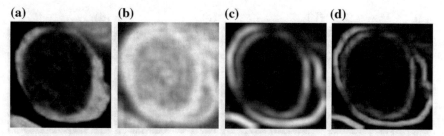

Fig. 7.16 Main stages of the segmentation algorithm for refining the boundaries of the simple cyst in the arterial phase after preprocessing: **a** input image; **b** entropy map; **c** local standard deviation map; **d** output image after final range filtration

(a)   **(b)**   **(c)**   **(d)**

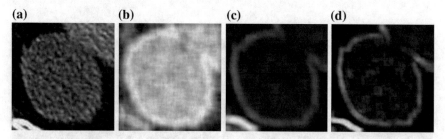

Fig. 7.17 Main stages of the segmentation algorithm for refining the boundaries of the simple cyst in the native phase without preprocessing: **a** input image; **b** entropy map; **c** local standard deviation map; **d** output image after final range filtration

accurate boundaries of the cyst inside the kidney area. The segmented area is not shown alone for better representation of the outer boundaries in relation to each other.

The same segmentation procedure is run over the enhanced image consisting of the same ROI with the cyst. The results are presented in Fig. 7.16. The order and the type of the different stages is the same as in Fig. 7.15.

In Fig. 7.17 the segmentation results are given for the native phase containing the single cyst from the ROI selected in the earlier selection. At first again the non-filtered image is passed through. The entropy map, the local standard deviation map and the output image after the final range filtration are shown. Similar to previous results from the arterial phase the more precisely located boundaries of the object of interest are underlined at the end.

The same boundaries but in a more smooth fashion and thus closer to their exact position are fixed by applying the texture segmentation along with the morphology described when using the filtered ROI in native phase. Visually the results are presented in Fig. 7.18.

The last test is done over images from the venous phase. Figure 7.19 reveals the same stages visual representation.

(a)   **(b)**   **(c)**   **(d)**

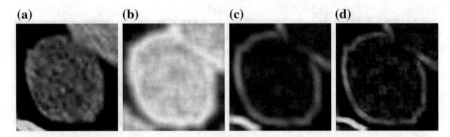

Fig. 7.18 Main stages of the segmentation algorithm for refining the boundaries of the simple cyst in the native phase after preprocessing: **a** input image; **b** entropy map; **c** local standard deviation map; **d** output image after final range filtration

(a)   **(b)**   **(c)**   **(d)**

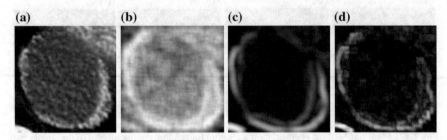

Fig. 7.19 Main stages of the segmentation algorithm for refining the boundaries of the simple cyst in the venous phase without preprocessing: **a** input image; **b** entropy map; **c** local standard deviation map; **d** output image after final range filtration

(a)   **(b)**   **(c)**   **(d)**

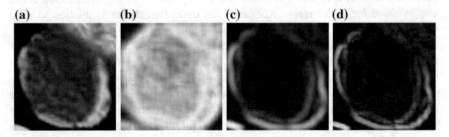

Fig. 7.20 Main stages of the segmentation algorithm for refining the boundaries of the simple cyst in the venous phase after preprocessing: **a** input image; **b** entropy map; **c** local standard deviation map; **d** output image after final range filtration

The experiments showed differences by the clusters from original and processed CT ROI images. They are illustrated in Fig. 7.20.

In Table 7.2 is given the average precision of finding the outer boundary of the single cyst for the three phases with and without the pre-processing stage.

Table 7.2 Single cyst contour detection accuracy in %

Phase	Original image	Pre-processed image
1. Arterial	93	98
2. Native	80	85
3. Venous	75	78

Fig. 7.21 Contour detection accuracy for single cysts from 3 images with and without pre-processing

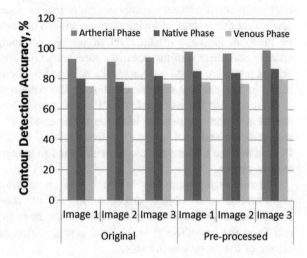

Figure 7.21 shows some extended results about the accuracy of the contour detection of single cysts from 3 CT images from the three phases. Results are similar to those presented in Table 7.2.

We have analysed some statistical parameters about the accuracy of the contour detection of single cysts in all the three phases in the axial plane.

The calculated standard deviation is respectively 0.03 for the original and 0.02 for the pre-processed images in the arterial phase. The mean value is respectively 92.67 % for the original and 98 % for the pre-processed images. In the native phase is the standard deviation respectively 0.04 for the original and 0.03 for the pre-processed images. The calculated mean values are 80 and 85.33 % respectively. The results are similar for the venous phase: standard deviation is 0.03 in both original and pre-processed images and the mean values are respectively 75.33 and 78.33 %. We can assume that the best results for accuracy of the contour detection of single cysts are obtained in the arterial phase of the pre-processed images.

7.5   Discussion

The implemented studying and obtained experimental results has shown that the processed images can better present contours and characteristic features of the kidney cysts. This can be very important for diagnosis and monitoring of this disease. The images, which are obtained in arterial phase of the axial plane, can visually better present the contours of the simple cyst and the differences between structures of the surrounding organs. Contrast enhancement, which is based on CLAHE can enhance not only the contrast of the image but reduce the noise in the homogenous areas, too. The proposed effective approach for noise reduction and contrast increasing based on combination of median filter and WPD can be adaptive applied for every stage of image preprocessing. The complementary adjustment can be made in the case of the level of wavelet shrinkage decomposition and the sparsity parameter α of the penalized threshold. Our investigations show that in the arterial phase the most accurate results emerge from better contrast in a global scale in the image and best separated texture in the same time.

Starting with the entropy feature the first map returned has most sharpness for the arterial phase which explains the better results in the later stages of texture segmentation and final boundary detection. Most of the original non-noisy details from inside the textures are cleaner subtracted in comparison to the native and the venous phases.

We can assume that the enhanced version of the image assures better final segmentation and object detection in all of the phases of axial plane. Non-informative details are removed from the homogenous areas inside the cyst and from the non-affected area of the organ, which additionally proves the efficiency of the preprocessing stage.

7.6 Conclusion

In the chapter is proposed a new and effective multistage approach for simple kidney cysts segmentation in CT images. This segmentation algorithm is very useful to extract objective information about the contours and structures of simple kidney cysts from CT scans. It can be implemented by parallel processing of the different phase of the selected plane. This approach can be very useful by easy obtaining of more precise diagnosis or in monitoring the disease progression. The implemented algorithm provides a basis for further investigations in several directions:

- Segmentation of example of critical case, when two simple kidney cysts are just one next to the other. This makes the separation of the two cysts and the delineation of their boundaries a challenging task.
- Another direction is the application of the method in case of autosomal dominant polycystic kidney disease (ADPKD).
- Analysis of tissue's structures of different type of kidney cysts is one more aspect of valuable application.
- It could be used as a method for segmentation and analysis of cysts in other organs.

Due to the reduced dimensionality of features used in the segmentation process it is faster in comparison with previously developed approaches. While preserving high accuracy of the cysts' detection different phases from the CT imaging could be used as a diversified mean for improving the results and as an input to more specialized analysing tools for more precise diagnosis.

Acknowledgement   The authors gratefully thank Professor Dr. V. Hadjidekov and Dr. Genov at the Department of Image Diagnostic on the Medical Academy in Sofia for the images and advices by the investigations.

References

1. Eknoyan, G.: A clinical view of simple and complex renal cysts. J. Am. Soc. Nephrol. **20**, 1874–1876 (2009)
2. Glassberg, K.I.: Renal dysgenesis and cystic disease of the kidney. In: Wein, A.J., Kavousi, L. R., Novick, A.C., Partin, W.A., Peters, C.A. (eds.) Campbell-Walsh Urology, 9th edn, pp. 3343–3348. W.B. Saunders, Philadelphia (2007)
3. Jose, J.S., Sivakami, R., Uma Maheswari, N., Venkatesh, R.: An efficient diagnosis of kidney images using association rules. Int. J. Comput. Technol. Electron. Eng. (IJCTEE), **2**, 14–20 (2013). ISSN: 2249-6343
4. Chapman, A., Guay-Woodford, L., Granrham, J., et al.: Renal structure in early autosomal-dominant polycystic kidney disease (ADPKD). Consort. Radiol. Imaging Stud. Polycystic Kidney Dis. (CRISP) Cohort. Kidney Int. **64**, 1035–1045 (2003)
5. Battiato, S., Farinella, G.M., Gallo, G., Garretto, O., Privitera, C.: Objective Analysis of Simple Kidney Cysts from CT Images, IEEE International Workshop on Medical Measurements and Applications, pp. 146–149, Certalo, Italy (2009). ISBN: 978-1-4244-3598-2
6. Mekhaldi, N., Benyttou, M.: Detection of renal cysts by method of SVM classification. Ann. Comput. Sci. Series. 11th Tome 2nd Fasc. 67–71 (2013)
7. Pham, D., Xu, C., Prince, J.: Current methods in medical image segmentation. Annu. Rev. Biomed. Eng. **2**(1), 315–337 (2000)
8. Freixenet, J., Muñoz, X., Raba, D., Martí, J., Cufí, X.: Yet another survey on image segmentation: region and boundary information integration. In: Computer Vision—ECCV, pp. 408–422. Springer, Berlin (2002)
9. van Ginneken, B., Katsuragawa, S., ter Haar Romeny, B., Viergever, M.: Automatic detection of abnormalities in chest radiographs using local texture analysis. IEEE Trans. Med. Imaging **21**(2), 139–149 (2002)
10. Felipe, J., Traina, A., Traina Jr., C.: Retrieval by content of medical images using texture for tissue identification. In: Proceedings of the 16th IEEE Symposium on Computer-Based Medical Systems, pp. 175–180 (2003)
11. Castellano, G., Bonilha, L., Li, L., Cendes, F.: Texture analysis of medical images. Clin. Radiol. **59**(12), 1061–1069 (2004)
12. Glatard, T., Montagnat, J., Magnin, I.: Texture based medical image indexing and retrieval: application to cardiac imaging. In: Proceedings of the 6th ACM SIGMM International Workshop on Multimedia Information Retrieval, pp. 135–142 (2004)
13. Pham, M., Susomboon, R., Disney, T., Raicu, D., Furst, J.: A comparison of texture models for automatic liver segmentation. In: Medical Imaging, International Society for Optics and Photonics, pp. 65124E–65124E (2007)

14. Haralick, R., Shanmugam, K., Dinstein, I.: Textural features for image classification. IEEE Trans. Syst. Man Cybern. **6**, 610–621 (1973)
15. Haralick, R.: Statistical and structural approaches to texture. Proc. IEEE **67**(5), 786–804 (1979)
16. Pal, N., Pal, S.: A review on image segmentation techniques. Pattern Recogn. **26**(9), 1277–1294 (1993)
17. Tuceryan, M., Jain, A.: Texture analysis. In: Chen, C., Pau, L., Wang, P. (eds.) The Handbook of Pattern Recognition and Computer Vision, 2nd edn. pp. 207–248. World Scientific Publishing Co (1998)
18. Malpica, N., Ortuño, J., Santos, A.: A multichannel watershed-based algorithm for supervised texture segmentation. Pattern Recogn. Lett. **24**(9), 1545–1554 (2003)
19. Tesar, L., Smutek, D., Shimizu, A., Kobatake, H.: 3D extension of haralick texture features for medical image analysis. In: SPPR 2007 Proceedings of the Fourth Conference on IASTED International Conference, pp. 350–355 (2007)
20. Tesař, L., Shimizu, A., Smutek, D., Kobatake, H., Nawano, S.: Medical image analysis of 3D CT images based on extension of haralick texture features. Comput. Med. Imaging Graph. **32**(6), 513–520 (2008)
21. Gueld, M., Keysers, D., Deselaers, T., Leisten, M., Schubert, H., Ney, H., Lehmann, T.: Comparison of global features for categorization of medical images. In: Medical Imaging, International Society for Optics and Photonics, pp. 211–222 (2004)
22. Chuctaya, H., Portugal, C., Beltran, C., Gutierrez, J., Lopez, C., Tupac, Y.: M-CBIR: a medical content-based image retrieval system using metric Data-Structures. In: SCCC, pp. 135–141 (2011)
23. El-Dahshana, E.-S.A., Hosnyb, T., Salem, A.-B.M.: Hybrid intelligent techniques for MRI brain images classification. Elsevier, Digital Signal Proc. **20**, 433–441 (2010)
24. Chaddad, A., Tanougast, C., Dandache, A., Bouridane, A.: Extracted Haralick's texture features and morphological parameters from segmented multispectrale texture Bio-Images for classification of colon cancer cells. WSEAS Trans. Biol. Biomed. **8**, 39–50 (2011)
25. Chen, E., Chung, P., Chen, C., Tsai, H., Chang, C.: An automatic diagnostic system for CT Liver image classification. IEEE Trans. Biomed. Eng. **45**(6), 783–794 (1998)
26. Howarth, P., Ruger, S.: Robust texture features for still-image retrieval. IEE Proc. Vis. Image Signal Process. **152**(6), 868–874 (2005)
27. Boas, F., Fleischmann, D.: CT artifacts: causes and reduction techniques. Imaging Med. **4**(2), 229–240 (2012)
28. Chhabra, Tarandeep, Dua, Geetika, Malhotra, Tripti: Comparative analysis of methods to denoise CT Scan Images. Int. Adv. Res. Electrical, Electronics and Instrumentation Eng. **2**(7), 3363–3369 (2013)
29. Georgieva, V., Kountchev, R., Draganov, I.: An adaptive approach for noise reduction in sequences of CT images. Adv. Int. Comput. Technol. Decis. Support Syst, Springer **486**, 43–52 (2014)
30. Zeyong, S., Aviyente, S.: Image denoising based on the wavelet co-occurrence matrix. IEEE Trans. Image Process. **9**(9), 1522–1531(2000)
31. Coifmann, R., Wickerhauser, M.: Entropy based algorithms for best basis selection. IEEE Trans. Inf. Theory **38**, 713–718 (1992)
32. MATLAB User's Guide. www.mathworks.com
33. Xu, Z., Lin, X., Chen, X.: For removal from video sequences using contrast limited histogram equalization. In: Proceedings of the International Conference on Computational Intelligence and Software Engineering, pp.1–4. IEEE Xplore Press, Wuhan (2009)
34. Faruquzzaman, A.B.M., Paiker, N.R., Arafat, J., Karim, Z., Ameer Ali, M.: Object segmentation based on split and merge algorithm. In: IEEE Region 10 Conference TENCON 2008, pp.1–4 (2008)
35. Rakesh, M., Ravi, T.: Image segmentation and detection of tumor objects in MR brain images using FUZZY C-MEANS (FCM) algorithm. Int. J. Eng. Res. Appl. (IJERA), **2**(3), 2088–2094 (2012). ISSN: 2248-9622

36. El-Dahshan, E.-S.A., Mohsen, H.M., Revett, K., Salem, A.-B.M.: Computer-aided diagnosis of human brain tumor through MRI: a survey and a new algorithm. Elsevier, Expert Syst. Appl. **41,** pp. 5526–5545 (2014)
37. Leung, C.-K., Lam, F.-K.: Image segmentation using maximum entropy method. In: International Symposium on. IEEE Speech, Image Processing and Neural Networks, Proceedings, ISSIPNN'94, pp. 29–32 (1994)

Chapter 8
Audio Visual Attention Models in the Mobile Robots Navigation

Snejana Pleshkova and Alexander Bekiarski

Abstract The mobile robots are equipped with sensitive audio visual sensors, usually microphone arrays and video cameras. They are the main sources of audio visual information to perform suitable mobile robots navigation tasks, modeling the human audio visual perception. The results from the audio and visual perception algorithms are widely used, separate or in conjunction (audio visual perception) in the mobile robots navigation, for example to control mobile robots motion in applications like people and objects tracking, surveillance systems, etc. The effectiveness and precision of the audio visual perception methods in the mobile robots navigation can be enhanced combining audio visual perception with audio visual attention. Sufficient relative knowledge exists, describing the phenomena of human audio and visual attention. Such approaches are usually based on a lot of physiological, psychological, medical and technical experimental investigations relating the human audio and visual attention, with the human audio and visual perception with the leading role of the brain activity. Of course, the results from these investigations are very important, but not sufficient for the mobile robots audio visual attention modeling, mainly because of brain missing in mobile robots audio visual perception systems. Therefore, in this chapter is proposed to use the existing definitions and models for human audio and visual attention, adapting them to the models of mobile robots audio and visual attention and combining with the results from the mobile robots audio and visual perception in the mobile robots navigation tasks.

Keywords Audio visual attention · Audio visual perception · Mobile robot navigation

S. Pleshkova (✉) · A. Bekiarski
Department of Telecommunications, Technical University of Sofia,
8 Kl. Ohridski Blvd., 1000 Sofia, Bulgaria
e-mail: snegpl@tu-sofia.bg

A. Bekiarski
e-mail: aabbv@tu-sofia.bg

© Springer International Publishing Switzerland 2016
R. Kountchev and K. Nakamatsu (eds.), *New Approaches in Intelligent Image Analysis*, Intelligent Systems Reference Library 108,
DOI 10.1007/978-3-319-32192-9_8

253

8.1 Introduction

Mobile robots are usually equipped with an appropriate audio visual system [1–4]. In most cases these systems are working as an appropriate and simple model of the human audio visual perception system [5–8]. The human audio visual perception is dedicated to model the mobile robot navigation in real indoor or outdoor environments in situation to avoid obstacles [9–12], objects [13–16], for people tracking [17–19], etc., in a variety of actual mobile robots applications like surveillance [20–23], military, police and rescue operations [24–26], home service, guiding robots [27–29], medical robots [30, 31], etc.

The effectiveness and precision of using the human audio visual perception in the mobile robot navigation and motion control can be extended and enhanced by combining the mobile robot audio visual perception with the associated to the human perception human audio visual attention [32–35]. To apply the human audio visual attention in the mobile robot navigation and motion control is necessary to describe the human audio visual attention in qualitative and quantitative terms suitable to be applied in the mobile robot audio visual systems and the corresponding algorithms, focusing the mobile robot audio visual attention mainly on tracking target objects or speaking persons. A lot of research works exist, and also the proposed methods [36, 37] which describe the phenomena of the human audio and visual attention, but they are usually based on a lot of physiological, psychological, medical and technical experimental investigations related to the human audio and visual attention with human audio and visual perception with the leading role of the brain activity. The results of these investigations are very important for the audio visual attention understanding, but they are not sufficient and entirely applicable for mobile robots audio visual attention modeling, because of brain missing in mobile robots audio visual perception systems. Therefore, to establish these circumstances, here is presented a brief description of the basic definitions of the human audio visual attention with the corresponding definitions of the human audio visual perception, and the importance of the human brain activity to control the human audio visual attention. The missing of brain or of something like mental ability in the mobile robot audio visual systems is the main stimulus for the researchers [38, 39] to propose a theoretical (mathematical) representation of the human audio visual attention model, where to avoid and replace the leading role of the human brain with a probabilistic representation, or to model the brain functions so that to focus and control attention in the general cases of computer vision.

8.2 Related Work

A lot of models of perception and attention as general probabilistic model are already proposed [40–44]. Most of them are specified separately to visual or audio perception and attention, but in [40] and [43] the described probabilistic models of

Fig. 8.1 The probabilistic model [44] obtained using visual geometry, assuming the defined initial probability density function in the observation area

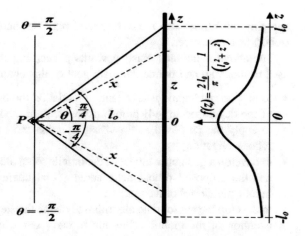

the audio and visual perception and attention are regarded as joined parts of the conceptual cognition foundation. This allows to use some of the probabilistic models related to audio perception and attention only, applying them after appropriate modifications to video perception and attention, or as joined audio and visual perception and attention probabilistic models. Therefore, the basic probabilistic vision model [44] can be considered as a start point for the development of the visual attention model for the mobile robot navigation, and to extend this model for the audio attention, in order to achieve a joined audio visual probabilistic model suitable for mobile robot navigation, as presented in this chapter. The existing probabilistic model [44] is obtained by using the visual geometry (Fig. 8.1), assuming also the definition of a chosen initial probability density function in the observation area:

$$f(z) = \frac{2l_0}{\pi} \frac{1}{(l_0^2 + z_0^2)} \tag{8.1}$$

It is assumed in the probabilistic attention model proposed in [44], that there is equal visual attention to all locations on the vertical plane z in the first instance of attention. This eliminates the importance of human brains in the initial step of attention. Therefore, the initial probability of getting attention at each point on the plane z is the same and respectively the corresponding probability density function of attention—uniformly distributed. The uniformly defined probability density function of the initial probabilistic attention model can be combined with the Bayesian attention models [42] and [43], suitable for description in each current step of the navigation (object tracking) for sharpening the mobile robot attention in a chosen direction, i.e. the current probability density function of attention is not uniformly distributed and should be calculated in each current step by using attention for the purposes like mobile robot navigation. The developed probabilistic attention model proposed in [44] is studied mainly considering the human as an observer and this model is then shown briefly in terms of the possibilities for applying this model also for architectural purposes and for using the virtual agent (robot) as an observer.

The above mentioned existing works concerning the area of attention can be used to motivate and define the main goal of this chapter in the following way:

- to propose an audio visual attention model on the base and on the modification of the developed in [44] probabilistic attention model;
- to apply the proposed audio visual attention model in the area of the mobile robot navigation;
- to develop a geometric model of the mobile robot area of observation, suitable to test the proposed probabilistic audio visual attention model in situations of indoor mobile robot navigation;
- to develop the appropriate algorithm to define the steps of the initial audio visual attention of the mobile robot and to carry out with this algorithm the corresponding experiments to demonstrate the main advantages achieved in the mobile robot indoor navigation, applying the proposed audio visual attention model in comparison with same mobile robot navigation tasks, but without information for the audio visual attention, i.e. by using the information from audio visual perception only.

8.3 The Basic Definitions of the Human Audio Visual Attention

The basic definitions of the human audio visual attention can be derived from our most general idea and conceptions for human audio and visual perception, since vision and audio are the major source of information in our comprehension of the environment [40]. This can be said also for the computer audio and video perception and especially for the mobile robots audio and visual perception in particular, but without the existence of brain and the related important functions in the human perception. Therefore, it is possible to present the following basic definitions considering the human audio visual attention in general aspects and addressing them to the computer audio vision attention applicable to mobile robot motion control and navigation:

- **perception** refers to the way in which humans interpret the information gathered and processed by human senses;
- **audio visual perception** is the process of acquiring knowledge about environment and events extracting information from the sounds and/or the light they emit or reflect;
- **early human audio and visual perception** is dedicated to build the first initial and comprehensive audio and visual information of the observed environment;

- **human mind of audio and visual perception** defines the dependence of the audio visual perception from our mind, i.e. it involves not only our ears and eyes but also the brains and their functions in formation of our knowledge about the audio visual information about the objects and thinks in the area of observation;
- **human audio and visual perception** means that perceiving something (sounds, objects, people, etc.) it is possible to recall their relevant properties;
- **human audio and visual attention** is associated and closely connected with the human audio and visual perception and represents the human ability to focus the perception on the important things, sound of the speaking person, music, noise, objects or people, etc. (selected by human mind).

It can be noted, that all of the above mentioned basic definitions, which consider the human audio visual perception and attention, are verbal and are proposed and accepted from philosophic and psychological point of view. Therefore, they are not directly applicable in areas like computer vision, robotic, virtual reality, etc., where it is obligatory to use audio visual perception and attention, but in corresponding mathematical or algorithmic sense and description.

8.4 General Probabilistic Model of the Mobile Robot Audio Visual Attention

In the existing general case studies [41–44], for the presentation of the human audio visual perception and attention as mathematical models, are used probabilistic processes to obtain and interpret the audio visual information from the environment. These models are also used in the mathematical definitions of the audio visual attention, openness or perceptivity, the visual openness measurement for design, scene description by perception, virtual reality and virtual agents and other characteristics of perceived audio visual information. There are also ideas [45–48] to apply these models in robotics, virtual reality, etc.

From these models, two cases of audio visual attention are defined in this chapter:

- prior, preliminary or initial audio visual attention;
- posterior or current audio visual attention.

The definition of the preliminary or initial audio visual mobile robot attention model refers to the first human audio visual impression of unknown environment and can be based on probabilistic definition with uniform distribution of the probability density function. The current audio visual mobile robot attention model, when mobile robot localized sound sources, speakers, observe and tracking the objects or people, can be also considered in the same way, using the probabilistic definition. But in the concrete situation the mobile robot audio visual attention depends on the goal of the mobile robot moving, the sound sources localization, the observation and the tracking task, which is similar to the human audio visual attention. The only difference is that humans use their brains to define the audio visual attention. Nevertheless, it is possible to assume that the initial and current audio visual mobile robot attention model can be modeled by using probabilistic uniform (for the initial model) and non uniform (for the current model) functions. This means that the general (the initial and the current) description of the audio visual mobile robot attention model in unknown environment can be defined by using the definition of the following probabilistic model:

$$P(E_A, E_V, A_i, V_i) = P(A_i, V_i | E_A, E_V) P(E_A, E_V), \qquad (8.2)$$

where

E_A and E_V are respectively the initial or current audio visual estimations (the results from the audio and video processing for feature extraction, object detection, scene analysis, etc.) of the audio visual information (sound signals, images);

A_i and V_i initial or current audio visual information as sound signals and images captured respectively by the microphone array and the video camera, mounted on the mobile robot.

The arguments in Eq. 8.2 used for the audio visual estimations E_A and E_V, can be defined and explained more precisely, especially in cases of using information from the audio visual attention in the current steps of the mobile robot navigation tasks. The arguments (estimations E_A and E_V) of the probabilities in Eq. 8.2 are the descriptions of the real audio visual objects SP_A, OB_V, for example the coordinates of the speakers SP_A, the visual objects OB_V, etc., defined after the execution of the algorithms for sound source localization, the visual objects detection, etc., which are the objects of previous investigations of the authors of this chapter, given in [46] and [49]. An example of applying these arguments from Eq. 8.2 in the current step of the mobile robot navigation is given as a probabilistic posterior audio visual attention model:

$$P(SP_A, D_A, OB_V, LC_V, A_i, V_i) =$$
$$= P(SP_A, D_A) P(OB_V, LC_V) P(A_i, V_i | SP_A, D_A, OB_V, LC_V), \qquad (8.3)$$

where

SP_A, OB_V are the speakers and the visual objects, respectively determined as results of the execution of the appropriate and the well known existing speaker localization and the visual objects detection algorithms, using the perceived and the processed audio and visual information captured by the mobile robot audio and the visual perception system (microphone array and video camera, respectively);

D_A and LC_V the directions of speakers, determined by using the sound localization algorithms and the visual objects locations, calculated by the
objects detection algorithms, respectively.

Equations 8.2 and 8.3 can be considered as definitive representation of the
general probabilistic audio visual attention model suitable for the following tasks
described in this chapter:

- applying Eqs. 8.2 and 8.3 with calculated arguments in concrete situations of
  the mobile robot navigation by using the probabilistic audio visual attention
  model;
- to arrange the various situations of the mobile robot navigation for applying and
  testing the proposed probabilistic audio visual attention model is created an
  indoor geometric model of the mobile robot area of observation;
- for the initial probabilistic audio visual attention model, it is sufficient to calculate and substitute in Eq. 8.2 the arguments E_A and E_V (feature extraction,
  object detection, scene analysis, etc.) and the arguments A_i, V_i as captured sound
  signals and images from the mobile robot microphone array and the video
  camera, respectively;
- for each current probabilistic audio visual attention model it is necessary to
  calculate the arguments in Eq. 8.3 SP_A, OB_V as the speakers and visual objects,
  and D_A, LC_V as directions of speakers, and of the visual objects locations;
- all arguments mentioned in Eqs. 8.2 and 8.3 can be the subject of calculation as
  results from the numerous executions of the existing tested and well working
  audio visual perception algorithms [46, 49] as developed methods and algorithms for audio-visual mobile robot perception, which are not the object here
  and therefore, it is not necessary to mention them in this chapter.

The proposed general probabilistic model of the mobile robots audio visual
attention and the related perception can be explained by means of the perceptual
geometry, presenting acoustic and visual environment in the field of the mobile
robot action. The mobile robot environment or the areas of action are usually
defined as indoor and outdoor. Here, on Fig. 8.2 is considered the case of the indoor
mobile robot observation environment (for example a room). The audio visual
sensors (for example the microphone array and the video camera), placed on the
mobile robot platform, have a chosen initial position in the room and observe the
environment in front of the robot in a maximal angle of observation chosen to be
$\theta = \pm\pi/4$ (Border of space observation on Fig. 8.2).

To achieve and explain the probabilistic characteristics of the mobile robot audio
visual perception and attention is used an imaginary perception line (on Fig. 8.2 as
the more frequently used 2D geometric model of the room) or imaginary plane (in
the case of the 3D geometric model of the room).

Therefore, in the area of observation ($\theta = \pm\pi/4$), i.e. of the audio visual perception and attention, it is possible to define the following distances from the audio
visual sensors initial position to the imaginary perception line (plane):

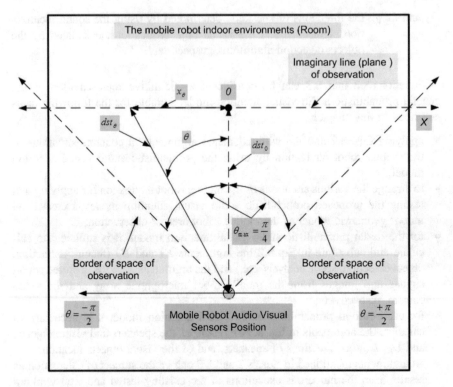

Fig. 8.2 The explanation of the proposed probabilistic model of the audio visual attention and related perception by means of the perceptual geometry, presenting the acoustic and the visual environment in the field of the mobile robot action for the case of indoor environments

- the distance dst_0 from the audio visual sensors initial position to the imaginary perception line (plane) in the initial direction of observation, defined by the angle $\theta = 0$;
- the distance dst_θ from the audio visual sensors initial position to the imaginary perception line (plane) in the arbitrary direction of observation, defined by the angle $\theta \neq 0$ in the range of $\theta = \pm \pi/4$.

Using the defined preliminary or initial probabilistic mobile robot audio visual attention and perception model (Eq. 8.2) in unknown environment (on Fig. 8.2—the room), it is possible to suppose, that all the arbitrary directions of observation (for $\theta \neq 0$) are assumed to have equal probability of observation. This means that the probability density function f_θ, which belongs to the angle θ, is uniformly distributed:

$$f_\theta = \frac{1}{\pi/2} \tag{8.4}$$

From the probabilistic characteristic of the directions of observation (angle θ) defined by Eq. 8.4, it is reasonable to assume the probabilistic basis of the distance dst_θ from the audio visual sensors initial position to the imaginary perception line (plane) in an arbitrary direction of observation, and the distance x_θ from the origin "0" on the imaginary perception plane on Fig. 8.2. Thus, by using these assumptions it is possible to calculate the corresponding probability density function of the distance dst_θ or of the distance x_θ. For example the following equation defines the probability density function of the distance x_θ:

$$f(x_\theta) = \frac{2dst_0}{\pi(dst_0^2 + x_\theta^2)} \tag{8.5}$$

The precision derivation expressed by the Eq. 8.5 can be found in the existing probabilistic theory [41] and [42] of the human perception. Here is presented the way of making the derivation of the Eq. 8.5 by means of the perceptual geometry for applications in the mobile robot audio visual attention only. It is assumed that the direction θ is a random variable and x_θ is a function of the direction θ:

$$x_\theta = \frac{dst_0}{\cos(\theta)} = f(\theta) \tag{8.6}$$

Therefore, the derivation of the probability density function of the distance x_θ can be calculated through applying the theory of the functions of the random variables as in Eq. 8.4. The random function $f(\theta)$ can be described by its real roots $\theta_1, \theta_2, \theta_3, \ldots$:

$$x_\theta = f(\theta_1) = f(\theta_2) = f(\theta_3) = \cdots \tag{8.7}$$

Following this way, two roots only could be detected in the range $\theta = \pm\pi/4$ of the random function, used to derivate the Eq. 8.5 of the random function $f(\theta)$. The random function $f(x_\theta)$ derived by Eq. 8.5 describes the probability density function of the initial audio visual attention of the mobile robot as a function of the observation directions (angle θ) and of the distance dst_θ from the audio visual sensors initial position to the imaginary perception line (plane) on Fig. 8.2. The probabilistic presentation through the random function $f(x_\theta)$ of the mobile robot audio visual attention is deeply related to the possibility to define the probabilistic characteristics of the mobile robot audio visual perception also. To justify this, the probabilistic characteristics of the audio visual attention are defined as a density in a very small range, or angle sector, $\Delta\theta$ region (Fig. 8.3).

If the probability density function $f(x_\theta)$ is integrated in the finite interval (x_{max} and x_{min} in Fig. 8.3) of the variable x, the result can be considered as the probabilistic definition of the initial mobile robot audio visual perception P_{AV}:

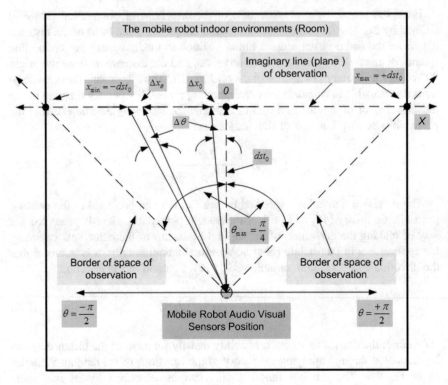

Fig. 8.3 The definition of mobile robot audio visual attention probabilistic characteristics as a density in very small interval of angle sector $\Delta\theta$

$$P_{AV}(x,\theta) = \int_{x_{min}}^{x_{max}} f(x_\theta)dx, \qquad (8.8)$$

where

$x_{max} = +dst_0$ and $x_{min} = -dts_0$ if the mobile robot audio visual perception and attention are considered in the area of observation ($\theta = \pm\pi/4$).

Therefore, it is possible to summarize the following from Eqs. (8.4–8.8):

- from the existing probabilistic description of the human audio and visual perception and attention are derived the mobile robot audio visual perception and attention probabilistic descriptions;
- the initial audio visual attention and perception of the mobile robot can be described as probabilistic functions with corresponding probabilistic characteristics;
- the initial audio visual attention and perception of the mobile robot are related together with their probabilistic characteristics;

- the initial audio visual attention of the mobile robot is defined by the probability density function from Eq. 8.5 as the random function $f(x_\theta)$ of the directions of the mobile robot observation (angle θ) and also of the distance dst_θ from the audio visual sensors initial position to the imaginary perception line (plane) on Fig. 8.2;
- the initial audio visual perception of the mobile robot can also be defined as the probabilistic function $P_{AV}(x, \theta)$ by Eq. 8.8 derived as the integrated function of the initial audio visual attention of the mobile robot in the finite interval (x_{max} and x_{min} in Fig. 8.3);
- by the proposed probabilistic mobile robot audio visual attention and perception description is allowed to avoid (absorb) the important participation and the role of the brain existing in the human audio visual attention and perception.

In the next part of this chapter is proposed to use the mobile robot probabilistic audio visual attention and perception model, presented above, in the development and testing the examples of the models of the mobile robots audio and visual attention combining and comparing them with the results achieved from mobile robots audio and visual perception only in the mobile robots navigation tasks.

8.5 Audio Visual Attention Model Applied in the Audio Visual Mobile Robot System

In the general case it is very difficult to model the phenomena of the mobile robots audio visual attention. This is because in the existing audio visual mobile robot systems, there is not a knowledge system, similar to the human brain knowledge system. Therefore, it is possible to try to model and to simulate some special cases of mobile robots audio visual attention situations only. In this chapter is proposed to build the concrete model of the audio visual attention of the mobile robots for the case of indoor mobile robot motion control and navigation. The proposed probabilistic audio visual attention model is applied in the audio visual mobile robot system placed in the room environment, shown on Fig. 8.4 as a simple horizontal representation space model of the room environment, describing the area of the mobile robot movements.

8.5.1 Room Environment Model for Description of Indoor Initial Audio Visual Attention

The audio visual attention model, shown on Fig. 8.4, is based on the audio visual mobile robot system for the case of indoor mobile robot motion control and navigation, and can be described by the following characteristics and the corresponding

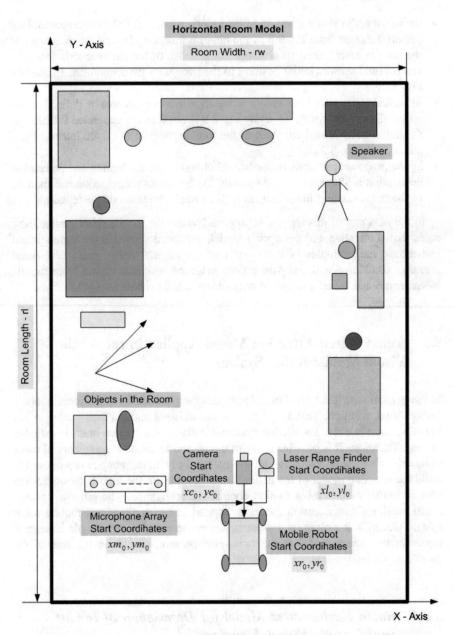

Fig. 8.4 The proposed audio visual attention model based on the audio visual mobile robot system for the case of indoor mobile robot motion control and navigation as a simple horizontal representation space model of a room environment describing the area of the mobile robot movements

coordinates needed for the initial precise description of the audio visual attention model:

- the room dimensions in the presented on Fig. 8.4 horizontal space model: length —rl and width—rw;
- the room coordinates X and Y along the room length and width, respectively, with the origin chosen in the bottom left corner on Fig. 8.4;
- the mobile robot platform start coordinates xr_0 and yr_0;
- the start room coordinates xc_0 and yc_0 of the video camera mounted on the mobile robot platform;
- the start room coordinates xm_0 and ym_0 of the microphone array mounted on the mobile robot platform;
- the start room coordinates xl_0 and yl_0 laser range finder, mounted on the mobile robot platform;
- all presented in the general case objects in the room like tables, chairs or other obstacles for the mobile robot motion in the room environment can be described by the corresponding room coordinates $xobj_i$ and $yobj_i$ for $i = 1, 2, \ldots n$ in case that in the room there are "n" number of objects, and for a simple case it is possible to accept, that all objects in the room, shown as a model in Fig. 8.4, are stationary, i.e. their room coordinates are fixed ($xobj_i = const$ and $yobj_i = const$), when using the horizontal space model, shown on Fig. 8.4, in the development of the audio visual attention algorithms;
- the speaker room start coordinates xsp and ysp, necessary for the audio visual attention algorithms to define the start position of the speaking person to focus the mobile robot audio attention if the speaking person sends the voice commands to the mobile robot motion control in the room.

All coordinates, mentioned above, can be changed, if necessary, in any current instance of the audio visual attention algorithms execution.

In the same way it is possible to present also the simple vertical space model of a room, shown on Fig. 8.5, as the area of the mobile robot movements.

The simple vertical part of the space model, shown on Fig. 8.5, defines the coordinates x and z only for the mobile robot (xr_0, zr_0), the microphone array (xm_0, zm_0) and the laser range finder (xl_0, zl_0) in the vertical $Z–X$ plane, if it is supposed that they are located in the mobile robot platform in such way, that their y—coordinate is equal, i.e. $yr_0 = ym_0 = yl_0 = const$.

The definition of all other the objects presented on Fig. 8.4, and the speaker, can be defined and shown in a similar way, but this is more convenient to make this definition direct in each developed algorithm, because of the concrete and arbitrary space position in the room and in the general case, shown on Fig. 8.4.

Therefore, using the information from Figs. 8.4 and 8.5, it is possible to summarize in the following way the general definitions of the space positions, applying the room space coordinate system X, Y, and Z of all existing and presented on Fig. 8.4 objects in the room, speaker, mobile robot, laser range finder and microphone array, for the chosen situation, suitable of modeling mobile robot attention:

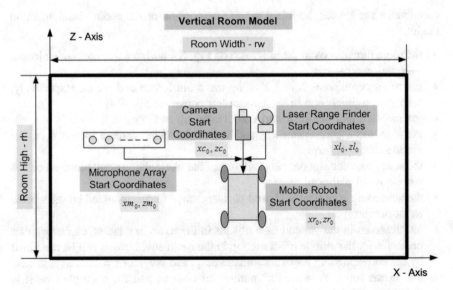

Fig. 8.5 The proposed audio visual attention model based on the audio visual mobile robot system for the case of indoor mobile robot motion control and navigation as a simple vertical representation space model of a room environment describing the area of the mobile robot movements

- room dimensions length, width and high (rl, rw, rh);
- mobile robot platform start coordinates (xr_0, yr_0, zr_0);
- video camera start room coordinates (xc_0, yc_0, zc_0);
- microphone array start room coordinates (xm_0, ym_0, zm_0);
- laser range finder start room coordinates (xl_0, yl_0, zl_0);
- objects in the room coordinates ($xobj_i$, $yobj_i$, $zobj_i$) for $i = 1, 2, \ldots n$;
- speaker room start coordinates (xsp, ysp, zsp).

8.5.2 Development of the Algorithm for Definition of the Mobile Robot Initial Audio Visual Attention Model

The general definitions presented above are applied in the proposed algorithm, shown on Fig. 8.6, to define the mobile robot initial audio visual attention.

The steps of the proposed algorithm, presented on Fig. 8.6, are arranged as a necessary sequence of initial and calculation operations in order to achieve a suitable audio visual mobile robot attention in the beginning of its motion in the room from the starting position.

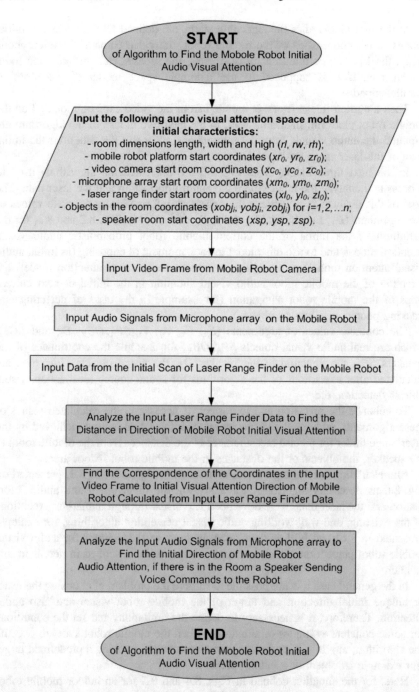

Fig. 8.6 The steps in the proposed algorithm to define the initial audio visual attention of the mobile robot

In the first block, after the algorithm starts, are defined all the necessary initial values as the coordinates of the mobile robot, the video camera, the microphone array, the laser range finder, a lot of objects, the speaker etc., and also the room dimensions, if it is supposed to use a simple space model of the room as parallelepiped.

Then a current image frame is inputted from the video camera, mounted on the mobile robot platform. In similar way, in the next two blocks of the algorithm are inputted the audio signals from the microphone array and the data after the initial scan of the laser range finder.

In the next three blocks from Fig. 8.6 are presented the algorithms used to process and analyze the input video, the audio information and the laser data. The goal of this analysis is to achieve the necessary information as concrete values of the arguments (E_A, E_V, A_i, V_i, SP_A, OB_V, D_A and LC_V) in Eqs. 8.2 and 8.3, for the definition of the initial or the current mobile robot probabilistic audio visual attention model, and to use this model in each moment of choosing (as initial audio visual attention model) or changing (as current audio visual attention model) the direction of the mobile robot audio visual attention in the initial or next current steps of the mobile robot navigation (for example in the tasks of detecting and tracking people, speakers or objects).

The concrete values of arguments (E_A, E_V, A_i, V_i, SP_A, OB_V, D_A and LC_V), which are real audio visual objects SP_A, OB_V, for example the coordinates of the speakers SP_A, the direction to the speaker D_A, the visual objects OB_V, etc., are calculated after executions of the algorithms for sound source localization, visual objects detection, etc.

To enhance the precision and to correct (if necessary) the calculated values of these arguments, additional and precise space information is used, achieved by the laser range finder for precise measurement of the distances from the mobile robot to the speaker, the objects or the obstacles in the mobile robot indoor area.

All calculations of the arguments, mentioned in Eqs. 8.2 and 8.3, presented on Fig. 8.6 as necessary algorithms for analyzing the input video and audio information and the laser data, were developed and tested through numerous executions of the existing, and well working audio visual perception algorithms, for example presented in [46] and [49] as developed methods and algorithms for audio visual mobile robot perception, which are not necessary to be described in details in this chapter.

In the general case it is not possible to resolve the problem of choosing the exact or unique initial direction and target of the mobile robot visual and also audio attention. Therefore, it is necessary to define the limitations and set the conditions for some concrete examples or situations, when the mobile robot starts to execute the algorithm, moving in the indoor environments and tracking a predefined target (for example an object or speaking person).

Here, for the situation defined in Figs. 8.4 and 8.5 for an indoor mobile robot movement, is proposed to apply the following limitations and conditions for the initial positions of the video camera, the microphone array and the laser range finder, mounted on mobile robot platform:

- the room coordinates Y and Z of the video camera, the microphone array and the laser range finder, mounted on the mobile robot platform, are chosen and defined to be the same and equal to the room coordinates Y and Z of the mobile robot initial position, i.e.:

$$yc_0 = ym_0 = yl_0 = yr_0; \qquad (8.9)$$

$$zc_0 = zm_0 = zl_0 = zr_0; \qquad (8.10)$$

- the room coordinate X of the video camera is also defined to be the same or equal to the coordinate X of the mobile robot, but the coordinate X of the microphone array and of the laser range finder, mounted on the mobile robot platform, are chosen to be different from each other and from the room coordinate X of the mobile robot initial position, i.e.:

$$xc_0 = xr_0 \neq xm_0 \neq xl_0. \qquad (8.11)$$

The conditions defined by the Eqs. 8.9, 8.10 and 8.11 allow to analyze in the next steps of the algorithm, shown on Fig. 8.6, the incoming from laser range finder data, to find their correspondence to the coordinates of the input video image and to analyze the microphone array audio signals to define the initial direction of the voice commands (if they exist) sent from a speaker to the mobile robot in their initial positions. For the analysis of the incoming data from laser range finder, in order to find their correspondence with the coordinates of the input video image is necessary to extract the information about the distance, which exists in the incoming laser range finder data in the format, shown on Table 8.1 as an example of a part of the data collecting after scan from the laser range finder.

In Table 8.1 are applied the following names defined for the used laser range finder model URG-04LX-UG01 [50]:

Table 8.1 Laser range finder data format

Index	Length	Angle	Coordinate X	Coordinate Y
379	197	−0.030	196.907	−6.042
380	197	−0.024	196.941	−4.834
381	197	−0.018	196.967	−3.626
382	197	−0.012	196.985	−2.417
83	197	−0.006	196.996	−1.208
384	197	0	197	0
385	195	0.006	194.996	1.196
386	195	0.012	194.985	2.392
387	195	0.018	194.967	3.589
388	193	0.024	192.942	4.736
389	193	0.030	192.909	5.920

- **Index** is the number of the current laser scan in direction depending on the current angle value in radians;
- **Length** is the calculated distance in centimeters from the laser range finder current position to an object (obstacle) in each moment of the current laser scan in direction depending on the current value of the angle in radians;
- **Angle** is the current value of the angle in radians in the appropriate current direction of the laser scan (the range of a laser scan is set default -120 to $+120$ degree, with step 0.3310 degree);
- **Coordinate X** and **Coordinate Y** are the current values of the local coordinates, with origin in the initial laser range finder position, calculated in the current laser scan direction defined by the current value of the angle in radians.

Analyzing the incoming data, given in Table 8.1, collected after scan by the laser range finder, is possible to find for the value zero of the angle (Angle = 0 in Table 8.1) the existence of equality between Length and Coordinate X:

$$Length(Index) = Coordinate\,X(Index)\,for\,Index(Angle)\,and\,Angle = 0 \quad (8.12)$$

Equation 8.12, given in general form, has the following interpretation analyzing the incoming data in Table 8.1, collected by the laser range finder after an example scan:

$$for\,Angle = 0 \rightarrow Index = 384 \rightarrow Length(384) = 197\,cm \rightarrow$$
$$\rightarrow Coordinate\,X(384) = Length(384) = 197\,cm \rightarrow Coordinate\,Y(384) = 0$$
$$(8.13)$$

The result from this analysis is that it is possible to propose to use the information after the laser range finder scan to determine the distance between the laser range finder and the objects (as the value of the Length or Coordinate X for Angle = 0), respectively between the mobile robot and these objects.

Analyzing in similar way the incoming data after the initial scan of the laser range finder, mounted on the mobile robot platform and using Eqs. 8.12 and 8.13, it is possible to determine the initial mobile robot audio visual attention in front of the mobile robot, i.e. in the direction defined by Angle = 0 for the initial local laser range finder position and orientation. If the conditions defined by Eqs. 8.9 and 8.10 for the initial position of the laser range finder, mounted on mobile robot platform are satisfied, then the mobile robot direction of the mobile robot audio visual attention matches with the initial local laser range finder position and orientation defined for Angle = 0.

The initial definition of mobile robot audio visual attention, derived from the analysis of the incoming data from the laser range finder scan in the mobile robot environment, can be connected with the local coordinates of the input video frame from the mobile robot camera, shown in the second step of algorithm defining the initial audio visual attention for the mobile robot initial position (Fig. 8.6). As it is mentioned with the Eqs. 8.9, 8.10 and 8.11, only the room coordinate X of the video camera, the microphone array and the laser range finder, mounted on the mobile robot platform, differ from each other and from the room coordinate X of the mobile robot initial position. This can be used in the next step of the algorithm shown in Fig. 8.6, to find the correspondence between the points in the input video frame and the points for which is already determined the initial mobile robot audio visual attention in front of the mobile robot in direction (Angle = 0) achieved from the analysis of incoming data from the laser range finder of the mobile robot environment. This possibility is explained with Fig. 8.7, where is shown an example of the room view of the proposed (on Fig. 8.4) audio visual attention model based on the audio visual mobile robot system for the case of indoor mobile robot motion control and navigation. From Fig. 8.7 can be seen that the location of the laser range finder, mounted on the mobile robot platform, can scan the mobile robot environments in a horizontal plane placed at a distance from the floor of the room and defined by the value z_{pl} of the coordinate Z, equal to the coordinate Z of the laser range finder, the video camera and the microphone array:

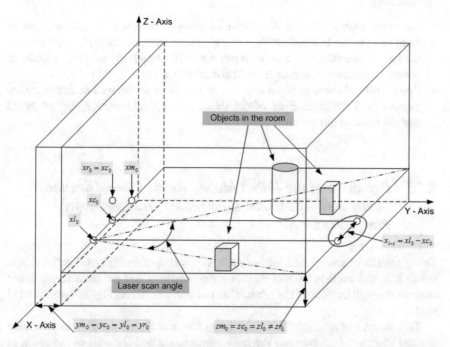

Fig. 8.7 The view of a room as an example to explain the correspondence between points of inputted video frame and the incoming data from laser range finder scan of mobile robot environment for the case of indoor mobile robot motion control and navigation using audio visual attention

$$z_{pl} = zl_0 = zc_0 = zm_0 \qquad (8.14)$$

Therefore, the incoming data from the scan of the laser range finder, according to Table 8.1, contains information as length (or distance), angle and coordinates X, Y of the points in the horizontal plane, which crosses the objects, speakers (if they are in the room) sending commands to the robot and also the walls of the room.

In the same time, these points exist also in the corresponding line of the video frame captured from the video camera, which also observes the mobile robot environments. The information for these points, in the inputted flat image as video frame, is brightness and/or colors, but as is proposed in the algorithm from Fig. 8.6 finding the correspondence of these points to the points and their data captured from the laser range finder, it is possible to add the existing information in the laser range finder about the distance and the values of the room coordinates X and Y to the points in the image video frame belonging to the line crossing the horizontal plane in which is made laser range finder scan. The reason for this is that the points from the laser range finder scan have their projections on the line, mentioned above, in the image frame.

All of these statements can be related with the initial mobile robot audio visual attention, defining the following conditions for the chosen mobile robot indoor environment:

- the initial mobile robot audio visual attention is chosen to be directed just in front of the mobile robot location at the motion start, i.e. the video camera and the laser range finder observe with priority (with attention) the space, objects or people (speakers) placed in front of the mobile robot;
- there can be chosen, for each case, an object, place in the wall or active and/or passive land marks in front of the robot as initial observation, i.e. as initial mobile robot audio visual attention.

8.5.3 Definition of the Initial Mobile Robot Video Attention Model with Additional Information from the Laser Range Finder Scan

These conditions are applied for the room views shown on Figs. 8.4, 8.5 and 8.7, where it is chosen that the mobile robot video camera and the laser range finder observe the wall in front of the robot. This place is marked on Fig. 8.7 with the red circle.

This means that after the initial laser range finder scan, it is possible to determine the real distance dst_{lw} between the laser range finder and the wall of the room in front of the robot, using Eqs. 8.12 and 8.13, and extracting the value of the **Length** for **Angle** = 0 from the captured laser range finder data:

$$dst_{lw} = Length(Angle) \text{ for } Angle = 0 \qquad (8.15)$$

For the example given on Fig. 8.7 and using Table 8.1 with the laser range finder scan data, Eq. 8.15 gives the following value for the distance dst_{lw} between the laser range finder and the wall of the room (as a example) in front of the robot:

$$dst_{lw} = Length(Angle) = 197\,\text{cm for } Angle = 0 \qquad (8.16)$$

The distance dst_{lw}, determined from Eq. 8.15, can be used to find the corresponding point in the initial video frame captured from the mobile robot video camera. From Fig. 8.7 it is seen, that if the mobile robot video camera observes the same initial audio visual attention as the laser range finder, then it is possible to find a line in the image, which corresponds to the imaginary line where the laser range finder scan crosses the observed wall in front of the robot (see as an example the horizontal line in the red circle area on in Fig. 8.7) in the horizontal plane where the laser range finder scan is done. If at the start of the mobile robot motion in the room the optical system of the mobile robot camera is set and adjusted to observe and capture the entire wall in front of mobile robot, then the line (row) im_{row}^{inw} in the image im^{inw} corresponding to the imaginary line where the laser range finder scan crosses the observed wall in front of the robot can be determined by the following equations used in the corresponding step of the algorithm, shown in Fig. 8.6:

$$im_{row}^{inw}(1 \div x_{im}^{max}) = im^{inw}(y_{im}^{row}, 1 \div x_{im}^{max}), \qquad (8.17)$$

where

x_{im}^{max} is the maximal value of the local horizontal image coordinate X, equivalent to the horizontal image resolution N_x of the used model of the mobile robot video camera;

y_{im}^{row} the unknown value of the local vertical image coordinate Y, corresponding to the imaginary line where the laser range finder scan crosses the observed wall in front of the robot.

The unknown value y_{im}^{row} of the local vertical image coordinate Y can be determined by using its correspondence to the value z_{pl} of the horizontal plane (Fig. 8.7) or the value zl_0 of the room coordinates Z of the laser range finder (Eq. 8.13). In accordance with the maximal value y_{im}^{max} of the local vertical image coordinate Y (or the vertical image resolution N_y of the concrete model of mobile robot video camera) and using the value of the room high rh (equivalent to the maximal z_{max} room coordinate Z), it is possible to derive the value of the unknown local vertical image coordinate y_{im}^{row}, as follow:

$$y_{im}^{row} = \frac{z_{pl}}{rh} y_{im}^{max} = \frac{zl_0}{rh} y_{im}^{max} \qquad (8.18)$$

In similar way it is possible to determine also the value x_{im}^{col} of the local horizontal image coordinate X in the image line or row im_{row}^{inw} of the input image im^{inw} corresponding to the imaginary point with coordinate xl_0, where the laser range finder scan for **Angle** = 0 crosses the observed wall in front of the robot, but as it is seen from Fig. 8.7 with applying the following horizontal coordinate translation:

$$x_{l-c} = xl_0 - xc_0 \tag{8.19}$$

Therefore, using Eq. 8.19 it is possible to find the value x_{im}^{col} of the local horizontal image coordinate X in the image line or on row im_{row}^{inw}:

$$x_{im}^{col} = \frac{xc_0 + x_{l-c}}{rw} x_{im}^{max} = \frac{xl_0}{rw} x_{im}^{max}, \tag{8.20}$$

where

rw is the room width equivalent to the maximal x_{max} room coordinate X;

x_{im}^{max} is the maximal value of the local horizontal image coordinate X, equivalent to the horizontal image resolution N_x of the concrete model of the mobile robot video camera.

The Eqs. 8.18 and 8.20 give the location of the initial mobile robot video attention as the local vertical y_{im}^{row} and the horizontal x_{im}^{col} image coordinates together with the value of the distance dst_{lw} (Eq. 8.15) from the initial laser range finder scan data measured for **Angle** = 0, i.e. in front of the mobile robot. The values calculated from Eqs. 8.15, 8.18 and 8.20 are the final results from the analysis done in the corresponding step of the algorithm, shown on Fig. 8.6 to analyze the data incoming from the laser range finder and to find their correspondence to the coordinates of the input video image bringing to them an additional and very useful information for the distance dst_{lw} of the defined initial mobile robot video attention location.

8.5.4 Development of the Initial Mobile Robot Video Attention Model Localization with Additional Information from a Speaker to the Mobile Robot Initial Position

In the same way can be executed the next step of the algorithm, shown on Fig. 8.6, where it is necessary to define the initial mobile robot audio attention location analyzing the microphone array audio signals to define the initial direction of the voice commands (if they exist) sent from a speaker to the mobile robot initial position. For this purpose, Fig. 8.7 is modified to present in Fig. 8.8 a situation of a speaking person in the room, sending voice commands to the mobile robot.

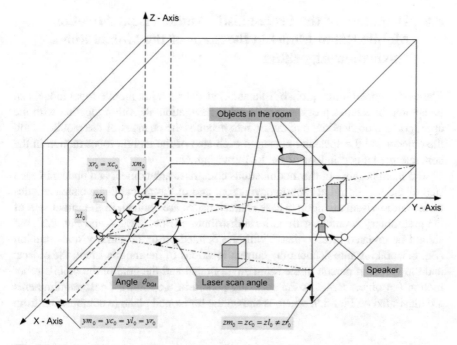

Fig. 8.8 The modified view of the room shown on Fig. 8.5 for the example of definition the initial mobile robot audio attention location analyzing the incoming microphone array audio signals to find the initial direction of the voice commands sent from a speaker to the mobile robot initial position

The analysis of the incoming microphone array audio signals in this step is executed using one famous and proved method [49] tested in real working applications, using sound source localization for the mobile robot audio visual motion control. The result, applying this method as a suitable algorithm for finding the initial mobile robot audio attention, in fact is the calculated direction of arrival (DOA), represented as the angle θ_{DOA} of the sound arrival from the speaker talking and sending the voice commands to the robot.

If the voice command send from the speaker is important (it depends on the embedded in the mobile robot motion control algorithm) for the motion of the mobile robot, then mobile robot platform performs a rotation in the direction defined by the calculated angle θ_{DOA}, i.e. the mobile robot audio attention is directed to the speaker (for example if the speaker sends a voice command to the robot "Go to me"). At same time the video camera and laser range finder mounted on the mobile robot platform, also change their direction on the angle θ_{DOA}, respectively as defined by Eqs. 8.15, 8.18 and 8.20 and the initial mobile robot video attention location must be redefined. Therefore, the new calculated values for the distance dst_{lw}, the local vertical y_{im}^{row} and horizontal x_{im}^{col} image coordinates, together with the calculated direction of arrival θ_{DOA} from the speaker can be considered as determinate joined initial audio visual attention.

8.6 Definition of the Probabilistic Audio Visual Attention Mobile Robot Model in the Steps of the Mobile Robot Navigation Algorithm

The so developed initial probabilistic audio visual attention mobile robot model can be applied in each step of the mobile robot navigation algorithm together with the appropriate modification, and taking into account the changes of the mobile robot, the objects and the speaking person in each step of the mobile robot motion in the concrete predefined indoor or outdoor environment.

The main purpose of the modifications comprises amendment and update in each step of the mobile robot motion navigation, and of the probabilistic characteristics of the audio visual attention. These characteristics are defined, in the initial step of the preliminary attention with uniform distribution (Eqs. 8.2, 8.4 and Fig. 8.2), but should be updated in accordance with the defined posterior audio visual attention (Eq. 8.3), taking into account the current locations of the mobile robot, the objects and the speaking person in the room environment and the goal of the mobile robot motion (an object target or the speaking person in the room). These requirements are illustrated on Fig. 8.9, where is shown the horizontal plane (a combination from

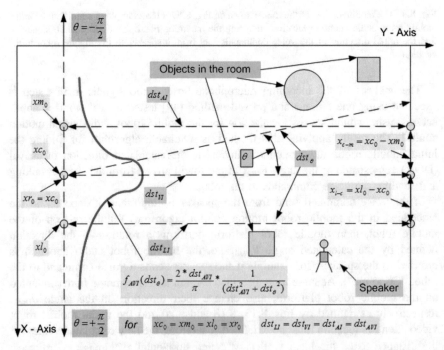

Fig. 8.9 Probabilistic characteristics of the audio visual attention mobile robot model applying the algorithm of the mobile robot motion navigation, based on horizontal planes for the defined initial visual and audio attention shown on Figs. 8.7 and 8.8, respectively

horizontal planes from Fig. 8.7 and from Fig. 8.8) of the initial audio visual mobile robot attention model additionally modified with the probabilistic features describing the mobile robot audio visual attention. On Fig. 8.9 is shown the probability density function $f_{AVI}(dst_\theta)$, defined by using the general Eq. 8.5, and the initial audio and visual attention model from Figs. 8.7 and 8.8, used to describe the probability characteristics:

$$f_{AVI}(dst_\theta) = \frac{2 \times dst_{AVI}}{\pi} \times \frac{1}{(dst_{AVI}^2 + dst_\theta^2)}, \tag{8.21}$$

where

dst_θ is the distance measured by laser range finder in the current direction (the angle θ on Fig. 8.7) from the initial mobile robot position to an object/speaker (if they exist in the room), or to the wall of the room in front of the mobile robot;

dst_{AVI} the initial distance measured by laser range finder in the direction defined by the angle $\theta = 0$ on Fig. 8.9 from the initial mobile robot position to an object/speaker (if they exist in the room) or to the wall of the room in front of the mobile robot. The value of the distance dst_{AVI} is equal to the value of distance, i.e.:

$$dst_{AVI} = dst_{lw} \tag{8.22}$$

The Eqs. 8.21 and 8.22 are valid only, if assumed the existence of the following condition for the equality of the coordinates X of the video camera, the microphone array and the laser range finder with coordinate X of the mobile robot:

$$xc_0 = xm_0 = xl_0 = xr_0, \tag{8.23}$$

from which follows the equality of the distances below:

$$dst_{LI} = dst_{VI} = dst_{AI} = dst_{AVI}, \tag{8.24}$$

shown on Fig. 8.9 for the general case of inequality of the distances $dst_{LI}, dst_{VI}, dst_{AI}, dst_{AVI}$ from Eq. 8.24.

Equation 8.21 gives the clarity of the mobile robot initial audio visual attention definition by applying the probability density function $f_{AVI}(dst_\theta)$ and thus solving the problem of lack in the mobile robot the system, similar to the human brain. Therefore, in other more complicated cases, when Eqs. 8.15, 8.23 and 8.24 are not satisfied, it is necessary to take into account the existence of the following differences, shown on Fig. 8.9, between coordinates X of the video camera, the microphone array and the laser range finder:

$$x_{l-c} = xl_0 - xc_0 \tag{8.25}$$

$$x_{c-m} = xc_0 - xm_0 \tag{8.26}$$

Thus, in case of existence of the differences defined by Eqs. 8.25 and 8.26, it is also necessary to calculate (applying the corresponding geometrical transformations) the different values of the distances dst_{LI}, dst_{VI} and dst_{AI}, i.e. to take into account the following condition:

$$dst_{LI} \neq dst_{VI} \neq dst_{AI} \neq dst_{AVI} \tag{8.27}$$

The condition defined by Eq. 8.27 shows the need to calculate the different probability density functions like the function $f_{AVI}(dst_\theta)$, but using in the following equations (similar to Eq. 8.21), the different values of the distances dst_{LI}, dst_{VI} and dst_{AI}, respectively:

$$f_{LI}(dst_{L\theta}) = \frac{2 \times dst_{LI}}{\pi} \times \frac{1}{(dst_{LI}^2 + dst_{L\theta}^2)} \tag{8.28}$$

$$f_{VI}(dst_{V\theta}) = \frac{2 \times dst_{VI}}{\pi} \times \frac{1}{(dst_{VI}^2 + dst_{V\theta}^2)} \tag{8.29}$$

$$f_{AI}(dst_{A\theta}) = \frac{2 \times dst_{AI}}{\pi} \times \frac{1}{(dst_{AI}^2 + dst_{A\theta}^2)} \tag{8.30}$$

In Eqs. 8.28, 8.29 and 8.30 are also used different indexes of the distance dst_θ measured by the laser range finder for the current direction defined in the general case as the angle θ from Fig. 8.9, but in each of these particular cases, related to the corresponding laser range finder $L\theta$, and the corresponding angle of the video camera $V\theta$ or of the microphone array $A\theta$. Therefore, the names of these differences dst_θ differ from each other only by the corresponding indexes distance: $dst_{L\theta}$ for the direction measured from the laser range finder local coordinate X; $dst_{V\theta}$ for the direction measured from the video camera local coordinate X; $dst_{A\theta}$ for the direction measured from the microphone array local coordinate X.

The initial probability density functions defined by Eq. 8.21 and also the corresponding Eqs. 8.28, 8.29 and 8.30 for the particular cases, can be used in the development of the appropriate algorithms for the continuous mobile robot audio visual motion control and navigation in concrete situations for indoor environments. It is evident that Eq. 8.21 is the simplest case (suitable for simulations) of the determination of the distance dst_θ measured by the laser range finder with the assumption for equal coordinates X for the video camera, the microphone array and the laser range finder, equivalent to the coordinate X of the initial mobile robot position. Therefore, it is proposed to apply Eq. 8.21 in the next simulations as the initial probability density function $f_{AVI}(dst_\theta)$ and the particularly corresponding

Eqs. 8.28, 8.29 and 8.30, modifying and updating them in each execution step of the algorithm in some examples of the mobile robot motion control and navigation in the room environments.

8.7   Experimental Results from the Simulations of the Mobile Robot Motion Navigation Algorithm Applying the Probabilistic Audio Visual Attention Model

The experiments for testing the properties of the proposed probabilistic audio visual attention model are carried out following the main block diagram, shown on Fig. 8.10. It is composed so that to present and explain all the necessary operations as appropriate blocks implementing the algorithms for testing the performance of the proposed probabilistic audio and visual attention model, applied in the tasks for mobile robot motion control and navigation in rooms, i.e. for indoor environments.

The audio information is collected as audio signals captured (for the general case) by N microphones (M1, M2, ... MN, shown on Fig. 8.10) arranged in a microphone array. In the concrete implementation the microphone array is arranged as a linear microphone array consisting of six microphones (N = 6). The microphone array is implemented as an appropriate microphone array module type Steval-MKI126v2 [51] with MEMS (Micro-Electro-Mechanical Systems), type of microphones MP34DB01 [52]. The video camera is Surveyor SRV-1 Blackfin

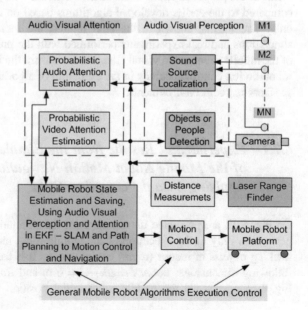

Fig. 8.10 Main block diagram used in the experiments carried out for testing the properties of the proposed probabilistic audio visual attention model

Camera [53] and [54] is mounted on the mobile robot platform module type
Surveyor SRV-1 Blackfin Robot [55]. The type of laser range finder used to scan
and measure the data for distances is URG-04LX-UG01 [50].

The audio and visual information collected by the microphone array and the
video camera, together with the data measured by the laser range finder after scan of
the room environments (Fig. 8.10) can be used both by the audio visual perception
blocks (Sound Source Localization and Objects or People Detection) or by the
audio visual attention blocks (Probabilistic Audio Attention Estimation and
Probabilistic Video Attention Estimation), which depend on the start of the concrete
algorithm of the block General Mobile Robot Execution Control (Fig. 8.10). The
main goal of using the results from the audio visual perception or the audio visual
attention blocks (depending from execution of concrete algorithm) is to estimate
and to save the current state or step of the motion control and navigation, applying
in the corresponding block on Fig. 8.10, the embedded algorithms for EKF–SLAM
and path planning [49, 56, 57]. The block diagram, shown on Fig. 8.10, is used in
all experiments, together with appropriate modifications, to test the proposed
probabilistic audio visual attention model in the algorithms for mobile robot motion
control and navigation.

The experiments for testing the proposed probabilistic model of audio visual
attention are based in part of the research, presented in [49] and also on previous
investigations of the authors, presented in publications [58–62]. In these resent
works is applied the audio and video perception only, without using the probabilistic model (proposed here) of the audio visual attention for implementing the
mobile robot motion control and navigation in indoor environments, i.e. rooms with
objects, people, etc. Therefore, this proposal is additionally proved by the experiments of the proposed audio visual attention model with probabilistic characteristics
compared to the earlier developed algorithms based on the audio visual perception
only. The results, shown in the next figures, are more important than multiple
simulations and real experiments performed with the proposed probabilistic model
of the mobile robot audio visual attention to confirm the improvement of the mobile
robot motion control accuracy in the execution of algorithms for tracking objects or
speakers in room environments.

8.7.1 Experimental Results from the Simulations of the Mobile Robot Motion Navigation Algorithm Applying Visual Perception Only

On Fig. 8.11 are presented the results from the simulation performed using visual
perception only in the execution of the mobile robot motion control algorithm for
tracking objects in indoor (room) environments. It is assumed, that the room is of
following dimensions: *Room Length—rl* = 6 m and *Room Width—rw* = 4 m. On
Fig. 8.11 are shown the mobile robot initial position in the room, and a lot of

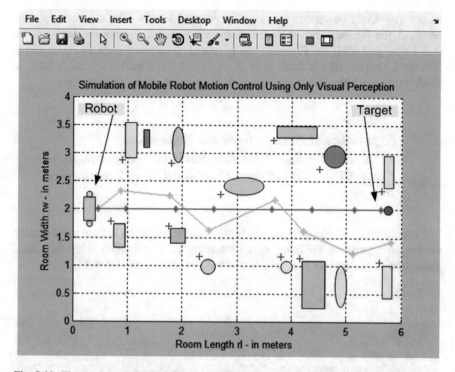

Fig. 8.11 The results as defined (desired) and real locations of mobile robot trajectory in each step of the execution of the mobile robot motion control algorithm in simulation performed using visual perception only

objects simulated as different types (tables, chairs, etc.), and placed arbitrary in the room environment.

With red line on Fig. 8.11 is defined the desired (ideal) trajectory of the mobile robot motion from the initial to the target position. The little blue crosses on Fig. 8.11 indicate the places near each object as the positions (coordinates X and Y) of each object measured by the laser range finder from the initial mobile robot position to the position of each object. It can be noted, that the objects positions are inputted in the simulation as necessary data about the distances and the coordinates X and Y of the objects obtained by the real measurements in the existing room with similar dimensions.

It is seen from Fig. 8.11, that there are objects in the room without little blue crosses placed near these objects. This is because they seem to be invisible for the laser range finder. With green color is presented on Fig. 8.11 the trajectory really achieved when an algorithm is executed using visual perception only to avoid the obstacles (the objects on Fig. 8.11) in each step of the mobile robot motion control from its initial position to the target. Analyzing the simulation results, on Fig. 8.11 can be noticed significant differences (errors) between the defined or desired (ideal) trajectory and the trajectory really achieved after the execution of the algorithm for mobile robot motion control using visual perception only. The simulation results are used for the preparation of the quantitative errors comparison between the simulation presented on Fig. 8.11 using visual perception only (without visual attention) and the results from next simulation when visual attention is used in combination with the visual perception in the mobile root motion control.

8.7.2 Experimental Results from the Simulations of the Mobile Robot Motion Navigation Algorithm Using Visual Attention in Combination with the Visual Perception

The experimental results achieved from the simulations of the mobile robot motion navigation algorithm, applying visual attention in combination with the visual perception in the mobile root motion control, are shown on Fig. 8.12.

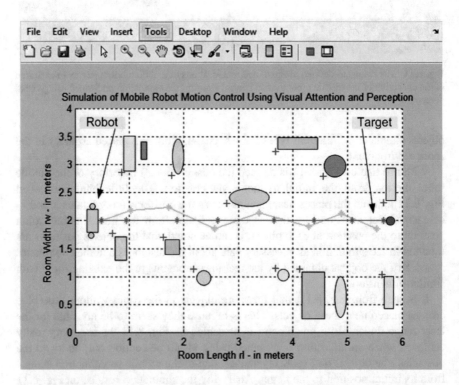

Fig. 8.12 The results as defined (desired) and the real locations of the mobile robot trajectory in each execution step of the mobile robot motion control algorithm in the simulation performed using visual attention in combination with visual perception

The visual analysis of the trajectory on Fig. 8.12, shown as a result of the execution of the algorithm using visual attention in combination with visual perception, achieves more precise motion control to the defined target in comparison with the same trajectory achieved after the execution of the algorithm using visual perception only (Fig. 8.11).

8.7.3   Quantitative Comparison of the Simulations Results Applying Visual Perception Only, and Visual Attention with Visual Perception

The visual analysis together with the visual comparison, presented above, give the qualitative estimation only of the existence of errors and differences between the trajectories of mobile robot motion to a chosen target applying in the presented two simulations (Figs. 8.11 and 8.12) and executing the algorithm of the mobile robot motion control and navigation using at first visual perception only (Fig. 8.11), and then—visual attention in combination with visual perception (Fig. 8.12). This visual analysis and comparison also show a more precise motion control in Fig. 8.12 to the defined target in comparison with same trajectory achieved in the execution of the algorithm using visual perception only (Fig. 8.11), but these results should be verified and confirmed also based on quantitative comparison. Therefore, the results from such quantitative comparison of the achieved trajectories in both simulations of the mobile robots motion control algorithms by using only visual perception (Fig. 8.11) and applying visual attention in combination with visual perception (Fig. 8.12), are presented in Table 8.2.

In Table 8.2 is presented briefly the quantitative comparison of a part of the numerous steps of mobile robots trajectories only, in both simulations as the current coordinates X and Y in some current steps in the execution of the algorithm for mobile robot motion control to a preliminary defined target, using in the first case visual perception only (Fig. 8.11) and in the second case—in combination with the visual attention (Fig. 8.12).

It can be mentioned, that the essential quantitative comparison of the trajectories variations and differences, in the cases with visual perception only (Fig. 8.11) and with visual attention in combination with visual perception (Fig. 8.12), can be considered only along the coordinates X, because in these two experiments is chosen a simple situation of the mobile robot motion to a target just in front of the robot, i.e. the coordinate Y is in direction of the mobile robot motion and is not important to estimate the error along the coordinates Y of the motion to the defined target.

As it is seen from Table 8.2, analysing the results, it is possible to confirm, that the error of the mobile robot motion to the target, when using the combination of visual attention and perception in the motion control, is smaller (the average error of the coordinate X is 0.0295), unlike the error of the mobile robot motion to the target,

Table 8.2 Quantitative comparison of the achieved trajectories in both mobile robot motion control algorithms by using visual perception only (Fig. 8.11) or visual perception combined with visual attention (Fig. 8.12)

Steps of the mobile robot motion to the target	Defined trajectory		Simulation using visual perception only (Fig. 8.9)		Simulation using visual attention and perception (Fig. 8.10)	
	Coordinates (m)					
	X	Y	X	Y	X	Y
1	2.0000	0.4562	2.0139	0.4167	2.0206	0.4650
2	2.0139	0.9539	2.3056	0.8889	2.1993	1.0115
3	2.0000	1.8664	2.2361	1.7639	2.2680	1.7182
4	2.0000	2.5438	1.5972	2.4722	1.8694	2.5774
5	1.9861	3.6221	2.1806	3.7083	2.1718	3.5811
6	2.0139	4.3410	1.6250	4.2222	1.9244	4.2818
7	2.0000	5.1429	1.2222	5.1389	2.1168	4.8776
8	1.9861	5.6129	1.4306	5.8056	1.8557	5.5704
Average error (m)			0.1736	0.0153	0.0295	0.0973
Deviation dx and dy from real target position (m)			dx 0.5555	dy 0.1927	dx 0.1304	dy 0.0425

when using visual perception only in the motion control (the average error of the coordinate X is higher -0.1736). It is seen from Table 8.2 that the variances in the coordinate Y are a little higher (the average error of the coordinate Y is 0.0973) in the case of using the combination of visual attention and perception in the motion control, when compared to the case of motion control using visual perception only (the average error of the coordinate Y is 0.0153), but this is not so important because the motion is in direction Y (just in front of the mobile robot) and this error indicates the existence of difference in each discrete step of the algorithm in the two simulations under comparison.

Additional means to quantitative comparison of trajectories variations and differences, in the case of two simulations using visual perception only (Fig. 8.11) and using visual attention in combination with visual perception (Fig. 8.12) in the mobile robot motion control, can be considered from the calculated and presented in Table 8.2 deviations dx and dy of the coordinates X and Y of the final mobile robot position from the coordinates X and Y of the real target position. It can also be seen from Table 8.2, that the deviations dx and dy are smaller ($dx = 0.1304$, $dy = 0.0425$) in the case of applying in motion control the combination of visual attention and perception, than the deviations dx and dy in the case of motion control using visual perception only ($dx = 0.5555$, $dy = 0.1927$). This result indicates, that in the case of motion control combining the visual attention with the visual perception, the final destination (the target in Fig. 8.12) is achieved more accurately in comparison with the case of motion control using visual perception only (the target in Fig. 8.11).

8.7.4   Experimental Results from Simulations Using Audio Visual Attention in Combination with Audio Visual Perception

On Fig. 8.13 are presented the results from the simulation of the mobile robot motion control algorithm using both audio and visual perception, but without using the audio and visual attention. The initial conditions of this simulation are similar to these shown on Fig. 8.11 (using visual perception only).

After the start of the mobile robot motion control algorithm, the mobile robot begins the motion tracking the target in front of the robot (the green line on Fig. 8.13) using the algorithm for visual perception only (as in Fig. 8.11). In this case the simulation differs from the simulation shown on Fig. 8.11, when the algorithm of audio perception is on, and the robot waits for the perception of a voice command, maybe from a speaker, in the mobile robot area. In the fourth step (marked with an arrow) of tracking to the target, the speaker (shown on Fig. 8.13) sends a voice command ("Go to me") to the robot. This command is perceived by the microphone array of the mobile robot and the execution of the audio perception algorithm for sound localization determines the direction of sound arrival from the

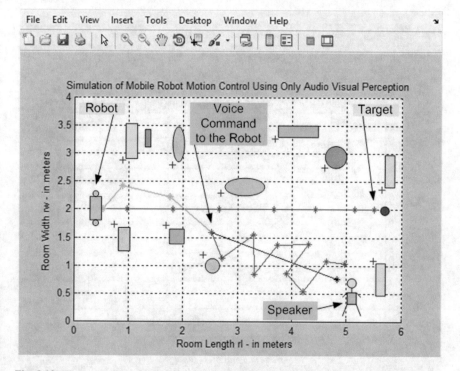

Fig. 8.13 The results from the simulation of the mobile robot motion control algorithm performed using both audio and visual perception, but without applying the audio and visual attention

speaker (the black line on Fig. 8.13). This calculated direction is used in the motion algorithm to change in the fourth step the way of the robot (the magenta line in Fig. 8.13), not to the target in front of it, but to the speaker. The next steps of the motion control algorithm use both audio perception (waiting for new voice commands from the speaker) and visual perception (tracking the body of the speaker) to arrive at the place of the speaker.

In the next simulation shown on Fig. 8.14 are presented the results from the simulation of the mobile robot motion control algorithm also, but applying additional audio and visual attention in combination with audio and visual perception (Fig. 8.13).

The simulation on Fig. 8.14 is carried out with similar initial conditions as these shown on Fig. 8.13 (using only audio visual perception). In the same way, after the start of the mobile robot motion control algorithm, the mobile robot begins the motion tracking the target in front of the robot (the green line on Fig. 8.14), but in this simulation (unlike the simulation in Fig. 8.13) are used both the algorithms for visual attention and visual perception (as in Fig. 8.12). As in the simulation presented on Fig. 8.13, in this simulation the robot waits to perceive a voice command, maybe from a speaker, in the mobile robot area. In the fourth step (marked with an arrow) of tracking to the target, the speaker (shown on Fig. 8.14) sends a voice

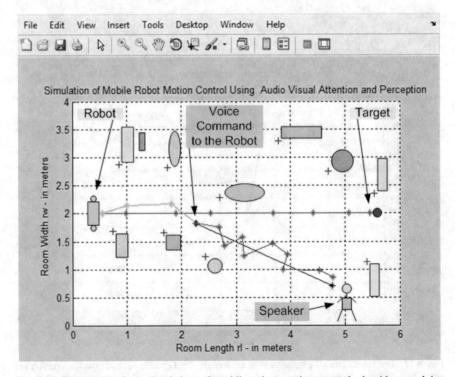

Fig. 8.14 The results from simulation of mobile robot motion control algorithm applying additional audio and visual attention in combination with audio and visual perception

command ("Go to me") to the robot. This voice command is perceived by the microphone array of the mobile robot and the execution of the audio perception algorithm for sound localization determines the direction of sound arrival from the speaker (the black line on Fig. 8.14). The so calculated direction is used in the motion algorithm to change in the fourth step the way of the robot (the magenta line on Fig. 8.14), not to the target in front of it, but to the speaker. In the next steps, the motion control algorithm uses audio attention and perception (waiting for new voice commands from the speaker) and also both visual attention and perception (tracking the body of the speaker) to arrive at the place of the speaker.

The visual analysis is prepared, to compare the trajectories from Figs. 8.13 and 8.12, achieved after the execution of the mobile robot motion control algorithm in two different ways: first (Fig. 8.13) using audio and visual perception only, without applying the audio and visual attention; and second (Fig. 8.14) using audio and visual perception with applying audio and visual attention also. From the visual comparison is determined that the mobile robot motion trajectory (the magenta line in Fig. 8.14) stands closer to the direction (the black line on Fig. 8.14) predefined by the sound localization algorithm, than the same mobile robot motion trajectory (the magenta line in Fig. 8.13) with greater variation around the direction (the black line on Fig. 8.14) predefined by the sound localization algorithm. Therefore, from this visual comparison is confirmed the main advantage, i.e.—the higher accuracy of the mobile robot motion control, in case of tracking the speaking person using the audio and visual attention in combination with the audio and visual perception.

8.7.5 Quantitative Comparison of the Results Achieved in Simulations Applying Audio Visual Perception Only, and Visual Attention Combined with Visual Perception

The properties of the above simulations of the mobile robots motion control algorithm using both the audio visual perception only, and the visual attention in combination with the visual perception, can be presented not only through the qualitative visual analysis and comparison of the trajectories on Figs. 8.13 and 8.14. They can be illustrated also as the results from the quantitative comparison (Table 8.3) of the mobile robot motion trajectories (the magenta line on Figs. 8.13 and 8.14) with the direction of the trajectory (the black line on Figs. 8.13 and 8.14) predefined by the sound localization algorithm.

In Table 8.3 is briefly presented the quantitative comparison of a part of the numerous steps of the mobile robots trajectories only, in case of tracking the speaking person, in both simulations; in Fig. 8.13 using audio and visual perception only without applying the audio and visual attention, and in Fig. 8.14—using audio and visual perception with applying audio and visual attention also. The results of the quantitative comparison presented in Table 8.3, are given as the values of the

Table 8.3 Quantitative comparison of the mobile robot trajectories (the magenta line in Figs. 8.13 and 8.14) with the direction or the trajectory predefined by the sound localization algorithm (the black line on Figs. 8.13 and 8.14)

Steps of the mobile robot motion control to the speaker	Direction determined by the sound localization algorithm		Simulation using audio visual perception only (Fig. 8.13)		Simulation using audio visual attention and perception (Fig. 8.14)	
	Coordinates (m)					
4	X	Y	X	Y	X	Y
5	1.8547	2.2811	1.5862	2.5081	1.8194	2.2448
6	1.6747	2.5991	1.1448	2.7021	1.7500	2.6882
7	1.5779	2.8618	1.5448	3.2702	1.4167	2.8129
8	1.4810	3.0553	0.8552	3.2979	1.5556	3.1039
9	1.4118	3.1935	1.3655	3.7275	1.2361	3.1316
10	1.2180	3.6359	1.3931	4.2818	1.4306	3.6721
11	1.1488	3.7880	0.8690	3.9076	1.2639	3.9492
12	1.0519	4.0092	0.5517	4.2263	1.0000	3.8661
13	0.8581	4.4654	1.0759	4.6420	1.0000	4.5450
14	0.7197	4.7834	1.0483	4.9746	0.8472	4.7806
Average error (m)			0.1562	0.2865	0.0323	0.0122
Deviations dx and dy from the real speaker position (m)			dx 0.3286	dy 0.1912	dx 0.1275	dy 0.0028

current coordinates X and Y in the corresponding execution steps of the algorithm for mobile robot motion control to the speaking person, using in the first case audio visual perception only (Fig. 8.13) and in the second case—audio visual attention in combination with audio visual perception (Fig. 8.14).

It is seen, analyzing the calculated average errors (Table 8.3), that the errors in the trajectory of the mobile robot motion to the speaking person, when using the combination of audio visual attention and perception in the motion control, are smaller (the average errors of the coordinates X and Y are 0.0323 and 0.0122, respectively), than the errors in the mobile robot motion trajectory to the speaking person, when in the motion control is used audio visual perception only (the average errors of the coordinates X and Y are 0.1562 and 0.2865, respectively).

For the two simulations, presented in Figs. 8.13 and 8.14, can also be applied the additional quantitative comparison of trajectories variations and differences, in cases of using audio visual perception only (Fig. 8.13) and using audio visual attention in combination with audio visual perception (Fig. 8.14). This is done considering (for both simulations) the calculated and presented in Table 8.3 deviations dx and dy of the coordinates X and Y of the mobile robot final position from the real speaker position. It can also be seen from Table 8.3, that the deviations dx and dy are smaller ($dx = 0.1275$, $dy = 0.0028$) in the case of applying the combination of audio visual attention and perception in motion control, than the deviations dx and dy in the case of motion control using audio visual perception only ($dx = 0.3286$, $dy = 0.1912$). This result indicates, that in the case of motion control combining the audio visual attention with the audio visual perception, the final destination (the speaking person in Fig. 8.14) more accurate than in the case of motion control using audio visual perception only (the speaking person in Fig. 8.13).

8.8   Conclusion

The simulations results for the mobile robot motion navigation algorithm applying the proposed probabilistic model of the audio visual mobile robot attention can be summarized as follows:

- it is confirmed from the qualitative and quantitative comparison (Figs. 8.11 and 8.12, and Table 8.2), that the average error in the real mobile robot trajectory to the target, when using the combination of visual attention and perception in the motion control, is lower (the average error of the coordinate X is 0.0295), than the error of the mobile robot motion to the target, when visual perception only is used in the motion control (the average error of coordinate X is higher -0.1736);
- this result indicates, that in the case of motion control combining the visual attention with the visual perception, each step of the mobile robot trajectory to the target (target in Fig. 8.12) is achieved more accurately than in the case of motion control using visual perception only (the target in Fig. 8.11);
- also the comparison of the deviations dx and dy of the coordinates X and Y (Table 8.2), shows that the deviations dx and dy are smaller ($dx = 0.1304$, $dy = 0.0425$) in the case of applying the combination of visual attention and perception in motion control, than the deviations dx and dy in the case of motion control using visual perception only ($dx = 0.5555$, $dy = 0.1927$);
- this result indicates, that in the case of motion control combining the visual attention with the visual perception, the final destination (the target in Fig. 8.12) is achieved more accurately than in the case of motion control using visual perception only (the target in Fig. 8.11);
- from Figs. 8.13 and 8.14, and Table 8.3 can be confirmed, that the average errors in the trajectory of the mobile robot motion to the speaking person, when using the combination of audio visual attention and perception in the motion control, are smaller (the average errors of the coordinates X and Y are 0.0323 and 0.0122, respectively), than the errors in the trajectory of the mobile robot motion to the speaking person, when audio visual perception only is used in the motion control (the average errors of the coordinates X and Y are 0.1562 and 0.2865, respectively);
- this result indicates, that in the case of motion control combining audio visual attention with audio visual perception, each step of the mobile robot trajectory to the speaking person (the speaker in Fig. 8.14) is achieved more accurately than

in the case of motion control using audio visual perception only (the speaker in Fig. 8.13);

- the comparison of the deviations dx and dy of the coordinates X and Y (Table 8.3) illustrates, that the deviations dx and dy are smaller ($dx = 0.1275$, $dy = 0.0028$) in the case of applying the combination of audio visual attention and perception in motion control, than the deviations dx and dy in the case of motion control using audio visual perception only ($dx = 0.3286$, $dy = 0.1912$);
- this result confirms, that in the case of motion control combining audio visual attention with audio visual perception, the final destination (the speaking person in Fig. 8.14) is achieved more accurately than in the case of motion control using audio visual perception only (the speaking person in Fig. 8.13);
- in the presented experimental results are included mainly the quantitative comparisons only of the results achieved in simulations of the existing and tested method, algorithm and model, using audio visual perception only presented in more details in [46] and [49], and the audio visual attention model in combination with visual perception, proposed in this chapter;
- the comparative test for speed of calculations of the proposed in this chapter audio visual attention model with the existing methods based on audio visual perception only, are not included, because the simulations are not prepared in real time and it is the goal in the future works to convert these simulations in the same programs means (for example C++ or C#) as the existing in [46] and [49] or other publication algorithms working in real time. Therefore after this conversion it is possible to make an appropriate comparison between the speed of calculations for the proposed and the existing methods;
- the comparison of the precision in the mobile robot navigation using the audio visual attention model (proposed in this chapter), and the existing methods based on audio visual perception only, are shown in the experimental results and it is seen that the precision achieved in the mobile robot navigation using the proposed audio visual attention is higher than the precision of the mobile robot navigation applying the existing methods based on audio visual perception only.

Finally could be declared that here is the proposed probabilistic audio visual attention model applicable in the audio visual mobile robot system for the purpose of mobile robot motion control and navigation tasks. This model is presented first as indoor initial audio visual attention model and corresponding algorithm, in combination with additional information, derived as incoming scan data from the laser range finder and additional information for the initial direction of arrival of the voice commands sent from a speaker to the mobile robot initial position. The initial audio visual attention model is then extended as a probabilistic audio visual attention mobile robot model applying it in each current step of the mobile robot navigation algorithm. To confirm the efficiency of the proposed probabilistic audio visual attention in the mobile robot motion control and navigation tasks, are conducted the simulations of the mobile robot motion navigation algorithm applying the defined probabilistic model of the audio visual mobile robot attention. The simulations are carried out in two ways: applying the visual attention and perception only in the mobile robot motion navigation algorithm for tracking a visual object, and applying the audio visual attention and perception in the mobile robot motion navigation algorithm for tracking the speaking person. The results from all simulations confirm the accuracy not only in each step of the mobile robot trajectory to the target or to the speaking person, but also the precision in reaching the final destination (the target or a speaking person).

This gives the reason to continue with future developments and publications to improve the proposed probabilistic audio visual attention model both in respect of the implementation of more precise probability functions focusing the mobile robot audio visual attention to important objects and speaking persons, as well as applying the audio visual attention model in different and more complex indoor and outdoor environments of the mobile robot motion activity. The attention will be directed also to create some mobile robot knowledge (modeling some human brain functions), accumulating the information of audio and visual objects in each step of the motion and using this information for focusing more accurately the mobile robot audio visual attention to the objects, targets or speaking persons.

References

1. Choset, H., Lynch, K., Hutchinson, S., Kantor, G., Burgard, W., Kavraki, L., Thrun, S.: Principles of Robot Motion: Theory, Algorithms, and Implementations. MIT Press (2005)
2. Okuno, H.G., Nakadai, K., Hidai, K.I., Mizoguchi, H., Kitano, H.: Human robot interaction through real time auditory and visual multiple talker tracking. In: Proceedings IEEE/RSJ International Conference on Intelligent Robots and Systems, pp. 1402–1409 (2001)
3. Brooks, R., Breazeal, C., Marjanovic, M., Scassellati, B., Williamson, M.: The Cog project: building a humanoid robot. In: Computation for Metaphors, Analogy, and Agents. Lecture Notes in Artificial Intelligence 1562. Springer, New York, pp. 52–87 (1999)
4. Filho, A.: Humanoid Robots. New Developments. Advanced Robotic Systems International and I-Tech, Vienna Austria (2010)
5. Bigun, J.: Vision with Direction. Springer (2006)
6. Jarvis, R.: Intelligent Robotics: past, present and future. Int. J. Comput. Sci. Appl. (Technomathematics Research Foundation) 5(3), 23–35 (2008)
7. Adams, B., Breazeal, C., Brooks, R., Scassellati, B.: Humanoid robots: a new kind of tool. IEEE Intell. Syst. Appl. 15(4), 25–31 (2000)
8. Eckmiller, R., Baruth, O., Neumann, D.: On human factors for interactive man-machine vision: requirements of the neural visual system to transform objects into percepts. In: Proceedings of IEEE World Congress on Computational Intelligence WCCI 2006—International Joint Conference on Neural Networks, Vancouver, Canada, 16–21 July, pp. 99–703 (2006)
9. Sezer, V., Gokasan, M.: A novel obstacle avoidance algorithm: "Follow the Gap Method". Robot. Auton. Syst. (2012)
10. Oroko, J., Ikua, B.: Obstacle avoidance and path planning schemes for autonomous navigation of a mobile robot: a review. Proc. Mech. Eng. Conf. Sustain. Res. Innov. 4, 314–318 (2012)
11. Kalmegh, S., Samra, D., Rasegaonkar, N.: Obstacle avoidance for a mobile exploration robot using a single ultrasonic range sensor. IEEE International Conference on Emerging Trends in Robotics and Communication Technologies INTERACT, pp. 8–11 (2010)

12. Zhu, Y., Zhang, T., Song, J., Li, X.: A new hybrid navigation algorithm for mobile robots in environments with incomplete knowledge. Knowl.-Based Syst. **27**, 302–313 (2012)
13. Sgorbissa, A., Zaccaria, R.: Planning and obstacle avoidance in mobile robotics. Robot. Auton. Syst. **60**(4), 628–638 (2012)
14. Kumari, C.: Building algorithm for obstacle detection and avoidance system for wheeled mobile robot. Glob. J. Res. Eng. 11–12 (2012)
15. Chen, K., Tsai, W.: Vision-based obstacle detection and avoidance for autonomous land vehicle navigation in outdoor roads. Autom. Constr. **10**(1), 1–25 (2000)
16. Jung, B., Sukhatme, G.: Detecting moving objects using a single camera on a mobile robot in an outdoor environment. In: International Conference on Intelligent Autonomous Systems, The Netherlands, pp. 980–987 (2004)
17. Beymer, D., Konolige, K.: Tracking people from a mobile platform. In: IJCAI-2001 Workshop on Reasoning with Uncertainty in Robotics, Seattle, WA, USA, pp. 99–116 (2001)
18. Chakravarty, P., Jarvis, R.: Panoramic vision and laser range finder fusion for multiple person tracking. In: Proceedings of IEEE/RSJ International Conference on Intelligent Robots and Systems (IROS), Beijing, China, pp. 2949–2954 (2006)
19. Bennewitz, M., Cielniak, G., Burgard, W.: Utilizing learned motion patterns to robustly track persons. In: Proceedings of IEEE International Workshop on VS-PETS, France, pp. 102–109 (2003)
20. Menegatti, E., Nori, F., Pagello, E., Pellizzari, C., Spagnoli, D.: Designing an omnidirectional vision system for a goalkeeper robot. In: Birk, A., Coradeschi, S., Tadokoro, S. (eds.) RoboCup 2001: Robot Soccer World Cup V. Springer, pp. 78–87 (2002)
21. Collins, R., Lipton, A., Kanade, T.: A system for video surveillance and monitoring. Technical report, Robotics Institute at Carnagie Mellon University (2000)
22. Gutchess, D., Jain, A., Wang, S.: Automatic surveillance using omnidirectional and active cameras. In: Asian Conference on Computer Vision (ACCV), pp. 916–920 (2000)
23. Wang, M., Lin, H.: Object recognition from omnidirectional visual sensing for mobile robot applications. In: IEEE International Conference on Systems, Man, and Cybernetics, San Antonio, Texas, USA, pp. 2010–2015 (2009)
24. http://www.irobot.com/For-Defense-and-Security.aspx#PublicSafety
25. Schilling, K., Driewer, F., Baier, H.: User interfaces for robots in rescue operations. In: Proceedings IFAC/IFIP/IFORS/IEA Symposium Analysis, Design and Evaluation of Human-Machine Systems, Atlanta, USA, pp. 79–84 (2004)
26. http://www.doc-center.robosoft.com/@api/deki/files/6211/=robuROC4_web.pdf
27. Huttenrauch, H., Eklundh, S.: To help or not to help a service robot: Bystander intervention as a resource in human-robot collaboration. Interact. Stud. **7**(3), 455–477 (2006)
28. Burgard, W., Cremers, A., Fox, D., Hahnel, D., Lakemeyer, G., Steiner, W., Thrun, S.: Experiences with an interactive museum tour-guide robot. Artif. Intell. **114**(1–2), 3–55 (1999)
29. http://www.robotnik.eu/services-robotic/mobile-robotics-applications/
30. Khan, M.: The development of a mobile medical robot using ER1 technology. IEEE Potentials **32**(4), 34–37 (2013)
31. Ben Robins, B., Dautenhahn, K., Ferrari, E., Kronreif, G., Prazak-Aram, B., Marti, P., Iacono, I., Gelderblom, G., Bernd, T., Caprino, F., et al.: Scenarios of robot-assisted play for children with cognitive and physical disabilities. Interact. Stud. **13**, 189–234 (2012)
32. Bundsen, C.: A theory of visual attention. Psychol. Rev. **97**(4), 523–547 (1990)
33. Itti, L., Koch, C., Niebur, E.: A model of saliency-based visual attention for rapid scene analysis. IEEE Trans. Pattern Anal. Mach. Intell. **20**(11), 1254–1259 (1998)
34. Posner, M., Petersen, S.: The attention system of the human brain. Annu. Rev. Neurosci. **13**, 25–39 (1990)
35. Tsotsos, J.: Motion understanding: task-directed attention and representations that link perception with action. Int. J. Comput. Vision **45**(3), 265–280 (2001)
36. Itti, L., Koch, C., Niebur, E.: Computational modeling of visual attention. Nat. Rev. Neurosci. 194–203 (2001)

37. Torralba, A., Oliva, A., Castelhano, M., Henderson, M.: Contextual guidance of eye movements and attention in real-world scenes: the role of global features in object search. Psychol. Rev. **113**(4), 766–786 (2006)
38. Treisman, M., Gelade, G.: A feature-integration theory of attention. Cogn. Psychol. **12**(1), 97–136 (1980)
39. Wang, Y.: On cognitive informatics. In: Proceedings of International Conference on Cognitive Informatics, pp. 34–42 (2002)
40. Foster, J.: The Nature of Perception. Oxford University Press (2000)
41. Chater, N., Oaksford, M.: The Probabilistic Mind: Prospects for Bayesian Cognitive Science. Oxford University Press, New York (2008)
42. Kersten, D., Yuille, A.: Bayesian models of object perception. Curr. Opin. Neurobiol. **13**, 1–9 (2003)
43. Chater, N., Tenenbaum, J., Yuille, A.: Probabilistic models of cognition: conceptual foundations. Trends Cogn. Sci. **10**, 287–291 (2006)
44. Ciftcioglu, Ö., Bittermann, M., Sariyildiz, I.: Towards computer-based perception by modeling visual perception: a probabilistic theory. In: Proceedings of IEEE International Conference on Systems, Man and CyberneticsTaipei, Taiwan, pp. 81–89 (2006)
45. Andersen, S., Tiippana, K., Lampinen, J., Sams, M.: Bayesian modeling of audiovisual speech perception in noise. In: Audiovisual Speech Perception Conference Proceedings, 34–41 (2001)
46. Bekiarski, A.: Visual mobile robots perception in motion control. In: Kountchev, R., Nakamatzu, K. (eds.) Advanced in Reasoning-Based Image Processing Intelligent Systems. Springer (2014)
47. Tarampi, M., Geuss, M., Stefanucci, J., Creem-Regehr, S.: A Preliminary Study on the Role of Movement Imagery in Spatial Perception. Spatial Cognition IX Springer International Publishing, pp. 383–395 (2014)
48. Creem-Regehr, S., Gagnon, S., Geuss, K., Stefanucci, M.: Relating spatial perspective taking to the perception of other's affordances: providing a foundation for predicting the future behavior of others. Front. Human Neurosci. **7**, 596–598 (2013)
49. Dehkharghani, Sh.: Development of methods and algorithms for audio-visual mobile robot motion control. Ph.D. thesis, conducted at The French Language Faculty of Electrical Engineering, Technical University of Sofia, Bulgaria (2014)
50. Laser scanning range finder (SOKUIKI sensor): URG-04LX-UG01 Hokuyo Corporation. http://www.hokuyo-aut.jp/02sensor/07scanner/urg_04lx_ug01.html
51. Steval-MKI126v2. Datasheet. http://www.st.com
52. MP34DB01. Datasheet. http://www.st.com
53. Surveyor SRV-1 Blackfin Camera. http://www.surveyor.com/blackfin/
54. OV9655 Camera Board. Datasheet. http://www.surveyor.com/blackfin/
55. Surveyor SRV-1 Blackfin Robot. http://www.surveyor.com/SRV_info.html
56. Frese, U.: An algorithm for simultaneous localization and mapping. In: Freksa, C. (ed.) Spatial Cognition IV. Springer, pp. 455–476 (2005)
57. Sariff, N., Buniyamin, N.: An overview of autonomous mobile robot path planning algorithms. In: Proceedings of 4th Student Conference on Research and Development (SCORED 2006), Shah Alam, Malaysia, pp. 183–188 (2006)
58. Dehkharghani, Sh., Bekiarski, Al., Pleshkova, Sn.: Application of probabilistic methods in mobile robots audio visual motion control combined with laser range finder distance measurements. In: Advances in Circuits, Systems, Automation and Mechanics, pp. 91–98 (2012)
59. Dehkharghani, Sh., Bekiarski, Al., Pleshkova, Sn.: Method and algorithm for precise estimation of joined audio visual robot control. In: Iran's Third International Conference on Industrial Automation, Tehran, 22–23 Jan 2013, pp. 143–149 (2013)

60. Venkov, P., Bekiarski, Al., Dehkharghani, Sh., Pleshkova, Sn.: Search and tracking of targets with mobile robot by using audio-visual information. In: Proceedings of the International Conference on Automation and Informatics (CAI'10), Sofia, pp. 63–469 (2010)
61. Bekiarski, Al., Pleshkova, Sn.: Microphone array beamforming for mobile robot. In: Proceeding CSECS'09 Proceedings of the 8th WSEAS International Conference on Circuits, Systems, Electronics, Control and Signal Processing, pp. 146–149 (2009)
62. Dehkharghani, Sh, Pleshkova, S.: Geometric thermal infrared camera calibration for target tracking by a mobile robot. Comptes rendus de l'Academie bulgare des Sciences **67**(1), 109–114 (2014)

Chapter 9
Local Adaptive Image Processing

Rumen Mironov

Abstract Three methods for two-dimensional local adaptive image processing are presented in this chapter. In the first one, the adaptation is based on the local information from the four neighborhood pixels of the processed image and the interpolation type is changed to zero or bilinear. An analysis of local characteristics of images in small areas is presented from which the optimal selection of thresholds for dividing into homogeneous and contour blocks is made and the interpolation type is changed adaptively. In the second one, the adaptive image halftoning is based on the generalized two-dimensional LMS error-diffusion filter for image quantization. The thresholds for comparing of input image levels are calculated from the gray values dividing the normalized histogram of the input halftone image into equal parts. The third one—the adaptive line prediction is based on two-dimensional LMS adaptation of coefficients of the linear prediction filter for image coding. An analysis of properties of 2D LMS filters in different directions was made. As a result of the performed mathematical description in the presented methods, three algorithms for local adaptive image processing was developed. The principal block schemes of the developed algorithms are presented. An evaluation of the quality of the processed images was made on the base of the calculated PSNR, SNR, MSE and the subjective observation. The given experimental results from the simulation in MATLAB environment for each of the developed algorithms, suggest that the effective use of local information contributes to minimize the processing error. The methods are extremely suitable for different types of images (for example: fingerprints, contour images, cartoons, medical signals, etc.). The developed algorithms have low computational complexity and are suitable for real-time applications.

Keywords Image interpolation · Local adaptation · Image processing · Image quantization · Error diffusion · Adaptive filtration · Image halftoning · Linear prediction · 2D LMS algorithm

R. Mironov (✉)
Department of Radio Communications and Video Technologies,
Technical University of Sofia, Boul. Kl. Ohridsky 8, 1000 Sofia, Bulgaria
e-mail: rmironov@tu-sofia.bg

© Springer International Publishing Switzerland 2016
R. Kountchev and K. Nakamatsu (eds.), *New Approaches in Intelligent Image Analysis*, Intelligent Systems Reference Library 108,
DOI 10.1007/978-3-319-32192-9_9

9.1   Introduction

Nowadays the medical industry, astronomy, physics, chemistry, forensics, remote monitoring, industrial and agricultural production, trade and defense are just some of the many areas that rely on digital images to store, reproduce and to provide information about the surrounding world. Systems for digital image processing are becoming more popular due to the easy accessibility to powerful personal computers, increased memory size of the various devices, graphics software, etc. Because of the need to continuously improve the quality of use in everyday life video-information, you need to develop new methods and algorithms for image processing, to meet the new, increased demands from consumers. Best results would be obtained with the use of locally-adaptive image processing.

Three methods for two-dimensional local adaptive image processing are presented in this chapter: 2D image interpolation, 2D image halftoning and 2D line prediction of images.

The basic methods for 2D image interpolation are separated in two groups: non-adaptive (zero, bilinear or cubic interpolation) [1–5] and adaptive interpolation [6–17]. A specific characteristic for the non-adaptive methods is that when the interpolation order increases, the brightness transitions sharpness decreases. On the other side, in result of the interpolation order decreasing, artefacts ("false" contours) in the homogeneous areas start to appear. To reduce them more sophisticated adaptive image interpolation methods were proposed in the recent years [13–17], etc. These methods are based on edge patterns prediction in the local area (minimum 4×4) and on adaptive high-order (bicubic or spline) interpolation with contour filtration. The main insufficiency of these methods is that the analysis is very complicated and the image processing requires too much time.

The linear filtering is related to the common methods for image processing and is separated into the two basic types—non-adaptive and adaptive [18, 19]. In the first group the filter parameters are obtained by the principles of the optimal (Winner) filtering, which minimizes the mean square error of signal transform and assumes the presence of the priory information for image statistical model. The model inaccuracy and the calculation complexity required for their description might be avoided by adaptive estimation of image parameters and by iteration minimization of the mean-square error of the transform.

Depending on the processing method, the adaptation is divided into global and local. The global adaptation algorithms refer mainly to the basic characteristics of the images, while the local ones are connected to adaptation in each pixel of the processed image based on the selected pixel neighborhood.

The coefficients of the filters, used by the other two local adaptive image processing methods (for image halftoning and linear prediction), are adapted with the help of generalized two-dimensional Least Mean Square (LMS) algorithm [20–24].

This chapter is arranged as follows: Sect. 9.2 introduces the mathematical description of the new adaptive 2D interpolation; Sect. 9.3 introduces the mathematical description of the new adaptive 2D error-diffusion filter; Sect. 9.4 introduces the mathematical description of the new adaptive 2D line prediction filter; Sect. 9.5 gives some experimental results and Sect. 9.6 concludes this chapter.

9.2   Method for Local Adaptive Image Interpolation

9.2.1   Mathematical Description of Adaptive 2D Interpolation

The input halftone image of size M × N with m-brightness levels and the interpolated output image of size pM × qN can be presented as follows:

$$\begin{aligned} \mathbf{A}_{M \times N} &= \left\{ \mathbf{a}(i, j) \, / \, i = \overline{0, M-1}; \; j = \overline{0, N-1} \right\}, \\ \mathbf{A}_{pM \times qN}^* &= \left\{ \mathbf{a}^*(k, l) \, / \, k = \overline{0, pM-1}; \; l = \overline{0, qN-1} \right\}, \end{aligned} \tag{9.1}$$

where: $a(i, j)$ and $a^*(k, l)$ are the current image elements in input and output images respectively; q and p are the interpolation's coefficients in horizontal and vertical direction [7, 8, 9, 25, 26].

The differences between any two adjacent elements of the image in a local neighborhood of size 2 × 2, as shown on Fig. 9.1, can be described by the expressions:

$$\begin{aligned} \Delta_{2m+1} &= |a(i+m, j) - a(i+m, j+1)|, \text{ for } m = 0, 1; \\ \Delta_{2n+2} &= |a(i, j+n) - a(i+1, j+n)|, \text{ for } n = 0, 1. \end{aligned} \tag{9.2}$$

These image elements are used as supporting statements in image interpolation.

Here are introduced four logic variables f_1, f_2, f_3 and f_4, which depend on the values of the differences of the thresholds for horizontal θ_m and vertical θ_n direction in accordance with the expressions:

$$f_{2m+1} = \begin{cases} 1, \text{ if } : \Delta_{2m+1} \geq \theta_m \\ 0, \text{ if } : \Delta_{2m+1} < \theta_m \end{cases}; \quad f_{2n+2} = \begin{cases} 1, \text{ if } : \Delta_{2n+2} \geq \theta_n \\ 0, \text{ if } : \Delta_{2n+2} < \theta_n \end{cases}. \tag{9.3}$$

Fig. 9.1 Structure of the supporting image elements

a(i,j)	a(i,j+1)
a(i+1,j)	a(i+1,j+1)

Then each element of the interpolated image can be represented as a linear combination of the four supporting elements from the original image:

$$a^*(k, l) = \sum_{m=0}^{1} \sum_{n=0}^{1} w_{m,n}(r, t) a(i + m, j + n),\tag{9.4}$$

for $r = \overline{0, p}$; $t = \overline{0, q}$. The interpolation coefficients:

$$w_{m,n}(r, t) = F \cdot ZR_{m,n}(r, t) + \overline{F} \cdot BL_{m,n}(r, t),\tag{9.5}$$

depend on the difference of the logical function F, which specifies the type of interpolation (zero or bilinear): $F = f_1 f_3 \cup f_2 f_4$. The coefficients of the zero (ZR) and the bilinear (BL) interpolation are determined by the following relations:

$$ZR_{m,n}(r, t) = \frac{1}{4}[1 - (-1)^m \text{sign}(2r - p)][1 - (-1)^n \text{sign}(2t - q)]\tag{9.6}$$

$$BL_{m,n}(r, t) = (-1)^{m+n}\left[1 - m - \frac{r}{p}\right]\left[1 - n - \frac{t}{q}\right]\tag{9.7}$$

The dependence of the function F upon the variables f_1, f_2, f_3 and f_4, defining the type of luminance transition in a local window with size 2×2, is shown in Table 9.1. In the image for homogeneous areas (F = 0) the bilinear interpolation is used, and in the non homogeneous areas (F = 1)—the zero interpolation is used.

9.2.2 Analysis of the Characteristics of the Filter for Two-Dimensional Adaptive Interpolation

The two-dimensional interpolation process can be characterized by the following generalized block diagram shown in Fig. 9.2:

In the unit for the secondary sampling, the frequencies f_{sr} and f_{st} were increased p and q times in vertical and horizontal direction. Accordingly, the elements a(i, j) of the input image are complemented with zeros to obtain the elements b(k, l) using the following expression:

$$b(k, l) = \begin{cases} a(k/p, l/q), & \text{for } k = \overline{0, \pm(M-1)p}, \ l = \overline{0, \pm(N-1)q}, \\ 0, & \text{otherwise.} \end{cases}\tag{9.8}$$

The resulting image is processed by a two-dimensional digital filter with transfer function $H(z_k, z_l)$ and the resulting output are the elements a*(k, l) of the interpolated image. In this case the expression (9.4) can be presented as follows:

Table 9.1 The dependence of function F of the variables f_1, f_2, f_3 and f_4

№	f_1	f_2	f_3	f_4	F	Transitions	№	f_1	f_2	f_3	f_4	F	Transitions
0	0	0	0	0	0		8	1	0	0	0	0	
1	0	0	0	1	0		9	1	0	0	1	0	
2	0	0	1	0	0		A	1	0	1	0	1	
3	0	0	1	1	0		B	1	0	1	1	1	
4	0	1	0	0	0		C	1	1	0	0	0	
5	0	1	0	1	1		D	1	1	0	1	1	
6	0	1	1	0	0		E	1	1	1	0	1	
7	0	1	1	1	1		F	1	1	1	1	1	

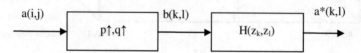

Fig. 9.2 Generalized block diagram of the 2D interpolator

$$a^*(k, l) = \sum_{m=0}^{1}\sum_{n=0}^{1} w_{m,n}(r, t) \cdot b\left(\left\lfloor\frac{k}{p}\right\rfloor + pm, \left\lfloor\frac{l}{q}\right\rfloor + qn\right), \quad (9.9)$$

where with the operation $\lfloor x \rfloor$ indicates the greatest integer not exceeding x.

Since the interpolation coefficients are repeated periodically, the analysis can be performed on one block of the image, as shown in Fig. 9.3. With the red line are marked the values for the output image elements by the bilinear interpolation and with the green—the corresponding values by zero interpolation. With black arrows are marked the four supporting image elements in the input image.

Then the relationship between image elements from the input block and output block can be represented as follows:

$$y(r, t) = \sum_{m=0}^{1}\sum_{n=0}^{1} w_{m,n}(r, t) \cdot x(pm, qn), \quad (9.10)$$

where $x(pm, qn)$ are the supporting elements in the current block $b(k, l)$ and $y(r, t)$ are the interpolated elements from the output block $a^*(k, l)$.

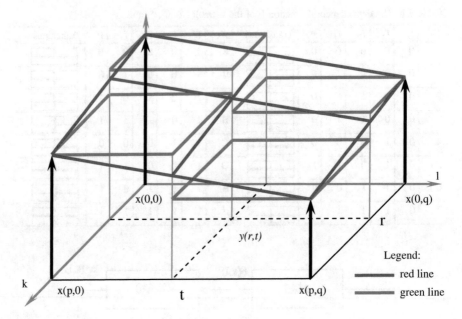

Fig. 9.3 2D interpolation scheme for one block of the image

9.2.2.1 Characteristics of the Bilinear Interpolation

From Eqs. (9.7) and (9.10) can be derived the following recurrent relation for bilinear interpolation of four neighboring image elements:

$$y(r, t) = \sum_{n=0}^{1} (-1)^n \left(1 - n - \frac{t}{q}\right) v(r, nq), \qquad (9.11)$$

where with $v(r, nq)$ is indicated the interpolation in vertical direction, which can similarly be represented as follows:

$$v(r, nq) = v(r - 1, nq) - \frac{1}{p} \sum_{m=0}^{1} (-1)^m x(mp, nq). \qquad (9.12)$$

Finally from the expression (9.11) we can derive:

$$y(r, t) = y(r, t - 1) - \frac{1}{q} \sum_{n=0}^{1} (-1)^n v(r, nq). \qquad (9.13)$$

Particular cases are when: $v(0, 0) = x(0, 0)$, $v(0, q) = x(0, q)$, $y(r, 0) = v(r, 0)$.

After completing the two-dimensional $\mathbf{Z}$ transformation of expressions (9.11) and (9.12), we obtain the final $\mathbf{Z}$ image of the expression (9.9) between the input and output statements in one block of the interpolated picture for bilinear interpolation:

$$Y(z_r, z_t) = \frac{1}{pq} X(z_r, z_t) \left[\frac{(1 - z_r^{-p})}{(1 - z_r^{-1})} \right]^2 \left[\frac{(1 - z_t^{-q})}{(1 - z_t^{-1})} \right]^2 . \tag{9.14}$$

Finally for the transfer function for the two-dimensional bilinear interpolation we get:

$$H_{BL}(z_r, z_t) = \frac{Y(z_r, z_t)}{X(z_r, z_t)} = \frac{1}{pq} \left[\frac{(1 - z_r^{-p})}{(1 - z_r^{-1})} \right]^2 \left[\frac{(1 - z_t^{-q})}{(1 - z_t^{-1})} \right]^2 \tag{9.15}$$

From the expression (9.15) for the amplitude-frequency response of the bilinear interpolator we obtain:

$$M_{BL}(\omega_r, \omega_t) = \frac{\sin^2 \left(\frac{p\omega_r}{2} \right) \sin^2 \left(\frac{q\omega_t}{2} \right)}{p.\sin^2 \left(\frac{\omega_r}{2} \right) q.\sin^2 \left(\frac{\omega_t}{2} \right)} = M_{BL}(\omega_r) . M_{BL}(\omega_t), \tag{9.16}$$

where: $\omega_r = 2\pi f_r / f_{sr}$ and $\omega_t = 2\pi f_t / f_{st}$ are the normalized circular frequencies respectively in vertical and horizontal direction.

Similarly, for the phase-frequency response we obtain:

$$\Phi_{BL}(\omega_r, \omega_t) = (1 - p)\omega_r + (1 - q)\omega_t = \Phi_{BL}(\omega_r) + \Phi_{BL}(\omega_t). \tag{9.17}$$

9.2.2.2 Characteristics of the Zero Interpolation

Similarly to the previous expression for bilinear interpolator from Eqs. (9.6) and (9.10) the transfer characteristic in the Z domain at zero interpolation of four neighboring supporting image elements can be derived:

$$H_{ZR}(z_r, z_t) = \frac{Y(z_r, z_t)}{X(z_r, z_t)} = \left[\frac{1 - z_r^{-p}}{1 - z_r^{-1}} \right] \left[\frac{1 - z_t^{-q}}{1 - z_t^{-1}} \right] . \tag{9.18}$$

After substitution of $z_r = \exp(j\omega_r)$ and $z_t = \exp(j\omega_t)$ in (9.18) the following relations are obtained for the amplitude-frequency response:

$$M_{ZR}(\omega_r, \omega_t) = \frac{\sin \left(\frac{p\omega_r}{2} \right) \sin \left(\frac{q\omega_t}{2} \right)}{\sin \left(\frac{\omega_r}{2} \right) \sin \left(\frac{\omega_t}{2} \right)} = M_{ZR}(\omega_r) M_{ZR}(\omega_t) \tag{9.19}$$

and phase-frequency response of the zero interpolator:

$$\Phi_{ZR}(\omega_r, \omega_t) = \frac{p-1}{2}\omega_r + \frac{q-1}{2}\omega_t = \Phi_{ZR}(\omega_r) + \Phi_{ZR}(\omega_t). \tag{9.20}$$

9.2.3 Evaluation of the Error of the Adaptive 2D Interpolation

In assessing the error of the interpolation we assume that the input image has a uniform spectrum in the intervals: $-\pi \le \omega_r \le \pi$ and $-\pi \le \omega_t \le \pi$. As a main criterion for evaluating the distortion, the definition of mean square error (MSE) described in [23] can be used:

$$\overline{\varepsilon^2} = \frac{1}{\pi^2}\int_0^\pi \int_0^\pi [1 - M_n(\omega_r, \omega_t)]^2 \partial\omega_r \partial\omega_t, \tag{9.21}$$

where:

$$M_n(\omega_r, \omega_t) = \frac{M(\omega_r, \omega_t)}{M(0,0)} = \frac{1}{pq}M(\omega_r, \omega_t), \tag{9.22}$$

is the normalized amplitude frequency characteristic of the adaptive interpolator and:

$$M(\omega_r, \omega_t) = \begin{cases} M_{BL}(\omega_r, \omega_t), & \text{for } F = 0; \\ M_{ZR}(\omega_r, \omega_t), & \text{for } F = 1. \end{cases} \tag{9.23}$$

are the expressions for the characteristics in bilinear and zero interpolation. Since amplitude frequency characteristics are separable functions on both directions, the equality (9.21) is transformed as follows:

$$\overline{\varepsilon^2} = 1 - \frac{2}{\pi^2}\int_0^\pi M_n(\omega_r)\,\partial\omega_r \int_0^\pi M_n(\omega_t)\,\partial\omega_t + \frac{1}{\pi^2}\int_0^\pi [M_n(\omega_r)]^2\partial\omega_r \int_0^\pi [M_n(\omega_t)]^2\partial\omega_t. \tag{9.24}$$

The examination of the integrals of Eq. (9.24) can be performed separately for two cases of interpolation.

- *First case*. The zero interpolation F = 1.

$$M_n(\omega_r, \omega_t) = \frac{M_{ZR}(\omega_r, \omega_t)}{M_{ZR}(0,0)} = \frac{1}{pq} M_{ZR}(\omega_r, \omega_t), \qquad (9.25)$$

as the amplitude frequency characteristic in the point (0, 0) is expressed by the following border transition:

$$M_{ZR}(0,0) = \lim_{\omega_r \to 0} M_{ZR}(\omega_r) \cdot \lim_{\omega_t \to 0} M_{ZR}(\omega_t) = pq. \qquad (9.26)$$

From the expression (9.24) for the separate integrals for zero interpolation, the following relations can the given:

$$\int_0^\pi M_n(\omega_r)\,\partial\omega_r = \frac{1}{p}\int_0^\pi \frac{\sin(p\omega_r/2)}{\sin(\omega_r/2)}\,\partial\omega_r = \frac{2}{p}\int_0^{\pi/2} \frac{\sin(p\omega_r)}{\sin(\omega_r)}\,\partial\omega_r = \frac{2}{p}\Im(p),$$

$$\int_0^\pi M_n(\omega_t)\,\partial\omega_t = \frac{1}{q}\int_0^\pi \frac{\sin(q\omega_t/2)}{\sin(\omega_t/2)}\,\partial\omega_t = \frac{2}{q}\int_0^{\pi/2} \frac{\sin(q\omega_t)}{\sin(\omega_t)}\,\partial\omega_t = \frac{2}{q}\Im(q),$$

$$\qquad (9.27)$$

where $\Im(p)$ and $\Im(q)$ are integrals of the kind:

$$\Im(u) = \Im(u - 2) + \frac{2}{u-1}\sin[(u-1)v]\big|_0^{\pi/2} = \Im(u-2) + \frac{2\sin[(u-1)\pi/2]}{u-1}.$$

$$\qquad (9.28)$$

After transforming the definite integral $\Im(u)$ can be calculated from the condition:

$$\Im(u) = \begin{cases} \frac{\pi}{2}, & \text{for } u = 2k+1 \\ 2\sum_{s=1}^{k} \frac{(-1)^{s-1}}{2S-1}, & \text{for } u = 2k \end{cases}. \qquad (9.29)$$

For the second group of integrals we obtain the expressions:

$$\int_0^\pi [M_n(\omega_r)]^2\partial\omega_r = \frac{1}{p^2}\int_0^\pi \frac{\sin^2(p\omega_r/2)}{\sin^2(\omega_r/2)}\,\partial\omega_r = \frac{2}{p^2}\int_0^{\pi/2} \frac{\sin^2(p\omega_r)}{\sin^2(\omega_r)}\,\partial\omega_r = \frac{2\Im_2(p)}{p^2}$$

$$\int_0^\pi [M_n(\omega_t)]^2\partial\omega_t = \frac{1}{q^2}\int_0^\pi \frac{\sin^2(q\omega_t/2)}{\sin^2(\omega_t/2)}\,\partial\omega_t = \frac{2}{q^2}\int_0^{\pi/2} \frac{\sin^2(q\omega_t)}{\sin^2(\omega_t)}\,\partial\omega_t = \frac{2\Im_2(q)}{q^2}$$

$$\qquad (9.30)$$

where $\Im_2(p)$ and $\Im_2(q)$ are integrals of the kind:

$$\Im_2(u) = \frac{u}{2}[\Im(2u+1) + \Im(2u-1)] = \frac{u\pi}{2}. \tag{9.31}$$

Finally from the transformations (9.27), (9.28), (9.30) and (9.31) it follows that in expression (9.24) the error of zero interpolation is as follows:

$$\overline{\varepsilon^2} = 1 - \frac{2}{\pi^2}\frac{4}{pq}\Im(p)\Im(q) + \frac{1}{\pi^2}\frac{4}{p^2q^2}\Im_2(p)\Im_2(q) = 1 - \frac{1}{pq}\left[\frac{8\Im(p)\Im(q)}{\pi^2} - 1\right]. \tag{9.32}$$

• *Second case.* Bilinear interpolation at F = 0.

$$M_n(\omega_r, \omega_t) = \frac{M_{BL}(\omega_r, \omega_t)}{M_{BL}(0,0)} = \frac{1}{pq}M_{BL}(\omega_r, \omega_t), \tag{9.33}$$

as the amplitude frequency characteristic in the point (0, 0) is expressed with the following border transition:

$$M_{BL}(0,0) = \lim_{\omega_r \to 0} M_{BL}(\omega_r) \cdot \lim_{\omega_t \to 0} M_{BL}(\omega_t) = pq. \tag{9.34}$$

From the expression (9.24) for the individual integrals for bilinear interpolation, the following relations can the described:

$$\int_0^\pi M_n(\omega_r)\,\partial\omega_r = \frac{1}{p^2}\int_0^\pi \frac{\sin^2(p\omega_r/2)}{\sin^2(\omega_r/2)}\,\partial\omega_r = \frac{2}{p^2}\int_0^{\pi/2}\frac{\sin^2(p\omega_r)}{\sin^2(\omega_r)}\,\partial\omega_r = \frac{2\Im_2(p)}{p^2}$$

$$\int_0^\pi M_n(\omega_t)\,\partial\omega_t = \frac{1}{q^2}\int_0^\pi \frac{\sin^2(q\omega_t/2)}{\sin^2(\omega_t/2)}\,\partial\omega_t = \frac{2}{q^2}\int_0^{\pi/2}\frac{\sin^2(q\omega_t)}{\sin^2(\omega_t)}\,\partial\omega_t = \frac{2\Im_2(q)}{q^2}$$

$$\tag{9.35}$$

where $\Im_2(p)$ and $\Im_2(q)$ are calculated in accordance to the expression (9.31).

For the second group of integrals we can derive the following expressions:

$$\int_0^\pi [M_n(\omega_r)]^2\partial\omega_r = \frac{1}{p^4}\int_0^\pi \frac{\sin^4(p\omega_r/2)}{\sin^4(\omega_r/2)}\,\partial\omega_r = \frac{2}{p^4}\int_0^{\pi/2}\frac{\sin^4(p\omega_r)}{\sin^4(\omega_r)}\,\partial\omega_r = \frac{2\Im_4(p)}{p^4}$$

$$\int_0^\pi [M_n(\omega_t)]^2\partial\omega_t = \frac{1}{q^4}\int_0^\pi \frac{\sin^4(q\omega_t/2)}{\sin^4(\omega_t/2)}\,\partial\omega_t = \frac{2}{q^4}\int_0^{\pi/2}\frac{\sin^4(q\omega_t)}{\sin^4(\omega_t)}\,\partial\omega_t = \frac{2\Im_4(q)}{q^4}$$

$$\tag{9.36}$$

where with $\Im_4(p)$ and $\Im_4(q)$ are described integrals of the kind:

$$\Im_4(u) = \int_0^{\pi/2} \frac{\sin^4(uv)}{\sin^4(v)} \partial v = \frac{u\pi}{6}\left[2u^2 + 1\right]. \tag{9.37}$$

Finally from the transformations (9.34), (9.35), (9.36) and (9.37) follows that in expression (9.24) the error at zero interpolation is as follows:

$$\overline{\varepsilon^2} = 1 - \frac{1}{pq}\left[2 - \frac{(2p^2 + 1)(2q^2 + 1)}{9p^2q^2}\right]. \tag{9.38}$$

In conclusion, from expressions (9.24), (9.32) and (9.38) is obtained the expression for MSE at the adaptive interpolator:

$$\overline{\varepsilon^2}(p,q) = \begin{cases} 1 - \frac{1}{pq}\left[\frac{8\Im(p)\Im(q)}{\pi^2} - 1\right], & \text{for } F = 1, \\ 1 - \frac{1}{pq}\left[2 - \frac{(2p^2 + 1)(2q^2 + 1)}{9p^2q^2}\right], & \text{for } F = 0. \end{cases} \tag{9.39}$$

9.2.4 Functional Scheme of the 2D Adaptive Interpolator

As a result of the discussion in paragraph 3 and from Eqs. (9.5), (9.6), (9.7), (9.10) and (9.13) can be synthesized the functional scheme of two-dimensional adaptive interpolator, developed for grayscale images, shown on Fig. 9.4.

The used modules in the scheme are as follows:

- MBI—module for bilinear interpolation;
- MZI—module for zero interpolation;
- MCA—module for adaptive control;
- AC—accumulator;
- LUT—lookup table;
- COM—digital comparer;
- z^{-k}—delay block for κ elements;
- $|\ .\ |$—block to determine the module of the current value;
- MUX—multiplexor;
- $\Delta_1, \Delta_2, \Delta_3, \Delta_4$—are the differences, calculated on the base of Eqs. (9.2).

The table LUT-1 was synthesized on the basis of Eq. (9.10) and the table LUT-2 is based on the equality for the logical function F. In the accumulators AC-1, AC-2 and AC-3 for starting values we introduce $x(0, 0)$, $x(0, q)$ and $v(r, 0)$ respectively. The values of the thresholds θ_m and θ_n are calculated based on descriptions, given in [11], depending on the level of noise in the input image and the type of difference histograms in vertical and horizontal direction.

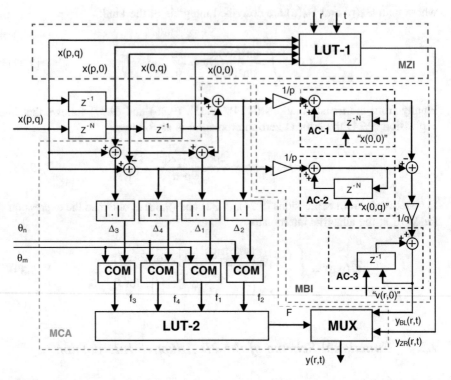

Fig. 9.4 Functional scheme of the 2D adaptive interpolator

For obtaining a high quality color image a divided processing of the signals for brightness and color is necessary. The proposed scheme applies only to the luminance signal processing. For color signals the use of adaptation is not necessary as only zero interpolation is sufficient.

9.3   Method for Adaptive 2D Error Diffusion Halftoning

9.3.1   *Mathematical Description of Adaptive 2D Error-Diffusion*

The input m-level halftone image and the output n-level ($2 \leq n \leq m/2$) image of dimensions M × N can be represented by the matrices:

$$
\begin{aligned}
C &= \left\{ \mathbf{c}(k, l) \,/\, k = \overline{0, M-1}; \ l = \overline{0, N-1} \right\}, \\
D &= \left\{ \mathbf{d}(k, l) \,/\, k = \overline{0, M-1}; \ l = \overline{0, N-1} \right\}.
\end{aligned}
\tag{9.40}
$$

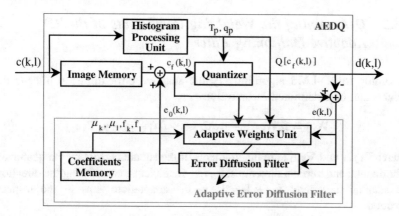

Fig. 9.5 Adaptive 2D error diffusion quantizer

Transformation of the image elements $\mathbf{c}(k, l)$ into $\mathbf{d}(k, l)$ is accomplished by the adaptive error diffusion quantizer (AEDQ) shown on Fig. 9.5.

The quantizer operation is described by the following equation:

$$\mathbf{d}(k, l) = Q[\mathbf{c}_f(k, l)] = \begin{cases} q_0, \text{ if } \mathbf{c}_f(k, l) < T_0 \\ q_p, \text{ if } T_{p-1} < \mathbf{c}_f(k, l) < T_p \ (p = \overline{1, n-2}) \\ q_{n-1}, \text{ if } \mathbf{c}_f(k, l) < T_0 \end{cases} \quad (9.41)$$

where $q_p \leq q_{p+1} \leq m \ (p = \overline{0, n-2})$ are the values of the function $Q[.]$.

Thresholds for comparison are calculated by the equation: $T_p = (C_p + C_{p+1})/2$, where C_p represents the gray values dividing the normalized histogram of the input halftone image $\mathbf{C}$ into n equal parts.

The value of the filtered element $\mathbf{c}_f(k, l)$ in Eq. (9.41) is:

$$\mathbf{c}_f(k, l) = \mathbf{c}(k, l) + \mathbf{e}_0(k, l) \quad (9.42)$$

The summarized error can be expressed as:

$$e_0(k, l) = \sum_{(r,t) \, \in \, W} \sum w_{k,l}(r, t) e(k - r, l - t) = \mathbf{W}_{k,l}^t \mathbf{E}_{k,l} \quad (9.43)$$

where $e(k, l) = c_f(k, l) - d(k, l)$ is the error of the current filtered element when its value is substituted by q_p; $w_{k,l}(r, t)$ are the filter weights defined in the certain causal two-dimensional window W; $\mathbf{W}_{k,l}$ and $\mathbf{E}_{k,l}$ are the vectors of the weights and their summarized errors, respectively.

9.3.2 Determining the Weighting Coefficients of the 2D Adaptive Halftoning Filter

According to 2D-LMS algorithm [7], the adaptive error diffusion filter (AEDF) weights can be determined recursively:

$$\mathbf{W}_{k,l} = f_k \mathbf{W}_{k,l-1} - \mu_k \nabla_{k,l-1} + f_l \mathbf{W}_{k-1,l} - \mu_l \nabla_{k-1,l} \qquad (9.44)$$

where: $\nabla_{k,l-1}$ and $\nabla_{k-1,l}$ are the gradients of the squared errors by the quantization in horizontal and vertical directions; f_k, f_l—coefficients, considering the direction of the adaptation, where: $f_k + f_l = 1$; μ_k, μ_l—adaptation steps in the respective direction.

According to [5], the convergence and the stability of the AEDF adaptation process are given by the following condition:

$$|f_k - \mu_k \lambda_i| + |f_l - \mu_l \lambda_i| < 1, \qquad (9.45)$$

where λ_i are the eigen values of the gray-tone image covariance matrix.

The sequence (9.44) is 2D LMS algorithm of Widrow summary from which the following two particular cases should hold:

- *First case.* If $f_k = 1, \mu_k = \mu$, $f_l = \mu_l = 0$, then the adaptive calculation of the weights is performed only in the horizontal direction:

$$\mathbf{W}_{k,l} = \mathbf{W}_{k,l-1} + \mu_k(-\nabla_{k,l-1}) = \mathbf{W}_{k,l-1} - \mu \frac{\partial \mathbf{e}^2(k,l-1)}{\partial \mathbf{W}_{k,l-1}} \qquad (9.46)$$

- *Second case.* If $f_l = 1, \mu_l = \mu$, $f_k = \mu_k = 0$ then the adaptive calculation is performed only in the vertical direction:

$$\mathbf{W}_{k,l} = \mathbf{W}_{k-1,l} + \mu_l(-\nabla_{k-1,l}) = \mathbf{W}_{k-1,l} - \mu \frac{\partial \mathbf{e}^2(k-1,l)}{\partial \mathbf{W}_{k-1,l}}. \qquad (9.47)$$

The derivatives of the quantization error in the respective directions are determined by the Eqs. (9.40), (9.41), (9.42), (9.43) and (9.44). The derivative in horizontal direction is obtained:

$$\frac{\partial \mathbf{e}^2(k, 1-1)}{\partial \mathbf{W}_{k,l-1}} = 2\mathbf{e}(k, 1-1)\mathbf{E}_{k,l-1}\left[1 - Q'_{c_f}(k, 1-1)\right], \qquad (9.48)$$

where the derivative of the quantization values of the filtered elements is:

$$Q'_{c_f}(k, 1-1) = \begin{cases} 0, & \text{if:} \quad \mathbf{c}_f(k, 1-1) \neq T_p \\ q_{p+1} - q_p, & \text{if:} \quad \mathbf{c}_f(k, 1-1) = T_p. \end{cases} \tag{9.49}$$

In the same way the derivative in the vertical direction from Eq. (9.47) is obtained from:

$$\frac{\partial \mathbf{e}^2(k-1,1)}{\partial \mathbf{W}_{k-1,1}} = 2\mathbf{e}(k-1,1) \, \mathbf{E}_{k-1,1} \left[1 - Q'_{c_f}(k-1,1)\right] \tag{9.50}$$

For the AIHF weights the condition must be hold:

$$\sum_{(r,t) \in W} \sum \mathbf{w}_{k,l}(r,t) = 1 \tag{9.51}$$

which guarantees that e(k, l) is not increased or decreased by its passing through the error filter.

Finally, from the Eqs. (9.44) and (9.46)–(9.50) the calculation of the weighs of the adaptive filter of $W_{k,l}$ is performed by the equation:

$$\mathbf{w}_{k,l}(r, t) = f_k \mathbf{w}_{k,l-1}(r, t) - 2\mu_k e(k, 1-1)e(k-r, 1-t-1)\left[1 - Q'_{c_f}(k, 1-1)\right]$$
$$+ f_l \mathbf{w}_{k-1,l}(r, t) - 2\mu_l e(k-1, l)e(k-r-1, 1-t)\left[1 - Q'_{c_f}(k-1, l)\right]. \tag{9.52}$$

9.3.3 Functional Scheme of 2D Adaptive Halftoning Filter

On the basis of the developed method for 2D adaptive error diffusion image halftoning the functional scheme of 2D adaptive halftoning filter is synthesized (Fig. 9.6).

An error diffusion filter with 4 coefficients has been used for the evaluation of the efficiency of the developed filter. The spatial disposition and the initial values of weights correspond to those in the Floyd-Steinberg filter [27].

The used units in the scheme on the Fig. 9.6 are as follows:

- LUT—lookup table for calculation of values of filtered image $c_f(k, l)$, according to the Eq. (9.42);
- COM—digital comparator, in which the current quantized elements are obtained according to the Eq. (9.49);
- $(N - 1)T$—delay unit for $N - 1$ elements of input image.

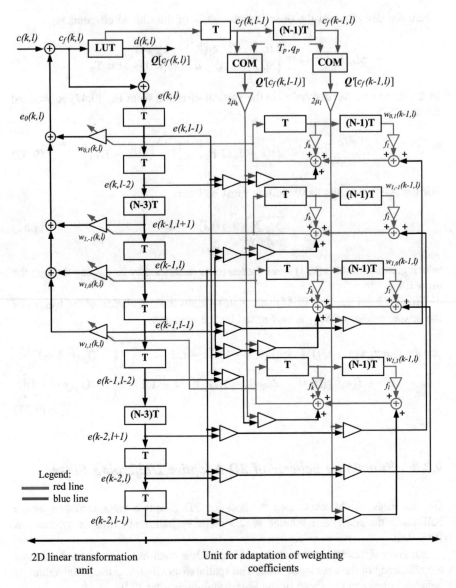

Fig. 9.6 Functional scheme of adaptive 2D error diffusion quantizer

With red color are denoted the units for iterative calculation of weights $w_{k,l}(r,t)$ of the adaptive halftoning filter, according to the Eq. (9.52). With blue color are denoted the units, for which the values of the coefficients, according to the 2D LMS algorithm are calculated.

The filter illustrated on Fig. 9.6 allows operation in four modes:

- first mode—adaptation of the coefficients in horizontal direction for $f_k = 1, f_l = 0$;
- second mode—adaptation of the coefficients in vertical direction for $f_k = 0, f_l = 1$;
- third mode—adaptation in horizontal and in vertical direction;
- fourth mode—non-adaptive for $f_k = 1, f_l = 0, \mu_k = \mu_l = 0$.

9.3.4 Analysis of the Characteristics of the 2D Adaptive Halftoning Filter

The analysis of the developed adaptive 2D error diffusion quantizer can be performed under the following assumptions—$f_k = 1, f_l = 0, \mu_k = \mu_l = 0$ and ignoring the effect of quantization. Then the proposed block scheme in Fig. 9.5 can be replaced by the equivalent scheme, shown on Fig. 9.7.

Here H_1 denotes the filter for calculation of the summarized error of transformation, INV denotes the unit for inverting the scanning direction, and the sign $\pm$ denotes the positive or negative value of the difference between the input and quantized image element. From the Eqs. (9.41)–(9.43) the filter transfer characteristic with four coefficients: $w_{k,l}(0,1) = w_1$, $w_{k,l}(1,0) = w_2$, $w_{k,l}(1,1) = w_3$ and $w_{k,l}(1,-1) = w_4$ can be obtained as follows:

$$H_1(z_k, z_l) = \frac{E_0(z_k, z_l)}{E(z_k, z_l)} = W_1 z_l^{-1} + W_2 z_k^{-1} + W_3 z_k^{-1} z_l^{-1} + W_4 z_k^{-1} z_l. \quad (9.53)$$

The expressions for the amplitude frequency response and phase frequency response of the filter for calculating of summarized error $e_0(k,l)$ are obtained from the Eq. (9.53) as follows:

$$M_1(\omega_k, \omega_l) = \sqrt{a^2 + b^2}, \quad \varphi_1(\omega_k, \omega_l) = -\text{arctg}\frac{b}{a}, \quad (9.54)$$

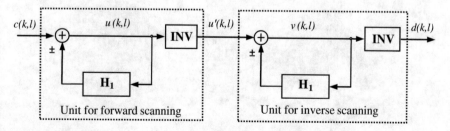

Fig. 9.7 Equivalent block scheme of adaptive 2D error diffusion quantizer

where the following assumptions are made:

$$a = W_1\cos(\omega_1) + W_2\cos(\omega_k) + W_3\cos(\omega_k + \omega_1) + W_4\cos(\omega_k - \omega_1)$$
$$b = W_1\sin(\omega_1) + W_2\sin(\omega_k) + W_3\sin(\omega_k + \omega_1) + W_4\sin(\omega_k - \omega_1).$$

$$(9.55)$$

The two-dimensional amplitude/phase frequency response characteristics are calculated in Matlab and shown on Fig. 9.8.

For equivalent error filter, from the scheme in Fig. 9.7, two transfer functions are obtained:

$$H_+(z_k, z_l) = \frac{1}{1 - H_1(z_k, z_l)}, \text{ for sign } "+" \text{ of } E(z_k, z_l)$$
$$H_-(z_k, z_l) = \frac{1}{1 - H_1(z_k, z_l)}, \text{ for sign } "-" \text{ of } E(z_k, z_l).$$

$$(9.56)$$

Then the amplitude and phase frequency response characteristics can be calculated:

$$M_+(\omega_k, \omega_1) = \frac{1}{\sqrt{(1-a)^2 + b^2}}, \quad \varphi_+(\omega_k, \omega_1) = \text{arctg}\frac{b}{1-a}$$
$$M_-(\omega_k, \omega_1) = \frac{1}{\sqrt{(1+a)^2 + b^2}}, \quad \varphi_-(\omega_k, \omega_1) = -\text{arctg}\frac{b}{1+a}.$$

$$(9.57)$$

The two-dimensional amplitude/phase frequency response characteristics for the filter for right scanning from Fig. 9.7 are computed by the Matlab and are shown on Fig. 9.9.

The values of the normalized round frequencies ω_k and ω_l are defined in an interval $[-\pi, +\pi]$, the amplitude frequency characteristics are shown on the left side of the figure and the phase frequency characteristics are shown on the right side.

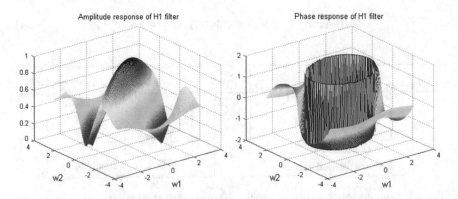

Fig. 9.8 Frequency response characteristics of the filter for calculating of $e_0(k, l)$

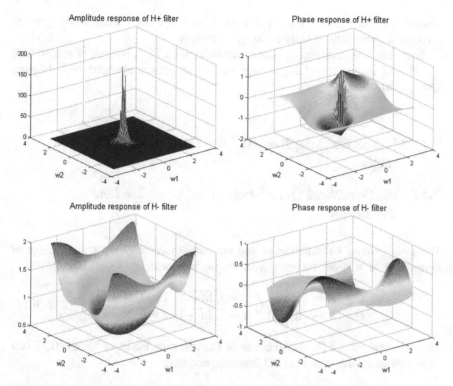

Fig. 9.9 Frequency response characteristics of H_+ and H_- filters

From the graphs is visible that the amplitude frequency characteristics are asymmetrical to zero frequency and phase frequency characteristics have nonlinear character. This leads to distortion of transitions, appearance of false structures in homogeneous areas and dependence of the transformed image from the mutual spatial arrangement of the scanning direction and the available transitions.

Based on this, an additional treatment in the opposite direction, bottom up and from right to left, designated as a reverse scanning unit of Fig. 9.7 is proposed. On the basis of equivalent scheme with the additional unit, the following relations for the common transfer function, amplitude and phase frequency characteristics, describing the operation of the filter in both directions of scanning, are obtained:

$$H(z_k, z_l) = H_1(-z_k, -z_l)H_1(z_k, z_l)$$
$$M(\omega_k, \omega_l) = M_1^2(\omega_k, \omega_l), \quad \varphi(\omega_k, \omega_l) = 0. \tag{9.58}$$

The resulting characteristics correspond to the distribution of the blue noise, which is recommended in the systems for pseudo-halftone conversion. Furthermore, amplitude frequency characteristics are better coordinated with spatial-frequency characteristic of human vision and zero phase frequency characteristics provide accurate reproduction of halftone image transitions.

Therefore, the discussed approach allows high quality playback of synthesized and natural images containing a large number of texture elements with different spatial orientation.

9.4 Method for Adaptive 2D Line Prediction of Halftone Images

9.4.1 Mathematical Description of Adaptive 2D Line Prediction

The input m-level halftone image and the output n-level ($2 \leq n \leq m/2$) image of dimensions M × N can be represented by the matrices:

$$\begin{aligned} C_{M \times N} &= \left\{ c(k, l) \ / \ k = \overline{0, M-1}; \ l = \overline{0, N-1} \right\}, \\ D_{M \times N} &= \left\{ d(k, l) \ / \ k = \overline{0, M-1}; \ l = \overline{0, N-1} \right\}. \end{aligned} \tag{9.59}$$

The predicted value $\hat{c}(k, l)$ of the current element from the input image $c(k, l)$ can be described by the following 2D linear dependence:

$$\hat{c}(k, l) = \sum_{(r,t) \in W} \sum w_{k,l}(r, t) c(k - r, l - t) = W_{k,l}^{t} C_{k,l}, \tag{9.60}$$

where $w_{k,l}(r, t)$ are the weighting coefficients of the filter for linear prediction, defined in given causal window W, for which the following condition must be fulfilled:

$$\sum_{(r,t) \in W} \sum w_{k,l}(r, t) = 1, \tag{9.61}$$

which guarantees that the error $e(k, l)$ of the prediction is not increased or decreased by its passing through the linear prediction filter. The coefficients $w_{k,l}(r, t)$ are real digits, determining parameters of the linear predictive filter (LPF), the output of which forms the predicted picture element (PE) $\hat{c}(k, l)$.

The choice of the prediction coefficients is performed in such a way as to allow a minimum value of the prediction error calculated as the difference between the current PE and the predicted one using the formula:

$$e(k, l) = c(k, l) - \hat{c}(k, l). \tag{9.62}$$

The prediction error is quantized unevenly through the use of optimal quantizer that provides or minimal mean square error of quantization or be reconciled with the properties of human visual perception. The quantized error value is determined by dependency:

$$e_q(k, l) = Q[e(k, l)], \tag{9.63}$$

where Q is an operator for quantization.

The coefficients of the adaptive linear prediction filter can be calculated recursively by introducing a two-dimensional generalization of the LMS Widrow algorithm, as is shown in paragraph 3.2 in horizontal and vertical directions as follows:

$$W_{k,l} = f_k W_{k,l-1} - \mu_k \nabla_{k,l-1} + f_l W_{k-1,l} - \mu_l \nabla_{k-1,l}, \tag{9.64}$$

where: $W_{k,l}$ is the vector of weight coefficients of adaptation; $\nabla_{k,l-1}$ and $\nabla_{k-1,l}$ are the gradients of the squared errors in horizontal and vertical directions; f_k and f_l are the coefficients, considering the direction of the adaptation, where $f_k + f_l = 1$; μ_k, μ_l—adaptation steps in the respective direction.

The conceptual block diagram of two-dimensional adaptive codec for encoding of halftone images using linear prediction, working accordingly to the described methodology, is shown on Fig. 9.10.

In the linear prediction unit, the value of the current predicted PE $\hat{c}(k, l)$, according to the Eq. (9.60) is calculated. The values of weight coefficients are calculated recursively according to the Eq. (9.64). In both summators, according to the Eq. (9.62), are calculated the prediction error $e(k, l)$ and the recovered value of the current PE $c'(k, l)$, that is used in the prediction of the next PE. In the quantization unit the output values of quantization error $e_q(k, l)$ are obtained, according to the Eq. (9.63).

With gray arrows are marked the input and output of the decoder, which is synthesized based on the following expressions:

Fig. 9.10 Block scheme of codec for 2D adaptive linear prediction

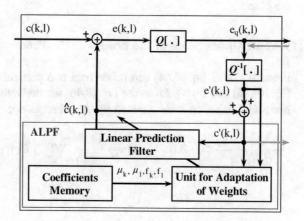

$$e'(k, l) = Q^{-1}[e_q(k, l)], \tag{9.65}$$

where: Q^{-1} is operator for de-quantization of the prediction error and the value of the recovered PE is calculated by the expression:

$$c'(k, l) = e'(k, l) + \hat{c}(k, l). \tag{9.66}$$

According to Eq. (9.45), the convergence and the stability of the adaptation process are given by the following condition:

$$|f_k - \mu_k \lambda_i| + |f_l - \mu_l \lambda_i| < 1, \tag{9.67}$$

where λ_i are the eigen values of the input gray-tone image covariance matrix.

9.4.2 Synthesis and Analysis of Adaptive 2D LMS Codec for Linear Prediction

From the theoretical point of view, it is not correct to divide the optimization of the processes of prediction and quantization, as there is a complicated relationship between the visibilities of quantization noise from the probability distribution of the values of the prediction error. Provided that as a criterion for evaluation the mean square error is used proves that the mean square value of the quantization noise is approximately proportional to the value of the mean square error of prediction, as shown by Musman, and has the form:

$$\bar{e}_q^2 = \left(\frac{9}{2}m^2\right)\bar{e}^2. \tag{9.68}$$

In this case, using this assumption it is permissible to carry out optimization of the parameters in the process of predicting on the one hand, and the quantization—on the other.

9.4.2.1 Optimization of the Process of Prediction

From 2D LMS Eq. (9.64) can be derived two particular cases described with the Eqs. (9.46) and (9.47). From the Eq. (9.46) the derivative of the prediction error in horizontal direction is determined by the relationship:

$$\frac{\partial e^2(k, l-1)}{\partial W_{k,l-1}} = -2e(k, l-1)\frac{\partial}{\partial W_{k,l-1}}\left[W_{k,l-1}^t C_{k,l-1}\right] = -2e(k, l-1) C_{k,l-1}, \tag{9.69}$$

where $C_{k,l-1} = \{c(k-r, 1-t-1)/(r, t) \in W\}$ is vector of input elements participating in the prediction area W in horizontal direction, defined by the Eq. (9.61).

Similarly, for the derivative of the prediction error in vertical direction from (9.47) is obtained:

$$\frac{\partial \, e^2(k-1, l)}{\partial \, W_{k-1,l}} = -2e(k-1, l) \, \frac{\partial}{\partial \, W_{k-1,l}} \left[W_{k-1,l}^t \, C_{k-1,l} \right] = -2e(k-1, l) \, C_{k-1,l},$$

(9.70)

where $C_{k-1,l} = \{c(k-r-1, 1-t)/(r, t) \in W\}$ is vector of input elements participating in the prediction area W in vertical direction, defined by the Eq. (9.61).

Finally, from the Eqs. (9.64), (9.69) and (9.70) the calculation of the weighs of the adaptive prediction filter is performed by the following recursive equation:

$$\begin{aligned}
w_{k,l}(r, t) = &f_k w_{k,l-1}(r, t) + 2\mu_k e(k, 1-1)c(k-r, 1-t-1) \\
&+ f_l w_{k-1,l}(r, t) + 2\mu_l e(k-1, l)c(k-r-1, 1-t).
\end{aligned}$$

(9.71)

Statistical studies described by Pratt [2], show that the value of each PE can be predicted with high accuracy by using the linear relationship from Eq. (9.60) for three coefficients located in the horizontal and vertical directions. In this case, using the Eq. (9.71) for 2D LMS recursive dependency for the calculation of values of the coefficients $w_{k,l}(r, t)$ is obtained:

$$\begin{aligned}
w_{k,l}(0, 1) = &f_k w_{k,l-1}(0, 1) + 2\mu_k e(k, 1-1)c(k, 1-2) \\
&+ f_l w_{k-1,l}(0, 1) + 2\mu_l e(k-1, l)c(k-1, 1-1), \\
w_{k,l}(1, 0) = &f_k w_{k,l-1}(1, 0) + 2\mu_k e(k, 1-1)c(k-1, 1-1) \\
&+ f_l w_{k-1,l}(1, 0) + 2\mu_l e(k-1, l)c(k-2, 1), \\
w_{k,l}(1, 1) = &f_k w_{k,l-1}(1, 1) + 2\mu_k e(k, 1-1)c(k-1, 1-2) \\
&+ f_l w_{k-1,l}(1, 1) + 2\mu_l e(k-1, l)c(k-2, 1-1).
\end{aligned}$$

(9.72)

The initial values of the weight coefficients are calculated from the dependencies:

$$w_{0,0}(0, 1) = \frac{R(0, 1)}{R(0, 0)}, \; w_{0,0}(1, 0) = \frac{R(1, 0)}{R(0, 0)} \text{ and } w_{0,0}(1, 1) = -\frac{R(1, 1)}{R(0, 0)},$$

(73)

where: $R(0, 0)$, $R(0, 1)$, $R(1, 0)$ and $R(1, 1)$ are the coefficients of autocorrelation function, calculated for the first two rows of the input image.

9.4.2.2 Optimization of the Process of Quantization

The error signal in the two-dimensional prediction has extremely irregular distribution of the probability density of their different values, as shown on Fig. 9.11.

Fig. 9.11 Distribution of
probability density of the
errors

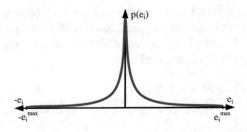

Fig. 9.12 Quantization scale

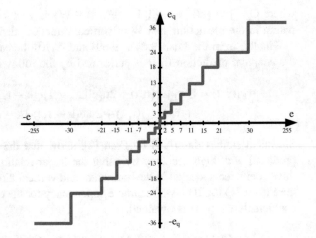

The highest probabilities have the smallest error values, i.e. those that have zero or close to zero values. Therefore, for the quantization of the difference signal the characteristic of quantizer must be unequal, as shown on Fig. 9.12.

A fundamental requirement for the quantization is the number of the levels to be set at a minimum for the given criterion of allegiance of the restored image, selected depending on the method of its use. The criteria for an optimal distribution of the quantization levels q_i $(i = 0, \ldots, n)$ are two: optimization by the minimization of mean square error and optimization in terms of subjective image quality.

The developed method of linear prediction uses the first criterion—optimization of mean square error by using the Max algorithm.

9.4.2.3 Functional Scheme of Adaptive 2D Codec for Linear Prediction

Based on the developed method for coding of halftone images by 2D adaptive linear prediction filter is synthesized functional scheme of codec for 2D adaptive linear prediction, shown on Fig. 9.13.

The used in the scheme functional units correspond to these, shown on Fig. 9.6. With red color are denoted the units for iterative calculation of weights $w_{k,l}(r, t)$ of

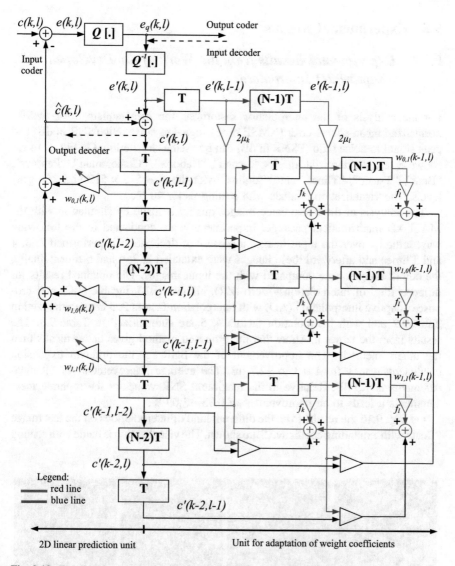

Fig. 9.13 Functional scheme of codec for adaptive 2D linear prediction

the adaptive prediction filter, according to the Eq. (9.72). With blue color are denoted the units, for which the values of the coefficients, according to the 2D LMS algorithm are calculated.

The shown in Fig. 9.13 filter allows adaptation of the coefficients in four modes: in horizontal direction, in vertical direction, in horizontal and in vertical direction, and non-adaptive operation.

9.5 Experimental Results

9.5.1 Experimental Results from the Work of the Developed Adaptive 2D Interpolator

For the analysis of the interpolation distortions the mean-square error (MSE), normalized mean-square error (NMSE in %), signal to noise ratio (SNR in dB) and peak signal to noise ratio (PSNR in dB) can be used as a criterion. On Fig. 9.14 are shown 8 test standardized images: "Lenna", "Baboon", "Cameraman", "Peppers", "Boat", "Tscale", "Tlines" and "Tdglins" with the size 512×512 and 256 gray levels. The visualization is made with scaling factor 40 %.

The analysis of the interpolated images quality is made by simulation with the MATLAB mathematical package. Experiments were conducted in the following way: initially, over the input images a process of decimation is performed 2, 3, 4 and 5 times and afterward these images were extended 2, 3, 4 and 5 times; finally, the output images were compared with the input images. The obtained results for different kind of interpolations—zero (ZR), bilinear (BL), bicubic (BC) and proposed adaptive interpolation (AD) with interpolation factors 2, 3 are summarized in Table 9.2 and with interpolation factors 4, 5 are summarized in Table 9.3. The results from the tables illustrate that the proposed method gives better results than the others methods. The improvement of the SNR for the different expansion coefficients varies from 0.3 to 12.7 dB. The average improvement for all measurements is 1.5 dB. Relative to the maximum SNR value for the separate measurements it leads to an improvement of 0.83–13.09 %.

On Fig. 9.15 the results from the different kind of interpolations on the test image "Boat" with expanding coefficient 5 are shown. The visualization is made with scaling

Fig. 9.14 Test images "Lenna", "Baboon", "Cameraman", "Peppers", "Boat", "Tscale", Tlines" and "Tdglins" with size 512×512, 256 gray levels

Table 9.2 Zero, bilinear, bicubic and adaptive interpolation with coefficients 2 and 3

Image, coefficient, kind of interpolation		Interpolation × 2				Interpolation × 3			
		ZR	BL	BC	AD	ZR	BL	BC	AD
Lenna	MSE	17.9432	16.3837	15.6725	14.3560	43.2827	10.3139	6.01325	5.2196
	NMSE, %	0.00018	0.00016	0.00016	0.00018	0.00056	0.00013	0.00008	0.00007
	SNR, dB	57.4469	57.8418	58.0345	57.3333	52.4915	58.7204	61.0636	61.7272
	PSNR, dB	35.6258	36.0207	36.2134	36.5945	31.8017	38.0306	40.3737	40.9884
Baboon	MSE	0.0000	65.8952	67.9308	8.46320	336.667	152.726	123.824	120.325
	NMSE, %	0.0000	0.00079	0.00082	0.00010	0.0041	0.0019	0.0015	0.0015
	SNR, dB	∞	50.9806	50.8484	59.8937	43.8502	47.2831	48.1942	48.3655
	PSNR, dB	∞	29.9763	29.8441	38.8895	22.8928	26.3257	27.2367	27.3613
Cameraman	MSE	0.02238	9.16620	7.45100	0.0014	32.6184	68.3241	84.2639	30.1430
	NMSE, %	3.09×10^{-7}	1.27×10^{-4}	1.03×10^{-4}	1.67×10^{-8}	0.0016	0.00095	0.0012	0.0004
	SNR, dB	85.0968	58.9744	59.8742	97.7640	47.9321	50.2056	49.2949	53.8045
	PSNR, dB	64.6653	38.5429	39.4427	76.7597	27.5456	29.8191	28.9084	33.3729
Peppers	MSE	0.04202	20.6692	20.4386	0.0382	45.2556	42.9575	35.8621	33.1673
	NMSE, %	5.66×10^{-7}	2.79×10^{-4}	2.76×10^{-4}	5.15×10^{-7}	0.00062	0.00059	0.00049	0.00045
	SNR, dB	82.4658	55.5475	55.5962	82.8800	52.0889	52.3152	53.0993	53.4936
	PSNR, dB	61.9300	35.0116	35.0603	62.3442	31.6081	31.8344	32.6184	32.9577
Boat	MSE	44.3498	25.6759	25.7716	24.3210	94.2767	46.2620	41.8361	43.7300
	NMSE, %	0.00046	0.00027	0.00027	0.00025	0.00099	0.00049	0.00044	0.00045
	SNR, dB	53.3644	55.7381	55.7219	55.9735	50.0391	53.1309	53.5677	53.4255
	PSNR, dB	31.6959	34.0695	34.0534	34.3050	28.4207	31.5126	31.9493	31.7570

(continued)

Table 9.2 (continued)

Image, coefficient, kind of interpolation		Interpolation × 2				Interpolation × 3			
		ZR	BL	BC	AD	ZR	BL	BC	AD
Tscale	MSE	128.000	128.000	126.008	124.326	0.00778	0.0000	0.0000	0.0000
	NMSE, %	0.0015	0.0015	0.0015	0.0015	9.46×10^{-8}	0.0000	0.0000	0.0000
	SNR, dB	48.1308	48.1308	48.1989	48.2573	90.2411	∞	∞	∞
	PSNR, dB	27.0927	27.0927	27.1608	27.2192	69.2539	∞	∞	∞
Tlines	MSE	122.070	120.125	122.070	119.613	0.0000	0.0311	0.0019	0.0000
	NMSE, %	0.0016	0.0015	0.0016	0.0015	0.0000	4.03×10^{-7}	2.52×10^{-8}	0.0000
	SNR, dB	48.0624	48.1322	48.0624	48.1507	∞	83.9456	95.9868	∞
	PSNR, dB	27.2987	27.3685	27.2987	27.3870	∞	63.2333	75.2745	∞
Tdglins	MSE	125.019	136.806	140.081	121.961	515.84	545.799	530.083	510.730
	NMSE, %	0.0015	0.0016	0.0017	0.0014	0.0062	0.0066	0.0064	0.0061
	SNR, dB	48.2670	47.8757	47.773	48.3746	42.0608	41.8157	41.9425	42.1549
	PSNR, dB	27.1950	26.8037	26.7010	27.3026	21.0396	20.7945	20.9214	21.0829

Table 9.3 Zero, bilinear, bicubic and adaptive interpolation with coefficients 4 and 5

Image, coefficient, kind of interpolation		Interpolation × 4				Interpolation × 5			
		ZR	BL	BC	AD	ZR	BL	BC	AD
Lenna	MSE	78.420	102.643	92.416	65.5730	100.837	74.621	53.8836	65.3470
	NMSE, %	0.0010	0.0013	0.0012	0.00084	0.0013	0.00097	0.0007	0.0008
	SNR, dB	49.9593	48.7903	49.2461	50.7363	48.8184	50.1260	51.5401	50.7513
	PSNR, dB	29.2205	28.0515	28.5074	29.9975	28.1286	29.4362	30.8502	30.0125
Baboon	MSE	627.282	169.598	87.1651	86.5573	718.837	584.649	599.836	695.709
	NMSE, %	0.0076	0.0021	0.0011	0.0010	0.0088	0.0072	0.0073	0.0084
	SNR, dB	41.1945	46.8749	49.7657	49.7961	40.5559	41.4533	41.3419	40.7448
	PSNR, dB	20.1902	25.8706	28.7614	28.7918	19.5985	20.4959	20.3845	19.7405
Cameraman	MSE	111.856	197.636	186.579	110.865	219.889	287.144	304.755	201.696
	NMSE, %	0.0015	0.0027	0.0026	0.0015	0.0031	0.0040	0.0043	0.0028
	SNR, dB	48.1097	45.6377	45.8877	48.1484	45.1292	43.9703	43.7118	45.5494
	PSNR, dB	27.6782	25.2062	25.4562	27.7168	24.7428	23.5838	23.3253	25.1178
Peppers	MSE	9.23876	24.5669	14.8640	8.6557	43.3513	155.314	186.963	42.0127
	NMSE, %	0.00012	0.00033	0.00020	0.00012	0.0006	0.0021	0.0026	0.0006
	SNR, dB	59.0445	54.7972	56.9793	59.3276	52.2756	46.7335	45.9281	52.4669
	PSNR, dB	38.5087	34.2613	36.4434	38.7918	45.9281	26.2527	25.4472	31.9310
Boat	MSE	195.795	197.656	205.692	194.127	6.0442	264.28	202.74	4.2013
	NMSE, %	0.0020	0.0021	0.0021	0.0020	0.00006	0.0028	0.0021	0.00004
	SNR, dB	46.9153	46.8742	46.7011	46.9525	61.9698	45.5625	46.7138	63.5995
	PSNR, dB	25.2468	25.2057	25.0326	25.2839	40.3514	23.9441	25.0954	41.9310

(continued)

Table 9.3 (continued)

Image, coefficient, kind of interpolation		Interpolation × 4				Interpolation × 5			
		ZR	BL	BC	AD	ZR	BL	BC	AD
Tscale	MSE	128.000	128.000	125.018	128.000	0.03113	0.03113	0.03113	0.03113
	NMSE, %	0.0015	0.0015	0.0015	0.0015	3.78×10^{-7}	3.78×10^{-7}	3.78×10^{-7}	3.78×10^{-7}
	SNR, dB	48.1308	48.1308	48.2332	48.1308	84.2205	84.2205	84.2205	84.2205
	PSNR, dB	27.0927	27.0927	27.1951	27.0927	63.2333	63.2333	63.2333	63.2333
Tlines	MSE	122.070	128.000	128.000	128.000	77.8198	84.1699	80.9637	77.8150
	NMSE, %	0.0016	0.0016	0.0016	0.0016	0.0010	0.0011	0.0010	0.0010
	SNR, dB	48.0624	47.8564	47.8564	47.8564	49.966	49.625	49.794	50.0179
	PSNR, dB	27.299	27.093	27.093	27.093	29.254	28.913	29.082	29.254
Tdglins	MSE	1137.06	1237.03	1214.90	1137.05	2059.16	2159.75	2063.76	2050.17
	NMSE, %	0.0136	0.0147	0.0145	0.0136	0.0248	0.0260	0.0249	0.0244
	SNR, dB	38.679	38.313	38.3913	38.679	36.049	35.842	36.039	36.1189
	PSNR, dB	17.607	17.2410	17.3194	17.607	15.028	14.821	15.018	15.0469

Fig. 9.15 Results of zero, bilinear, bicubic and adaptive interpolation with expanding coefficient 5 for the test image "Boat"

factor 20 %. The visual quality of the adaptive interpolation is higher than the other interpolations, which is clearly visible for the larger expansion coefficients.

9.5.2   Experimental Results from the Work of the Developed Adaptive 2D Halftoning Filter

An error diffusion filter with 4 coefficients has been used for the evaluation of the efficiency of the described in Sect. 9.3 adaptive 2D halftoning filter. On its base, an algorithm for adaptive 2D image halftoning is developed.

It has been tested on a set of five different types of images: "Lenna", "Baboon", "Cameraman", "Peppers" and "Boat", shown on Fig. 9.14, with size 512×512 and 256 gray levels. The analysis of 2D variation of peak signal to noise ratio (PSNR) depending on parameters f and μ ($f_k = f$, $f_l = 1 - f$, $\mu_k = \mu_l = \mu$.) is made in [23] and [24]. The examination of the function PSNR (f, μ) shows that the most proper mean values of the adaptation parameters f and μ are: $f_k = 0.7$, $f_l = 0.3$, $\mu = 1.67 \times 10^{-6}$. For quantification of the adaptive algorithm, a comparison with the non-adaptive algorithm was made and the results are shown in Table 9.4.

The visual results after adaptive 2D halftoning of the five test images with size 512×512 and 2 levels are shown in Fig. 9.16. For output values are selected the

Table 9.4 Results from halftoning of some test images

Test images		MSE	NMSE	SNR, dB	PSNR, dB
"Lenna"	Adaptive	5.4053e + 003	0.3106	5.0783	10.8366
	Non adaptive	1.2166e + 004	0.6990	1.5550	7.3133
"Babbon"	Adaptive	4.0277e + 003	0.2305	6.3737	12.1143
	Non adaptive	1.2775e + 004	0.7310	1.3607	7.1013
"Cameraman"	Adaptive	5.9302e + 003	0.3342	4.7599	10.4341
	Non adaptive	1.0727e + 004	0.6046	2.1857	7.8599
"Pappers"	Adaptive	6.8425e + 003	0.3993	3.9870	9.8126
	Non adaptive	1.1602e + 004	0.6770	1.6939	7.5196
"Boat"	Adaptive	6.4109e + 003	0.3014	5.2081	10.0956
	Non adaptive	1.1576e + 004	0.5443	2.6419	7.5294

Fig. 9.16 Results from halftoning of test images "Lenna", "Baboon", "Cameraman", "Peppers" and "Boat"

values $C_{p+1} = 0.95h_n$ and $C_p = 0.05h_n$, calculated from the histograms of the input gray images according to the dependencies given in [22].

From the conducted experiments can be concluded that the developed adaptive 2D image halftoning methods give better results than non-adaptive method, wherein the improvement of PSNR for the different images varies from 2.3 to 5 dB (the mean improvement for all measurements is 2.5 dB), as seen from Table 9.4. The visual quality of the adaptive method is higher, at the expense of some increase in the complexity of the calculations.

9.5.3   *Experimental Results from the Work of the Developed Codec for Adaptive 2D Linear Prediction*

The developed algorithms for encoding and decoding, with the use of adaptive two dimensional prediction, are tested on the images: "Lenna", "Baboon", "Cameraman", "Peppers" and "Boat", shown on Fig. 9.14, with size 512×512, 256 gray levels and values of adaptation parameters: $f_k = 0.7$, $f_l = 0.3$ and $\mu = 1.67 \times 10^{-6}$. Prediction is performed with three weight coefficients that are located spatially in the vicinity of the current image element. On Fig. 9.17 reconstructed test images after adaptive encoding with 2 bits/element and four times shrinking are shown and on Fig. 9.18 the corresponding error images from the reconstruction.

Fig. 9.17 Reconstructed test images after adaptive encoding with 2 bits/element

Fig. 9.18 Error images after adaptive encoding with 2 bits/element

Table 9.5 Results from encoding of the test images

Test images (2 bits/element)		MSE	NMSE	SNR, dB	PSNR, dB
"Lenna"	Adaptive	326.1565	0.0187	17.2723	23.0305
	Non adaptive	403.8962	0.0232	16.3438	22.1021
"Babbon"	Adaptive	370.0712	0.0212	16.7413	22.4819
	Non adaptive	394.1929	0.0226	16.4671	22.2077
"Cameraman"	Adaptive	690.5548	0.0389	14.0986	19.7728
	Non adaptive	790.2266	0.0445	13.5131	19.1873
"Pappers"	Adaptive	415.4637	0.0242	16.1538	21.9795
	Non adaptive	505.0852	0.0295	15.3055	21.1312
"Boat"	Adaptive	441.8626	0.0208	16.8244	21.7119
	Non adaptive	514.8935	0.0242	16.1601	21.0476

For quantification of the adaptive algorithm, a comparison with the non-adaptive classical DPCM algorithm with 2 bits/element was made and the results for the MSE, NMSE, SNR and PSNR are summarized in Table 9.5.

From the performed experiments can be concluded that the developed adaptive 2D image linear prediction methods gives better results than non-adaptive method, wherein the improvement of PSNR for the different images varies from 0.27 to 0.93 dB (the mean improvement for all measurements is 0.64 dB), as seen from Table 9.5. The visual quality of the adaptive method is higher, at the expense of some increase in the complexity of the calculations. When performing coding with 3 bits /element PSNR increases on average by 6.5 dB, and when performing coding with 4 bits /element the PSNR is close to 40 dB and the decoded image is virtually indistinguishable from the original.

9.6 Conclusion

Based on the performed experiments for 2D adaptive interpolation on halftone images, the following conclusions can be made:

- the use of optimal thresholds for selection of homogenous and contour blocks leads to the decreasing of mean-square error, normalized mean-square error and the increasing of signal to noise ratio and peak signal to noise ratio with about 7–10 %;
- by the changing of interpolation type, zero or bilinear, depending on the presence or lack of contours in the area of considerate fragment, better visual quality is achieved;
- the visual quality is better then the zero, bilinear and bicubic interpolation, which is shown for the biggest interpolation coefficients;
- the complexity of the adaptive interpolation is higher than zero and bilinear interpolations but is lower than other high-level interpolations;

- using smaller area for analysis and choice of optimal thresholds for image separation lead to decrease of calculation speed;
- the most effective interpolation for the local characteristic of images can be achieved by using coefficients p, q = 2, 3, 4.

The results for the quality of interpolated images show that the proposed method for adaptive interpolation can change the high-level interpolations, which are slower in systems, using digital image processing and visualization such as digital photography, videoconference systems, security systems and etc.

The developed generalized adaptive error-diffusion quantizer results in the following particular cases: the wide-spread non-adaptive error diffusion filter of Floyd and Steinberg (for n = 2, $f_k = 1, \mu_k = \mu_l = f_l = 0$); adaptive error diffusion using the weights only in the horizontal (from the same image row—$f_k = 1, f_l = 0$) or only in the vertical direction (from the previous image row—$f_l = 1, f_k = 0$). The adaptive filter provides minimum reconstruction error, uniform distribution of the arranged structures in the homogeneous areas and precise reproduction of edges in the output multilevel images. The coefficients f_k, f_l, μ_k, μ_l must be selected on the basis of PSNR analysis and keeping Eq. (9.44) as is done in [24]. The developed AEDQ is appropriate for realization on special VLSI circuit to accelerate calculation of image transform.

The presented error diffusion filter can be used for transformation of color palettes or brightness of pixels in multimedia systems, for printing color and halftone images and transmission by facsimile devices.

The developed adaptive 2D coder provides minimum processing error and lied to increase of PSNR with about 0.3 dB in comparison with 3 coefficients non-adaptive prediction coder.

The given experimental results from the simulation in MATLAB environment for each of the developed algorithms, suggest that the effective use of local information contributes to minimize the processing error. The methods are extremely suitable for different types of images (for example: fingerprints, contour images, cartoons, medical signals, etc.). The developed algorithms have low computational complexity and are suitable for real-time applications.

References

1. Gonzalez, R.C., Woods, R.E.: Digital Image Processing, 3rd edn. Pearson Prentice Hall, New Jersey (2008)
2. Pratt, W.K.: Digital Image Processing. Wiley, New York (2007)
3. Stucki, P.: Advances in Digital Image Processing. Plenum Press, New Jersey (1979)
4. Crochiere, R., Rabiner, L.: Interpolation and decimation of digital signals—a tutorial review. Proc. IEEE 69(3) (1981)
5. Chen, T.C., Figueiredo, R.P.: Image decimation and interpolation techniques based on frequency domain analysis. IEEE Trans. Commun. COM-32(4) (1984)
6. Wong, Y., Mitra, S.K.: Edge preserved image zooming, signal processing IV: theories and applications. EURASIP (1988)

7. Kountchev, R., Mironov, R.: Analysis of distortions of adaptive two dimensional interpolation of halftone images. In: XXV Science Session "Day of Radio'90", Sofia, Bulgaria, 7–8 May 1980 (in Bulgarian)
8. Kunchev, R., Mironov, R.: Adaptive interpolation block used in high quality videowall system. In: International Conference VIDEOCOMP'90, Varna, Bulgaria, 24–29 Sept 1990
9. Mironov, R.: Error estimation of adaptive 2D interpolation of images. In: XXXIX International Scientific Conference on Information, Communication and Energy Systems and Technologies (ICEST'2005), Niš, Serbia and Montenegro, June 29–July 1. Proc. Pap. 1, 326–329 (2005)
10. Mironov, R.: Analysis of quality of adaptive 2D halftone image interpolation. In: National Conference with Foreign Participation (Telecom'2007), International House of Scientists "F. J. Curie", "St. Constantine" Resort, Varna, Bulgaria, 11–12 Oct Proc. Pap. 1, 117–122 (2007)
11. Mironov, R., Kountchev, R.: Optimal thresholds selection for adaptive image interpolation. In: XLIII International Scientific Conference on Information, Communication and Energy Systems and Technologies (ICEST 2008), Niš, Serbia and Montenegro, 25–27 June. Proc. Pap. 1, 109–112 (2008)
12. Mironov, R., Kountchev R.: Adaptive contour image interpolation for sign language interpretations. In: The 8th IEEE International Symposium on Signal Processing and Information Technology (ISSPIT 2008), Sarajevo, Bosnia and Herzegovina, vol. 1, pp. 164–169, 16–19 Dec 2008
13. Carrato, S., Ramponi, G., Marsi, St.: A simple edge-sensitive image interpolation filter. Proc. ICIP'96 (1996)
14. Allebach, J., Wong, P.W.: Edge-directed interpolation. Proc. ICIP'96 (1996)
15. Zhang, X., Wu, X.: Image interpolation by adaptive 2-D autoregressive modeling and soft-decision estimation. IEEE Trans. Image Process. 17(6), 887–896 (2008)
16. Muresan, D.D., Parks, T.W.: Adaptively quadratic (aqua) image interpolation. IEEE Trans. Image Process. 13(5), 690–698 (2004)
17. Keys, R.G.: Cubic convolution interpolation for digital image processing. IEEE Trans. Acoust. Speech Signal Process. ASSP-29(6), 1153–1160 (1981)
18. Widrow, B., Stearns, S.D.: Adaptive Signal Processing. Englewood Cliffs, Prentice-Hall, Inc., New York (1985)
19. Sid-Ahmed, M.A.: Image Processing: Theory, Algorithms, and Architectures. McGraw-Hill Inc., New York (1995)
20. Hadhoud, M.M., Thomas, D.W.: The two-dimensional adaptive LMS (TDLMS) algorithm. IEEE Trans. Circuits Syst. CAS-35(5), 485–494 (1988)
21. Makoto, O., Hashiguchi, S.: Two-dimensional LMS adaptive filters. IEEE Trans. Consum. Electron. 37(1), 66–73 (1991)
22. Mironov, R., Kunchev, R.: Adaptive error-diffusion method for image quantization, electronics letters. IEEE Intr. Publ. Inst. Electr. Eng. 29(23), 2021–2023 (1993)
23. Mironov, R.: Algorithms for local adaptive image processing. In: XXXVII International Scientific Conference on Information, Communication and Energy Systems and Technologies (ICEST 2002), Niš, Yugoslavia. Proc. Pap. 1, 193–196 (2002)
24. Mironov, R.: Analysis of two-dimensional LMS error-diffusion adaptive filter. In: XXXIX International Scientific Conference on Information, Communication and Energy Systems and Technologies (ICEST 2004), Bitola, Macedonia. Proc. Pap. 1, 131–134 (2004)
25. Mironov, R.: Local adaptive interpolation of halftone images. In: Kountchev, R. (ed.) New Approaches in Intelligent Image Processing, Published by WSEAS Press. Chapter 13, 195–212 (2013)
26. Mironov, R.: Adaptive interpolation and halftoning for medical images. In: Iantovics, B., Kountchev, R. (eds.) Advanced Intelligent Computational Technologies and Decision Support Systems, Studies in Computational Intelligence, vol. 486, pp. 83–96. Springer International Publishing, Switzerland (2014)
27. Ulichney, R.A.: Dithering with blue noise. Proc. IEEE 76(1) 56–79 (1988)

Chapter 10
Machine Learning Techniques for Intelligent Access Control

Wael H. Khalifa, Mohamed I. Roushdy and Abdel-Badeeh M. Salem

Abstract Access control is a set of regulations that governs access to certain areas or information. By access we mean entering a specific area, or logging on a machine. The access regulated by a set of rules that specifies who is allowed to get access and what is the restrictions on such access. Across the years several access control systems have been developed. Due to the rapid advancement in technology over the past years, older systems are now easily by passed, thus the need to have new methods of access control. Biometrics is referred to as an authentication technique that relies on a computer system to electronically validate a measurable biological characteristic that is physically unique and cannot be duplicated. Biometrics has been used for ages as access control security system. In this chapter we will present several biometric techniques their usage, advantages and disadvantages.

Keywords Data protection · Privacy · Biometrics · Machine learning

10.1 Introduction

The term "biometrics" is derived from the Greek words "bios" (life) and "metrics" (to measure). Automated biometric systems have only become available over the last few decades, due to significant advances in the field of computer processing.

W.H. Khalifa (✉) · M.I. Roushdy · A.-B.M. Salem
Artificial Intelligence and Knowledge Engineering Research Labs,
Computer Science Department, Faculty of Computer and Information
sciences, Ain Shams University, Khalifa El-Maamon st, Abbasiya sq.,
Cairo 11566, Egypt
e-mail: wael.khalifa@cis.asu.edu.eg

M.I. Roushdy
e-mail: miroushdy@hotmail.com

A.-B.M. Salem
e-mail: abmsalem@yahoo.com

© Springer International Publishing Switzerland 2016
R. Kountchev and K. Nakamatsu (eds.), *New Approaches in Intelligent Image Analysis*, Intelligent Systems Reference Library 108,
DOI 10.1007/978-3-319-32192-9_10

Many of these new automated techniques, however, are based on ideas that were originally conceived hundreds, even thousands of years ago [1].

Human beings since the beginning of civilizations have been using biometrics to identify one another. People recognize each other's via their faces. Moreover, human beings identify each other with behavioral traits such as voice and gait. Computer biometric systems try to mimic the human mind by identifying individuals via physical or behavioral traits.

It is generally agreed by the relevant research community and industry that for a biometric system to be efficient it should meet as many as possible from the following features [2]:

(a) **Changeability**: The user must have the possibility to change his access features, take for example a finger print biometric, if the user finger print is compromised (someone stole it) and the system admin detected both the user and the imposter will be banned from the system with no means of allowing the user only to access the system.
(b) **Shoulder-surfing resistance**: The system should be resistant to shoulder surfing especially with the advances of cameras and recording devices.
(c) **Theft protection**: The theft protection covers physical theft of the authentication features or an easy guess for the features.
(d) **Protection from user non-compliance**: Users tend to bend the rules when they are in a tough situation, a user can give his password to a friend to get them something urgent from their office or pc. A biometric system should try to prevent user non-compliance.
(e) **Stable over time**: Some biometric features change over time for example Voice, ECG and EEG Signals. For example as the user gets older their voice changes slightly same happens with EEG and ECG. The system should handle these changes overtime.
(f) **Easy to Deploy**: Users always look for fast and secure way for authentication. Magnetic Resonance Imaging (MRI) maybe more accurate than fingerprint, but going through the process of putting the user in an MRI machine every time they require access is not a feasible way for security.
(g) **Liveness Detection**: the biometric system should try to interpret that the captured data is from a live user not a replica or a dead person.

This chapter is organized as follows; in Sect. 10.2 we discuss the machine learning methodology for intelligent access control. In Sect. 10.3 we present an introduction about the various user authentication techniques. In Sect. 10.4 we demonstrate some of the commonly used physiological biometric features. In Sect. 10.5 some of the commonly used behavioural biometrics are presented. Section 10.6 describes the fusion of multiple biometrics features to a single multimodal system. Section 10.7 shows some of the commonly used application for biometrics. Section 10.8 explores various machine learning techniques used in EEG biometrics systems. Section 10.9 contains the comparison of the various biometrics techniques and conclusion.

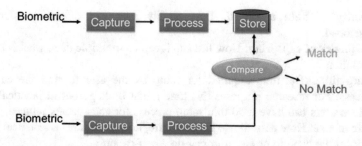

Fig. 10.1 Biometric processes

10.2 Machine Learning Methodology for Biometrics

Biometrics systems follow four main activities: capturing, enrollment, comparison, and decision (see Fig. 10.1).

- **Capturing**: is the use of sensors to capture the biometric features.
- **Enrollment**: Evolves using the captured biometric data and extracting a unique set of features for each user and string them in the system.
- **Comparison**: includes recapturing of the biometric features then running the same steps of the enrollment and comparing the generated signature to the stored signatures.
- **Decision**: Based on the signature comparison, the system decides whether or not to give the user access.

To put these processes in a computer science (informatics) point of view the steps are:

- **Signal Capturing**: signal capturing is using hardware sensors to capture the biometric features.
- **Feature Extraction**: is discovering unique information in the capture data. This information will be used as the user signature.
- **Classification**: is trying to find similarity between the generated signature at the enrollment phase and the verification phase.

From an initial look at the process it seems an easy task to develop a new biometric system but the truth is that every step has a set of challenges that will be covered in the following sections.

10.2.1 Signal Capturing

Signal capturing usually involves a capturing hardware. When developing a biometric the hardware type used affects the data captured. It is advised to consider the below items while selecting your hardware:

- **Quality of data captured**: Low quality data will affect the algorithms developed.
- **Frequency of capturing**: How fast can it capture multiple data, what is the data resolution.
- **Setup time**: The time required for setup by the user to start the capturing process. For research purposes this time might be large but in practical applications users can have a 30 min setup process for each access request.
- **Ease of Use**: How easy to use the hardware, does it requires a technical user to perform the capture or any user can do a self-capture.
- **Cost**: While expensive hardware usually provides better quality of data, the challenge is create a secure robust system with cheaper hardware. If a biometric system uses a few thousand dollars capturing device to have high quality data. It will not be practical to install this system on every door at the office building; users would rather have a few hundred dollars capturing device and more sophisticated software to handle the low quality data.

10.2.2 Feature Extraction

After the capturing phase, comes the feature extraction, which has the following challenges:

- **Data Cleaning**: The challenge in data cleaning is removing noise or useless information without destroying valuable information in the process.
- **Selection of Features**: There are many features that can be used depending on the type of biometrics you are using. Which features to use and why? Some understanding of the data captured should direct you to which features to extract.
- **Features combinations**: There are usually many features that can be extracted from the data, which feature combination is the best match for this signal. This will need a bit of try and error. The system need to run with several combinations and see which one would give the best results. Feature reduction techniques may as well be applied.

10.2.3 Classification

Some considerations that need to be taken care of when building a biometric classifier:

- **Type of Problem**: If the system will be using a verification or identification technique the type of classifier used might differ.

- **Change over Time**: How well does the classifier handle noise or slight changes in the features over time? The classifier should be developed to learn the new changes as the time passes.
- **Training Time**: Time required to train the system for a new user.
- **Matching Time**: Time required to match a user; while a classifier that has a 100 % accuracy can take 20 min to match the user, in real world it needs be done in seconds at most.
- **Accuracy**: The most important factor is accuracy and in the security system accuracy is divided to 2 parts. False acceptance rate where a user is allowed access and they do not have the privilege, this must be really near zero percent. As for the other measure it is the false rejection rate, that's when a privileged user is denied access. That is a problem as well but it is less serious than the false acceptance rate.

A biometric recognition system can run in two different modes: Identification or verification. In the identification case, the system is trained with the signatures of several persons. For each of the persons, a biometric signature is generated in the enrollment stage. A signature that is going to be identified is matched against every known signature, yielding either a score or a distance describing the similarity between the new signature and the stored ones. In the verification case, a person's identity is claimed a priori. The signature that is verified only is compared with the person's individual signature. Similar to identification, it is checked whether the similarity between new signature and stored ones is sufficient to provide access to the secured system or area.

10.3   User Authentication Techniques

The security field uses three different types of authentication [3]:

- Something you know: A password, PIN, or piece of personal information
- Something you have: A card key, smart card, or token (like a SecureID card)
- Something you are: A biometric.

Table 10.1 shows a comparison between various existing user authentication techniques [4].

Biometrics is a measurable physical characteristic or personal behavioral trait used to recognize the identity, or verify the claimed identity of an enrollee. Biometrics is divided to two types; namely Physiological biometrics and behavioural biometrics. Physiological Biometrics are related to the shape of the body. Example, fingerprint, face recognition, DNA, hand and palm geometry, iris recognition. While behavioral biometrics are related to the behavior of a person. Examples, typing rhythm, gait, voice, Electroencephalography (EEG), Electrocardiogram (ECG). Some researchers have coined the term behaviometrics for this class of biometrics. Figure 10.2 illustrates the biometrics taxonomy.

Table 10.1 Existing user authentication methods and techniques [4]

Method	Examples	Properties
What you know	User ID Password PIN	• Shared • Many passwords forgotten
What you have	Cards Badges Keys	• Shared • Can be duplicated • Lost or stolen
What you know and What you have	ATM Card + PIN	• Shared • Can be duplicated • Lost or stolen
What you are	Fingerprint Face Voice Iris EEG	• Not possible to share • Hard to Forge • Cannot be lost or stolen

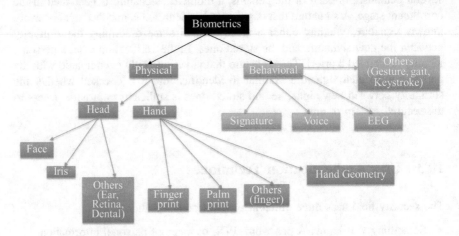

Fig. 10.2 Biometrics taxonomy

10.4 Physiological Biometrics Taxonomy

10.4.1 Finger Print

Fingerprint is the oldest biometric method in identity authentication and has been in use since 1896 especially for criminal identification. The main idea is based on fingertips that have corrugated skin with line like ridges flowing from one side of the finger to another. The flow of these ridges is non-continuous and it forms a pattern. The pattern of flow gives rise to a classification pattern such as arches, loops and whorls while the discontinuity in the ridge flow give rise to feature points, called minutiae as in Fig. 10.3 [5].

Fig. 10.3 Finger print
definition [22]

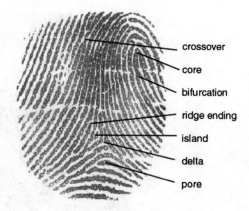

Fingerprint recognition can achieve good accuracy sufficient for both verification
and identification. Because of low cost and compactness it is popular consumer
product. On the other hand the sensor is not able to capture acceptable quality
fingerprint images for people with very dry or wet skin [5].

10.4.2   Face

Face recognition is the oldest biometric known to man since the start of history.
Human beings identified each other via the faces. With the spread in digital cameras, human face identification usage have grown. There are two main approaches
for face identification namely; Feature-based approach and Holistic approach.

Feature-based approach extracts distinctive facial features such as the eyes,
mouth, nose, etc., as well as other fiducial marks, and then compute the geometric
relationships among those facial points, thus reducing the input facial image to a
vector of geometric features. Standard statistical pattern recognition techniques are
then employed to match faces using these measurements. While Holistic approaches identify faces using descriptions based on the entire image rather than on local
features of the face [6] see Figs. 10.4 and 10.5.

10.4.3   Iris

The iris is a thin, circular structure in the eye. It controls the diameter and size of the
pupils and thus the amount of light reaching the retina. Eye colour is the colour of
the iris [7]. Upon imaging an iris, a 2D Gabor wavelet filters and maps the segments
of the iris into phasors (vectors). These phasors include information on the orientation and spatial frequency. This information is used to map the Iris Codes [7] see
Fig. 10.6.

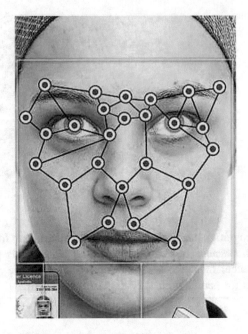

Fig. 10.4 Face detection feature-based approach [23]

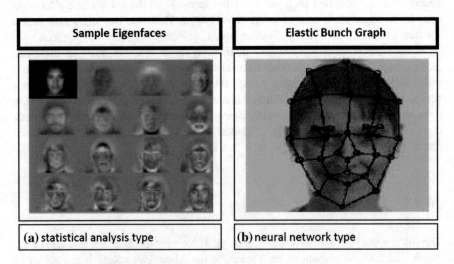

Fig. 10.5 Face detection holistic approach [5]

Fig. 10.6 Iris encoding [24]

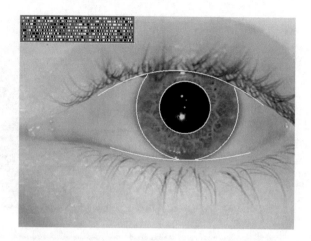

10.5   Behavioral Biometrics Taxonomy

10.5.1   Keystroke Dynamics

The idea behind Keystroke Dynamics has been around since World War II. It was well documented during the war that telegraph operators on many U.S. ships could recognize the sending operator. Known as the "Fist of the Sender," the uniqueness in the keying rhythm (even of Morse-code), could distinguish one operator from another [8].

Keystroke dynamics is the process of analyzing the way a user types on a keyboard to identify their typing rhythm. A user's typing pattern may be unique because similar neuro-physiological factors that make written signatures unique are also exhibited here Keystroke dynamics is a behavioral biometric Natural choice for computer login and network security [9]. The key features used are "flight time" the amount of time that a user spends "reaching" for a certain key and "dwell time" the amount of time a user spends pressing one key (See Fig. 10.7).

There are two modes of operation for keystroke dynamics systems; static verification and continuous verification. In static verification, the keystrokes are analyzed only at specific times e.g., during login. Static approaches provide more robust user verification than simple passwords, but static methods cannot detect substitution of the user after the initial verification. Continuous verification monitors the user's typing behavior all the time.

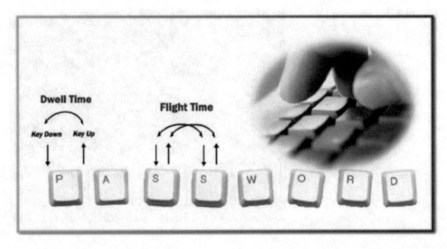

Fig. 10.7 Flight and dwell time [25]

10.5.2 Voice

Speaker, or voice, recognition uses a user's voice for recognition purposes. It is a different from "speech recognition", which recognizes words as they are articulated, which is not a biometric. The speaker recognition process relies on features influenced by both the physical structure of an individual's vocal tract and the behavioral characteristics of the individual [10].

There are two forms for speaker recognition, text dependent and text independent. In text dependent mode the user is required to say a specific word or phrase while in text independent the user can say anything. The speaker recognition system analyzes the frequency content of the speech and compares characteristics such as the quality, duration, intensity dynamics, and pitch of the signal.

10.5.3 EEG

EEG signals are brain activities (see Fig. 10.8 Sample EEG Signal) recorded from electrodes positioned on the scalp. The EEG signals can be used in biometrics due to the advances in its hardware devices; there are some EEG signal capturing devices that are equal in size to a mobile phone or computer headset, data can be acquired continuously. After capturing, the data is filtered to remove artefacts. After the EEG signals have been cleaned, they can be analyzed using a variety of signal processing approaches. Lastly, a wide range of machine learning algorithms have been applied to perform the classification process.

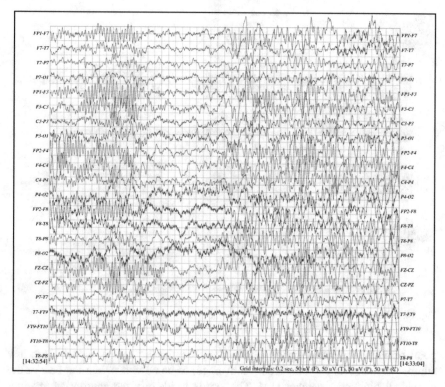

Fig. 10.8 Sample EEG signal

10.6   Multimodal Biometrics

Multimodal biometric systems are those which utilize, or have the capability of utilizing, more than one physiological or behavioral characteristic for enrollment, verification, or identification. Multimodal biometric systems combine multiple sources of biometric features. The integration of features is known as fusion. There are various levels of fusion namely raw data fusion, feature level fusion and score level fusion [11] (see Fig. 10.9 Fusion Levels).

Sensor level fusion refers to the consolidation of raw data obtained using multiple compatible sensors or multiple snapshots of a biometric using a single sensor [12].

In feature level fusion, feature sets are calculated from different biometric feature sources and combined to a new feature set. The combination process can be homogeneous or non- homogeneous [13]. Dimensionality reduction scheme like feature selection/transformation may be applied to obtain a minimal feature set.

In Decision level fusion, different biometric matchers provide scores indicating the degree of similarity between the input and stored signatures for a specific user. The combined result is used as the total score of the fusion. The combination can be done by various means such as weighted decision, and, or,.. etc. From theoretical

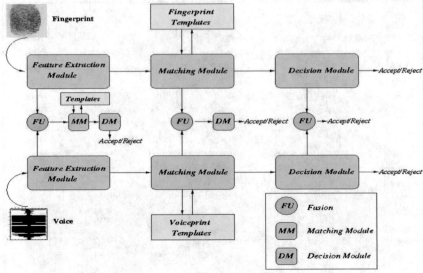

Ross, Jain, "Information Fusion in Biometrics", Pattern Recognition Letters, September 2003.

Fig. 10.9 Fusion levels

point of view the performance obtained by combining match scores from any number of matchers is guaranteed (on average) to be no worse than the best of the individual biometric matcher [14]. The key is to identify the appropriate method which combines the matching scores reliably and maximize the matching performance.

System Designers should take into consideration the following points while designing a multi modal system:

- Cost versus performance
- Throughput
- Verification versus Identification mode
- Choice and number of biometrics
- Level of fusion
- Fusion methodology
- Assigning weights to biometrics
- Multimodal databases

10.7   Applications

Depending on the application type, specific biometrics features will be best suited to be used. Below are some of the applications of using biometrics [10]:

- Access Control
- Criminals Detection
- Monitor human behavior
- Victim Detection
- Marketing (methods of biometrics are used to identify owners of loyal cards)
- Attendance systems at work, schools, etc.
- Voting system (during the functionality of voting system identification/ authentication of people, that take part in voting is demanded)
- Biometric identifiers are used for registration if immigrants and foreign workers.

10.8   Machine Learning Techniques for Biometrics

In this section we will explore several of the machine learning techniques used in biometrics and specially using EEG signals as a biometric feature.

10.8.1   Fisher's Discriminant Analysis

Riera et al. [15] have developed a multimodal authentication algorithm based on EEG and ECG signals. They conducted the test on 40 healthy subjects. Each subject was required to sit in a comfortable armchair, to relax, be quiet and close their eyes. Then 3 min takes are recorded to 32 subjects and four "3 min" takes are recorded to the 8 subjects. The 32 subject set are used as reference subject in the classification stage and the 8 subjects are the ones that are enrolled into the systems. Then several "1 min" takes are recorded afterwards to these enrolled subjects, in order to use them as authentication tests. Two electrodes were used to capture the EEG signals and 2 for the ECG. The data was divided to 4 s epochs.

The data acquisition module is the software that controls the ENOBIO [16] sensor in order to capture the raw data. Four channels are recorded: two EEG channels placed in the forehead, one ECG channel placed in the left wrist and one electrode placed in the right earlobe for referencing the data. At this point the data are separate in EEG data and ECG data and sent to two parallel but different biometric modules for EEG and ECG (Fig. 10.10).

Figure shows Riera et al. system flowchart steps after the raw EEG and ECG data is collected from the sensors. Each biometric feature is processed separately by a set of modules each feature has the following steps:

1. Signals are preprocessed
2. Features Extraction
3. Signature created and stored in DB
4. When the user is authenticated the same steps are repeated but the new signature is compared to the signature in DB.

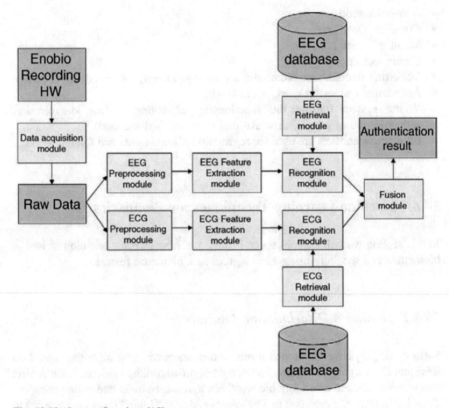

Fig. 10.10 System flowchart [15]

In Reira algorithm they worked with four different DFs:

- Linear: Fits a multivariate normal density to each group, with a pooled estimate of the covariance.
- Diagonal Linear: Same as "linear," except that the covariance matrices are assumed to be diagonal.
- Quadratic: Fits a multivariate normal density with covariance estimates stratified by group.
- Diagonal Quadratic: Same as "quadratic," except that the covariance matrices are assumed to be diagonal.

Two types of features were extracted from the 4 s epochs, one channel features (Auto regression, Fourier Transform) and Synchronicity features. Three features were selected from the Synchronicity features namely; Mutual information (measures the dependency degree between two random variables given in bits, when logarithms of base 2 are used in its computation), Coherence (quantizes the correlation between two time series at different frequencies), Correlation measures (measure of the similarity of two signals,). The classifier used in the authentication process is the classical Fisher's Discriminant Analysis, Four different discriminant functions were used (Linear, Diagonal Linear, quadratic, diagonal quadratic). The five best classifiers from the original 28 classifiers generated for each subject are selected during the enrollment and authentication of each subject.

The False Acceptance Rate (FAR) is computed taking into account both the intruder and the impostor cases (21.8 %). The True Acceptance Rate (TAR) only takes into account the legal cases. (71.9 %).

After combining the 2 signals (EEG and ECG) the TAR is 97.9 % and the FAR is 0.82 %.

10.8.2   Linear Discriminant Classifier

Palaniappan [17] proposed a multiple mental thought identification modal. The experiment was conducted on four subjects. The subjects were seated in an Industrial Acoustics Company sound controlled booth with dim lighting and noise-less fan (for ventilation). An Electro-Cap elastic electrode cap was used to record EEG signals from positions C3, C4, P3, P4, O1 and O2 defined by the 10–20 system of electrode placement. Each subject was requested to do up to five mental tasks. Signals were recorded for 10 s during each task and each task was repeated 10 times. Each recording was segmented into 20 segments, each 0.5 s length. The five mental tasks performed by the subjects are:

- **Baseline task**. The subjects were asked to relax and think of nothing in particular. This task was used as a control and as a baseline measure of the EEG signals.
- **Geometric Figure rotation task**. The subjects were given 30 s to study a particular three-dimensional block object, after which the drawing was removed and the subjects were asked to visualize the object being rotated about an axis. The EEG signals were recorded during the mental rotation period.
- **Math task**. The subjects were given nontrivial multiplication problems, such as 79 times 56 and were asked to solve them without vocalizing or making any other physical movements. The tasks were non-repeating and designed so that an immediate answer was not apparent. The subjects verified at the end of the task whether or not he/she arrived at the solution and no subject completed the task before the end of the 10 s recording session.
- **Mental letter composing task**. The subjects were asked to mentally compose a letter to a friend or a relative without vocalizing. Since the task was repeated for several times the subjects were told to continue with the letter from where they left off.
- **Visual counting task**. The subjects were asked to imagine a blackboard and to visualize numbers being written on the board sequentially, with the previous number being erased before the next number was written. The subjects were instructed not to verbalize the numbers but to visualize them. They were also told to resume counting from the previous task rather than starting over each time.

The captured data features were extracted using Auto Regression (AR) modeling. Six AR coefficients were obtained for each channel, giving a total of 36 feature vector for each EEG segment for a mental thought. When two mental thoughts were used, the size of the feature vector was 72 and so forth when more mental thoughts were used.

Linear Discriminant Classifier was used to classify the EEG feature vectors, LDC is a linear classification method that is computationally attractive as compared to other classifiers like artificial neural network. Various results were presented showing the error rate using 1, 2, ..., 5 five combination of the mental tasks. Using 1 task an average of error rate is 2.6 %, while using the 5 mental tasks, the error rate was 0.1 %.

10.8.3 LVQ Neural Net

Cempírek and Šťastný [18], proposed neural network classification technique for user identification. The algorithm was conducted on a datasets of 8 subjects. The subject sat is a dim and silent room, eyes kept closed. Then the EEG recordings were segmented (segment length 180 s, step 22.5 s); the single segments were centered. Linear magnitude spectra of the single segments were computed by Fast Fourier transform (Hamming window was used).

Learning Vector Quantisation (LVQ) is a supervised version of vector quantization, similar to Self organizing Maps (SOM) (see [13, 19] for a comprehensive overview). It can be applied to pattern recognition, multi-class classification and data compression tasks, e.g. speech recognition, image processing or customer classification. As supervised method, LVQ uses known target output classifications for each input pattern of the form.

LVQ algorithms do not approximate density functions of class samples like Vector Quantization or Probabilistic Neural Networks do, but directly define class boundaries based on prototypes, a nearest-neighbor rule and a winner-takes-it-all paradigm. The main idea is to cover the input space of samples with 'codebook vectors' (CVs), each representing a region labelled with a class. A CV can be seen as a prototype of a class member, localized in the center of a class or decision region in the input space. As a result, the space is partitioned by hyper planes perpendicular to the linking line of two CVs. A class can be represented by an arbitrarily number of CVs, but one CV represents one class only [19].

The LVQ neural network is a self–organizing neural network, with added second layer for vectors classification intended to be used with unlabeled training data. The first network layer detects subclasses. The second layer combines these subclasses into one single class. Actually, the first layer computes distance between input and stored patterns; the winning neuron is the one with minimum distance. Hence LVQ network is a kind of nearest-neighbor classifier; it does not make clusters, but the algorithm search through the weights of connections between input layer neurons and output map neurons. These represent classes [18].

10.8.4   Neural Networks

Sun [20] has developed a user identification system based on Neural Networks. The system was tested on 9 subjects. The task was to imagine moving his or her left or right index finger in response to a highly predictable visual cue. EEG signals were recorded with 59 electrodes mounted according to the international 10-10 system. Only Signals from 15 electrodes were used in the system. Totally 180 trials were recorded for each subject. Ninety trials with half labeled left and the other half right were used for training, and the other 90 trials were for testing. Each trial lasted 6 s with two important cues. The preparation cue appeared at 3.75 s indicating which hand movement should be imagined, and the execution cue appeared at 5.0 s indicating it was time to carry out the assigned response. The common spatial patterns (CSP) is employed to carry out energy feature extraction. As a result, each trial is modeled by an 8-dimensional vector (4 sources from each kind of mental task is assumed in this paper). Based on these features, neural network classifiers can be learned. Neural networks of one hidden layer and one output layer for experiments. The results showed that imagining left index finger movements is more appropriate for personal identification. Left index movement gave a classification accuracy of 95.6 % and right index accuracy gave 94.81 %. To summarize the above mentioned techniques', Table 10.2 presents a summary of these techniques.

Table 10.2 A summary of selected machine learning techniques

Technique	Channels	Subjects	Task	TAR	FAR	
Fishier discriminant analyses	2	40	Rest	79.2 %	21.8 %	[15]
Linear discriminant analysis	6	4	Rest, math, letter, count, rotation	–	0.1 % avg combination using 5 features	[17]
LVQ	–	8	Rest	80 %		[18]
Neural networks	15	9	Left/right hand movement	95.6 % (left) 94.81 % (right)		[20]

Table 10.3 Comparison of various biometric [21] (H = High, M = Medium, L = low)

Biometric	Universality	Distinction	Permanence	Collectability	Performance	Acceptability	Circumvention
Face	H	L	M	H	L	H	H
Fingerprint	M	H	H	M	H	M	M
Hand geometry	M	M	M	H	M	M	M
Iris	H	H	H	M	H	L	L
Keystroke	L	L	L	M	L	M	M
Signature	L	L	L	H	L	H	H
Voice	M	L	L	M	L	H	H
EEG	H	H	M	M	M	M	H

10.9 Conclusion

Table 10.3 shows a comparison of various biometric technologies. High, Medium, and Low are Denoted by H, M, and L, Respectively.

A brief comparison of some of the biometric identifiers based on seven factors is provided in Table 10.3. Universality (do all people have it?), distinctiveness (can people be distinguished based on an identifier?), permanence (how permanent is the identifier?), and collectability (how well can the identifier be captured and quantified?) are properties of biometric identifiers. Performance (speed and accuracy), acceptability (willingness of people to use), and circumvention (foolproof) are attributes of biometric systems [21].

Table 10.4 Shows the advantages and disadvantages of each biometric. There is no right or wrong about selecting which biometric to use to for you specific access control. The implementer has to consider these factors while selecting the best biometric or combination of biometrics for their application.

- Cost versus performance
- Throughput
- Verification versus Identification mode
- Choice and number of biometrics
- Level of fusion
- Fusion methodology
- Assigning weights to biometrics
- Multimodal databases

In this chapter we explained the different types of popular biometrics that are used for access control. We have demonstrated the methodology for developing machine learning techniques for intelligent access control. Moreover, we showed the challenges that arise during the development of each step of the methodology. Moreover we explored various famous biometric techniques that are widely used and the advantages and disadvantages of each technique. We also presented the benefits and challenges for developing a multimodal biometric system. We also demonstrated some of the machine learning techniques used in biometrics. Finally we present a comparison for the biometric features covered in the chapter as depicted in Table 10.3.

Table 10.4 Advantages and disadvantages of biometrics [10]

Biometric	Advantages	Disadvantages
Finger print	• Subjects have multiple fingers • Easy to use • Some systems require little space • Has proven effective in many large scale systems over years of use • Fingerprints are unique to each finger of each individual and the ridge arrangement remains permanent during one's lifetime	• Privacy concerns • Health or societal concerns with touching a sensor used by countless individuals • An individual's age and occupation may cause some sensors difficulty in capturing a complete and accurate • No aliveness detection • If a user fingerprint is copied and discovered by the admin, the user and the forger both will be denied access to the system
Face	• No contact required • Commonly available sensors (cameras) • Easy for humans to verify results	• Face can be obstructed by hair, glasses, hats, scarves, etc. • Sensitive to changes in lighting, expression, and pose • Faces change over time • Liveness detection, system can be fooled with pictures or 3d models.
Iris	• No contact required • Protected internal organ; less prone to injury • Believed to be highly stable over lifetime	• Difficult to capture for some individuals • Easily obscured by eyelashes, eyelids, lens and reflections from the cornea • Acquisition of an iris image requires more training and attentiveness than most biometrics • Cannot be verified by a human • Can be fooled by pictures
Keystroke	• Non-intrusive and wide user acceptance • Natural authentication mechanism for computer and network security • Continuous verification (monitoring) is possible • Minimal training • No additional hardware	• High false reject rate • Sensitive to changes in keyboard, user's physical condition (fatigue or illness) and other operational conditions • Narrow range of applications • Need to account for problems like typing errors
Voice	• Public acceptance • No contact required • Commonly available sensors (telephones, microphones)	• Difficult to control sensor and channel variances that significantly impact capabilities • Not sufficiently distinctive for identification over large databases • Easily by passed by recorders
EEG	• Prone to Forgery and Theft • Prone to Shoulder surfing • Can be changed • Protected from user non compliance • Stable over time • User must be alive	• High processing power • Lengthy enrollment process.

References

1. Biometrics History: Biometrics.gov. http://www.biometrics.gov/Documents/BioHistory.pdf (2014). Accessed 26 Apr 2014
2. Gupta, C.N., Palaniappan, R.: Biometric paradigm using visual evoked potential. In: Encyclopedia of Information Science and Technology, vol. 1, 2nd edn, pp. 362–368 (2009)
3. Liu, S., Silverman, M.: A practical guide to biometric security technology. IT Prof. 3(1), 27–32 (2001)
4. Ratha, N.K., Connell, J.H., Bolle, R.M.: Enhancing security and privacy in biometrics-based authentication systems. IBM Syst. J. 40(31), 614–634 (2001)
5. Yun, Y.W.: The '123' of biometric technology. In: Biometrics Working Group of Security and Privacy Standards Technical Committee, pp. 80–96 (2002)
6. Jafri, R., Arabnia, H.R.: A survey of face recognition techniques. J. Inf. Process. Syst. 5(2), 41–68 (2009)
7. Wildes, R.P.: Iris recognition: an emerging biometric technology. Proc. IEEE 85(9), 1348–1363 (1997)
8. Shen, P., Andrew, B., Jin, T., Shiang, Y.: A survey of keystroke dynamics biometrics. Sci. World J. 1–24 (2013)
9. Fabian, M., Aviel, R.: Keystroke dynamics as a biometric for authentication. Future Gener. Comput. Syst. 16, 351–359 (2000)
10. Biometrics.gov. http://www.biometrics.gov/ (2014). Accessed 26 Apr 2014
11. Arun, R., Anil, J.: Information fusion in biometrics. Pattern Recogn. Lett. 24, 2115–2125 (2003)
12. Ross, A., Jain, A.K.: Fusion techniques in multibiometric systems. In: Face Biometrics for Personal Identification, pp. 185–212. Springer (2007)
13. Ross, A., Nandakumar, K., Jain, A.K.: Handbook of Multibiometrics. Springer Science + Business Media, LLC (2006)
14. ISO/IEC TR 24722. Information technology. In: Biometrics: Multimodal and Other Multibiometric Fusion (2007)
15. Riera, A., Soria-Frisch, A., Caparrini, M., Cester, I., Ruffini, G.: Multimodal physiological biometrics authentication. In: Biometrics: Theory, Methods, and Applications, pp. 461–482. Wiley Press (2010)
16. ENOBIO. http://www.neuroelectrics.com/enobio (2014). Accessed 26 Apr 2014
17. Ramaswamy, P.: Multiple mental thought parametric classification: a new approach for individual identification. Proc. Int. J. Signal Process. 2, 222–226 (2006)
18. Cempírek, M., Šťastný, J.: The optimization of the EEG-based biometric classification. Appl. Electron. 25–28 (2007)
19. Forecasting with Artificial Neural Networks. http://www.neural-forecasting.com/lvq_neural_nets.htm (2014). Accessed 26 Apr 2014
20. Sun, S.: Multitask learning for EEG-based biometrics. In: Proceeding of International Conference on Pattern Recognition, pp. 51–55 (2008)
21. Jain, K.A., Pankanti, S., Prabhakar, S., Uludag, U.: Issues and challenges. Proc. IEEE 92(6), 948–960 (2004)
22. Finger Print Features. http://biometrics.derawi.com/wp-content/uploads/2011/01/fingerprint_definition.jpg (2014). Accessed 26 Apr 2014
23. Face Recognition Features. http://www.engineersgarage.com/sites/default/files/imagecache/Original/wysiwyg_imageupload/28714/Face-Recognition.jpg (2014). Accessed 26 Apr 2014
24. Iris Features. http://www.cl.cam.ac.uk/~jgd1000/iriscode.jpg (2014). Accessed 26 Apr 2014
25. KeyStroke Features. http://img.zdnet.com.cn/0/702/liIVtnoGEwL5o.jpg (2014). Accessed 26 Apr 2014

Chapter 11
Experimental Evaluation of Opportunity to Improve the Resolution of the Acoustic Maps

Volodymyr Kudriashov

Abstract The work is devoted to the generation of acoustic maps. The experimental work considers the possibility to increase the resolution of the maps. The work uses two-dimensional microphone array with randomly spaced elements to generate acoustic maps of sources located in its near-field region. In this region, the wavefront is not flat and phase of input signals depends on the direction of arrival and the range as well. The input signals are partially distorted by indoor multipath propagation and related interference of sources emissions. For acoustic mapping with the improved resolution, an algorithm in the frequency domain is proposed. The algorithm is based on the modified method of Capon. Acoustic maps of point-like noise sources are generated. The maps are compared with the maps generated using other standard methods including built-in equipment software. The resolution improvement is up to 2.7 times. The obtained results are valuable in the estimation of the direction of arrival for Noise Exposure Monitoring.

Keywords Beamforming · Modified method of Capon · Microphone array · Acoustic noise source localization

11.1 Introduction

The acoustic camera provides a fusion of optical camera image and acoustic map. It is aimed to locate and characterize sound sources in wide frequency range. The acoustic maps quality is limited by the system performance. The first acoustic camera system was described in the early 80-es when Billingsley proposed an acoustic telescope [1]. Two years later, Billingsley and Kinns developed a full-scale system for real-time sound source location [2].

V. Kudriashov (✉)
Mathematical Methods for Sensor Information Processing Department,
Institute of Information and Communication Technologies, Bulgarian
Academy of Sciences, Acad. G. Bonchev Str. Bl. 25 A, 1113 Sofia, Bulgaria
e-mail: KudriashovVladimir@Gmail.com

© Springer International Publishing Switzerland 2016   353
R. Kountchev and K. Nakamatsu (eds.), *New Approaches in Intelligent Image Analysis*, Intelligent Systems Reference Library 108,
DOI 10.1007/978-3-319-32192-9_11

Noise Exposure Monitoring uses acoustic maps [3]. Angular resolution is an essential feature of such maps [4]. The latter depends on the estimated signal frequency and aperture dimensions of the antenna of the equipment used for the mapping [5].

Dougherty [6] and Sarradj [7] enhanced acoustic maps performance by means of beamforming based on eigenvectors of a cross-spectrum matrix. Existing methods of spectral analysis, for example, the Capon method [8], allow increasing the resolution [9, 10]. The application of this method for estimation of signals direction of arrival may be found in [11, 12]. Near-field localization made by applying of time delays into incoming signals may be found in [13]. The current work is focused on the enhancement of the quality of acoustic maps generated with the particular two-dimensional random microphone array. The purpose of the work is reached via applying of the proposed beamforming method in the frequency domain. The experimental setup and details are given and the obtained results are analyzed.

11.2 Theoretical Part

Let us consider the generation of acoustic maps for one source of acoustic emission. The emission propagates towards an array of microphones, where it is transformed to electrical signals (Fig. 11.1). For further digital processing, these signals are supplied into multi-channel Analog-to-Digital Converter (ADC). Accumulation of the ADC output signals taking into account weighting factors allows generating an acoustic map. These weighting factors may compensate both time and phase shifts of signals affected by their propagation delays ("Phase shift 1", Fig. 11.1) to microphones of the array [3].

11.2.1 Signal Model Limitations

The distance between the source of the acoustic signal and the microphone array is not less than $0.62\sqrt{L_{Max}^3/\lambda}$, where L_{Max}, m is physical aperture size of the microphone array; $\lambda = v_S/f$, m is the wavelength of the highest frequency of the acoustical signal; v_S, m/s is the acoustic signal propagation velocity. The acoustic signal is stationary. The signal frequency bandwidth Δf satisfies the following condition: the difference (L_{Max}) of signal propagation distances to any two microphones does not exceed the signal coherence length. The latest is defined by relation $v_S/\Delta f$.

The intrinsic time delays of signals in acoustic camera equipment are affected by: (a) the initial phase mismatch of channels; (b) the temporal instability of the parameters of channels of the acoustic camera. We consider negligible the influence of both factors. The influence of multipath signal propagation is also not taken into

Fig. 11.1 Signal flow blocks diagram for an acoustic imaging

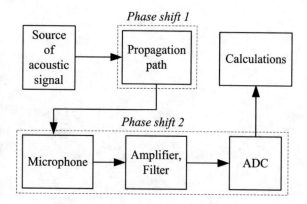

account. The intrinsic noise of acoustic camera equipment is considered as additive noise which power is much lower than acoustic signal power.

11.2.2 Signal Model

The spectrum of the acoustic signal is limited in frequency range $0 \leq f \leq f_S/2.56$, where f_S is sampling frequency of the ADC in Fig. 11.1. The ADC output signals spectrums are estimated using discrete Fourier transform. Let us denote frequencies of estimated discrete spectrum by $f(i)$ Hz, when $1 \leq i \leq N$. The quantity of these samples $N = f_S T$ is defined by the sampling frequency f_S and the emitter observation time T.

The point-like source of the acoustic signal is placed in a field of view of the acoustic camera—in front of its microphone array. The source position is constant during the observation time T. At time instance t, within the observation interval $0 \leq t \leq T$, the digitized acoustic signal $s(t)$ equals to:

$$s(t) = \sum_{i=1}^{N} A(i) \exp[j\, 2\, \pi f(i)t], \qquad (11.1)$$

where $A(i)$ denotes complex amplitude factors for discrete frequency counts $f(i)$; $j = \sqrt{-1}$ is the imaginary unit; N is the quantity of frequencies, noted above.

Let us suppose that a propagation path of the acoustic signal towards a microphone of the array is rectilinear and only one. The slant ranges $r_{SE}(k)$ between the source and microphones of the array are calculated as Euclidean distances in the three-dimensional Cartesian coordinate system (Fig. 11.2) as follows:

$$r_{SE}(k) = \sqrt{[x(k) - x_{SE}]^2 + [y(k) - y_{SE}]^2 + [z(k) - z_{SE}]^2},$$

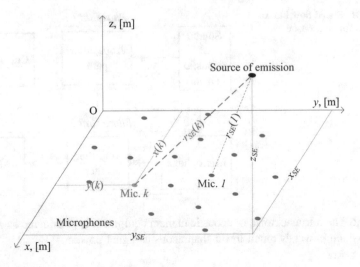

Fig. 11.2 Acoustic imaging geometry

where $k = [1, \ldots, K]$, $K = 18$ is quantity of the microphones in the array; $x(k)$, $y(k)$, $z(k)$ are these microphones coordinates; x_{SE}, y_{SE}, z_{SE} denote coordinates of the source.

Let us suppose that the speed of sound along the $r_{SE}(k)$ equals to v_S m/s for all frequencies of bandwidth Δf. The speed of sound in dry (0 % humidity) air is equal to: $v_S \approx 331.4 + 0.6\,T_C$, where T_C °C is the temperature at the time of an experiment [14]. For temperature $T_C = 21$ °C, the speed of sound equals to: $v_S \approx$ 344 m/s. The time of the acoustic signal to get to microphones of the array is expressed as: $\tau_{SE}(k) = r_{SE}(k)/v_S$, s. The phase delay is calculated correspondingly:

$$\varphi(k, i) = 2\,\pi f(i)\,\tau_{SE}(k), \tag{11.2}$$

where $\varphi(k, i)$ denotes the phase delay for the kth microphone. The frequencies values $f(i)$ depend on ADC sampling frequency f_S and the observation time T.

Acquired signals contain electronic noise from microphones and input modules of the acoustic camera [15]. The self-noise of individual channels is added to acquired signals. The noise is expressed as:

$$u(k, t) = \sum_{i=1}^{N} b(k, i)\,\exp[j\,2\,\pi f(i)\,t], \tag{11.3}$$

where $b(k, i)$ are random amplitude factors for $f(i)$ in K channels of the acoustic camera.

At time instance t, the ADC output signal is a mixture of the self-noise and acoustic signal with respect to its delay. The signal is expressed as:

$$s(k, t) = u(k, t) + \sum_{i=1}^{N} a(i) \, \exp[j \, 2 \, \pi f(i) \, t] \, \exp[j \, \varphi(k, i)], \qquad (11.4)$$

where $s(t, k)$ is a complex amplitude of the acoustic signal on the output of the channel k at time instance t; $a(i)$ denotes complex amplitude factors of the discrete spectrum of the signal. Thus, $A(i)$ were enclosed as an additive mix of self-noise of the acoustic camera with delayed signal of the source of emission. Let us denote the first channel as the reference channel. The phase differences of received signals and reference channel signal depend on differences of slant ranges: $r_{SE}(k) - r_{SE}(1)$ (Fig. 11.2).

The phase delays (11.2) can be compensated using corresponding delay units (phase shifters) on the outputs of K channels. The compensation enables summation of the channels signals in phase [3]. A microphone array scanning along angular coordinate can be performed in sequential or/and parallel handling of the delays. Each pixel of the acoustic map corresponds to a result of the summation.

Digital signal processing enables to generate acoustic maps in discrete nodes of angular coordinate grids. The grid is placed in the focal plane of the microphone array (in the azimuth-elevation plane). The range of the grid is a distance between the acoustic signal source and the microphone array. The angular dimensions of the grid are limited by the microphone array field of view.

11.2.3 Acoustic Mapping Methods

The focalization of the array is considered in three-dimensional Cartesian coordinate system using time delays $\tau_{SE}(k)$. The focalization in time domain includes introducing the time delays to output signals of channels. ADC digitizes those signals with sampling interval $1/f_S$, which acts as the time delay increment. For instance, if the sound source emits monochromatic input signal with frequency $f = 10$ kHz digitized by ADC with sampling frequency $f_S = 65.536$ kHz. One period of the signal will have $(1/f)/(1/f_S) \approx 6.5$ ADC counts. The latest corresponds to the time delay increment of $360°/6.5 \approx 55°$. The big increment value limits available angular positions of the microphone array beam, affects its sidelobe level and its pattern performance. The focalization in frequency domain enables to alleviate such limitation.

A channel output signal obtained during the observation time is denoted as a column with N samples (11.4). We obtain spectrum realization for this column by its Fourier transform. We denote by matrix S all the spectrums realizations obtained; the matrix S consists of N rows and K columns ($N \times K$). For each of N rows, we calculate the $K \times K$ matrix, using the product of a row conjugate transposed and the row. We obtain the realization of the cross-spectrum matrix as a result of the sum of all N matrices $K \times K$. We denote by matrix $F(\Delta f)$ the realization of cross-spectrum matrix obtained for defined bandwidth Δf. The acoustic camera enables operation in wide range of frequencies $f(i)$ from 10 to 25.6k Hz. The acoustic imaging approach requires defining center frequency and bandwidth around it. The work considers narrowband signals those bandwidth is not more than 10 % of the center frequency.

The time delays $\tau_{SE}(k)$ are calculated for the defined range of the acoustic map grid. The phase delays $\varphi_{SE}(k) = 2\pi f_C \tau_{SE}(k)$ are calculated for the defined: (a) acoustic signal center frequency f_C; (b) the grid nodes coordinates. The microphone array spatial (angular) scan delays are regarded as rows $X_{SE}(f_C)$ with K elements equal to: $\exp[-j \varphi_{SE}(k)]$. The beam scanning is realized via adjustment of delays $\tau_{SE}(k)$ of the row $X_{SE}(f_C)$ with respect to the defined range of the grid and beam angular position.

The delay-and-sum (DAS) is the mostly used beamforming method based on the sum of signals with respect to considered delays [15]. The matrix form representation of the method can be figured out in the frequency domain. Taking into account all considerations mention above, the output signal power $P_{SE}(\Delta f)$, in the particular node of acoustic map, is equal to:

$$P_{SE}(\Delta f) = X_{SE}(f_C)F(\Delta f)X_{SE}^H(f_C), \tag{11.5}$$

where superscript letter H denotes conjugate transpose.

The internal noises of microphones, preamplifiers, and the ADC dither are considered as the non-correlated additive noise of K channels of acoustic camera equipment. The noise main contribution is concentrated in the main diagonal elements of the cross-spectrum matrix [15]. The matrix diagonal elements removal increases the signal-to-noise ratio of generated acoustic maps, suppresses sidelobe level up to 0.6 dB [15]. We denote cross-spectrum matrix with nulled elements of main diagonal by $F_0(\Delta f)$. Thus, the power of the output signal $Q_{SE}(\Delta f)$, in the particular node of acoustic map, becomes:

$$Q_{SE}(\Delta f) = X_{SE}(f_C)F_0(\Delta f)X_{SE}^H(f_C). \tag{11.6}$$

A method for high-resolution spectrum analysis was proposed in 1969 by Capon [8]. Later the Capon method was described for estimation of the direction of arrival of a signal [11]. The method does not need any prior information about the number of sources in the field of view [11]. The method does not require prior knowledge about signal amplitude distribution. The method is limited by requirements to inversion of matrices and to signal-to-noise ratio more than 10 dB. The method applicability is extended by: (a) addition unit matrix of size $K \times K$ to the matrix under consideration; (b) modern methods of matrix pseudo-inverse. Signal processing can be done in the frequency domain. We use the modified pseudo inverse cross-spectrum matrix $F_M^{-1}(\Delta f)$, which calculation includes the unit matrix addition. The output signal power $R_{SE}(\Delta f)$ in a particular node of acoustic map is equal to:

$$R_{SE}(\Delta f) = \frac{1}{X_{SE}(f_C)F_M^{-1}(\Delta f)X_{SE}^H(f_C)}. \tag{11.7}$$

11.3 The Experimental Acoustic Camera Equipment

The acoustic camera is suitable for frequency, time and spatial analysis. The acoustic camera utilizes a fusion of optical image and acoustic map. Brüel&Kjaer (B&K) Sound and Vibration Measurement A/S manufactured both camera software and hardware. The camera equipment includes the microphone array with the optical camera; 6- and 12-channel input modules and the laptop with software for acoustic signals analysis (Fig. 11.3). The camera microphone array is B&K type WA-1558-W-021 [16]. It is two-dimensional randomly distributed microphone wheel array with diameter ≈0.33 m.

The quantity of the microphones is $K = 18$ (Fig. 11.4). The input modules are B&K type 3053-B-120 and type 3050-B-060. The signals used for acoustic map generation (11.5–11.7) are recorded using Time Data Recorder option of B&K Pulse LabShop software. The input modules dynamic range is up to 160 dB; the modules provide high phase stability and interchannel isolation. The above factors allow analysis of weak acoustic emissions.

The acoustic camera microphones B&K type 4958 have built-in preamplifiers and contain transducer electronic data sheets. These datasheets are aimed to transfer

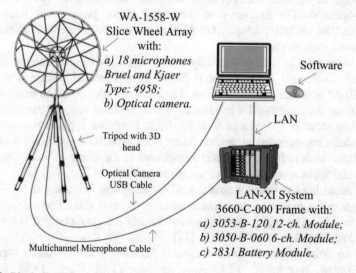

WA-1558-W
Slice Wheel Array
with:
a) 18 microphones
Bruel and Kjaer
Type: 4958;
b) Optical camera.

Software

Tripod with 3D head

LAN

Optical Camera
USB Cable

LAN-XI System
3660-C-000 Frame with:
a) 3053-B-120 12-ch. Module;
b) 3050-B-060 6-ch. Module;
c) 2831 Battery Module.

Multichannel Microphone Cable

Fig. 11.3 Block diagram of the acoustic camera

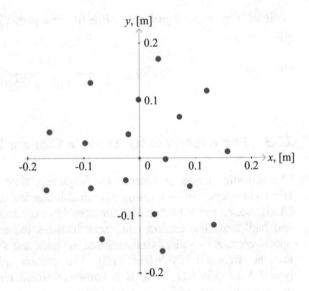

Fig. 11.4 The layout of microphones in the array

each microphone features to input modules. The microphones dimensions are: 34 mm × 7 mm. Their sensitivity is 11.2 mV/Pa. Their operation temperatures range is from −10 to +55 °C. Their dynamic range is from 28 to 140 dB.

The quality of acoustic signal amplitude estimation is provided by the acoustic camera amplitude calibration. The pistonphone calibrator B&K type 4228 corresponds to requirements of IEC 942 (1988) Class 1L or Class 0L (with external barometer) and ANSI S1.40-1984. The calibrator has the high stability of calibration signal level and frequency. The calibration signal accuracy is provided in a wide range of operation temperature, humidity, and pressure. The calibrator is battery-operated. The frequency of the signal of the pistonphone calibrator is 251.2 Hz. The calibrator adaptor DP-0775 is suitable for sequential calibration of the acoustic camera channels.

The input module with extended dynamic range is 6-channel. The second input module is 12-channel. Its resolution is 24 bit. Its interchannel leakage is not worse than −80 dB according to its datasheet. The input modules dynamic range depends on sampling frequency and a bandwidth. These modules support transducer electronic data sheet in order to provide the B&K Response Equalization technique. The modules are mounted in module frame B&K type 3660-C-000 for 5 modules. The battery module B&K type 2831 is mounted in the frame as well. The microphones and input modules features are given in Table 11.1.

The acoustic camera input modules ADCs sampling frequency is 65.536 kHz. The highest frequency of the camera channels is 25.6 kHz. The sampling ratio equals to 2.56. The input modules output signals are synchronized according to IEEE 1588 Precision Time Protocol [17].

The acoustic camera software includes: Acoustic Test Consultant type 7761; Beamforming type 8608; FFT Analysis type 7770; Time Data Recorder

Table 11.1 Frequency characteristics for modules of the Acoustic Camera

Module			
Name	Type	Frequency range, kHz	Features
Microphones	4958	0.01–20	–
Input modules	3053-B-120	0–25.6	1. Sampling rate: 65.5 k samples/s 2. Number of input channels: 12
	3050-B-060	0–51.2	1. Sampling rate: 131 k samples/s 2. Number of input channels: 6

Type 7708. The software provides generation of acoustic maps in defined frequency range, time and frequency representation of signals, their recording, etc.

The signals can be stored for further post-processing with third-party software using multi-buffer from Time Data Recorder option of B&K Pulse LabShop software. We use observation time equal to 0.25 s for each multi-buffer. The signals are stored to the hard disk drive with time stamps and the acoustic camera settings. The post-processing is done with developed scripts using MATLAB computing environment.

11.4   Experimental Results

The experiments on the estimation of angular coordinates of several point-like sources of acoustic signals were held in an office room. The room is not optimized in terms of multipath propagation of acoustic signals inside it. The above approach enables to compensate the narrowband signals delays including such from near-field region of the considered two-dimensional microphone array. Two sources of acoustic signals are placed on the range of 0.78 m which is comparable to the microphone array dimensions. The picture of the experimental setup is given on the Fig. 11.5.

The amplitude calibration is performed after the Acoustic Camera switching-on. The amplitude calibration uses the calibration option of B&K Pulse LabShop software and the pistonphone calibrator described above.

The source of the acoustic signal is placed on array boresight direction (PC speaker marked as "Source 1" in Fig. 11.5). The source emits acoustic noise signal. The sources of acoustic signals are 2 × 2 W stereo speakers; their switching-on increased the Acoustic Camera output signal (in the frequency range from 10 to 25.6 kHz) from 17 to 20 dB.

The experimental estimation of the two-dimensional microphone array pattern is done for several center frequencies. The signal bandwidth is 10 % of its center frequency. The chosen values of center frequencies belong to the frequency range from 0.1 to 18 kHz. The observation time is not less than 0.25 s. The acoustic signal source ("Source 1" in Fig. 11.5) position is constant during the observation time.

Fig. 11.5 Picture of imaging scenario: two point-like sources of the acoustic signal in front of the acoustic camera

The experiments are held for the microphone array field of view ±90° in azimuth and in elevation. The microphone array pattern (AP) parameters under investigation are: (a) beamwidth of the AP; (b) the AP sidelobes levels (SLL); (c) the AP highest null level; (d) angular position of highest sidelobe of the AP. The highest null and the highest SLL are considered as such with the highest values. All further estimations are held using normalized AP.

11.4.1 Microphone Array Patterns Generated with the Delay-and-Sum Beamforming Method

The experimental estimation of the two-dimensional microphone array pattern is done using DAS beamforming (11.5). The AP beamwidth is measured as the normalized AP half-power level (−3 dB level) regarding its peak. We denote the beamwidth as $\beta°$ and $\varepsilon°$ in azimuth and in elevation, correspondingly, where superscript symbol ° denotes that the value is given in degrees.

For center frequencies lower than 500 Hz the estimated AP lowest level in the field of view do not cross the −3 dB level thus, the AP beamwidth can not be estimated. For center frequency 500 Hz the $\beta° \approx 141°$; the elevation slice lowest level is −2.7 dB.

Table 11.2 The parameters of the microphone array pattern estimated indoor using delay-and-sum beamforming method

f_C, kHz	$\beta°$	$\varepsilon°$	Sidelobes levels, dB									
			No. 1		No. 2		No. 3		No. 4		No. 5	
			Az.	El.	Az.	El.	Az.	El.	Az.	El.	Az.	El.
1	61.5	59.3	−7.8	−8.2	–	–	–	–	–	–	–	–
3	17.8	20.5	−8.4	−11	−11.5	−9.4	–	–	–	–	–	–
10	7	7	−11.5	−16	−8.5	−16	−5.3	−7.8	−11.4	−8.7	−8.2	−9.8
18	3.4	3.4	−12.2	−15.8	−11.6	−12.1	−6.6	−9.2	−6.6	−8	−9.2	−10

Estimated microphone array patterns parameters are given in Table 11.2 for center frequencies from 1 to 18 kHz. The angular position of the highest sidelobe of the AP varies from 85° to 13°. The highest null level of the AP varies from −2.8 to −18 dB. The highest null level estimated on center frequency 3 kHz equals to −11 dB in azimuth and −12 dB in elevation.

The variation of the listed values along repeated measurements is inconsiderable. The repeatability depends on the experiment conditions and on parameter stability of B&K equipment.

11.4.2 Microphone Array Patterns Generated with the Christensen Beamforming Method

The sidelobes of the array pattern determine the level of penetration of unwanted acoustic signals from corresponding angular directions. It is known that the removal of main diagonal of the cross-spectrum matrix can diminish the sidelobes level [3, 15]. The considered above signals recordings were processed with Christensen beamforming method (11.6). The estimated AP parameters are given in Table 11.3.

For center frequencies lower than 500 Hz the AP beamwidth is about 10° narrower than estimated one by DAS beamforming method. For center frequency 500 Hz the $\beta° \approx 128°$ and the $\varepsilon° \approx 152°$. Thus, Christensen beamforming method enabled to estimate $\varepsilon°$ for center frequency 500 Hz unlike the DAS beamforming method.

For center frequency 1 kHz, the Christensen beamforming enabled to narrow the beamwidth $\beta°$ in 4.9° and $\varepsilon°$ in 1.5° comparable to DAS beamforming (Table 11.2). Christensen paper is focused on sidelobe level suppression, not on the main lobe narrowing [3, 15].

Let us compare the level of unwanted penetration via sidelobes in AP, estimated with DAS and Christensen beamforming methods (Tables 11.2, 11.3). The first sidelobe is considered. For center frequency 1 kHz, the sidelobe is suppressed on 3.1 and 3.7 dB compared to such levels estimated with DAS beamforming.

Table 11.3 The parameters of the microphone array pattern estimated indoor using Christensen beamforming method

f_C, kHz	$\beta°$	$\varepsilon°$	Sidelobes levels, dB										Highest null level, dB	
			No. 1		No. 2		No. 3		No. 4		No. 5			
			Az.	El.	Az.	El.	Az.	El.	Az.	El.	Az.	El.	Az.	El.
1	56.6	57.8	−10.9	−11.9	−	−	−	−	−	−	−	−	−	−
3	16.5	18.8	−10.4	−10.5	−10.7	−10.4	−	−	−	−	−	−	−11	−10.5
10	5.5	5	−12.1	−11.3	−11.4	−11.2	−6.3	−10.4	−11.2	−12.5	−12.2	−11.4	−12.5	−12.2
18	3.2	3.2	−12	−12.1	−8	−11.8	−12.4	−12.3	−12.3	−12.4	−7.4	−12.4	−13	−12.4

Table 11.4 Beamwidth difference of the microphone array patterns estimated with delay-and-sum beamforming method and Christensen beamforming method for center frequency 10 kHz

Parameter	DAS	Christensen	Difference
β°	7	5.5	2.5
ε°	7	5	2

Such comparison for center frequency 3 kHz shows suppression from 1 to 2 dB for the second sidelobe in elevation and the first sidelobe in azimuth, correspondingly. The second sidelobe in azimuth is 0.8 dB higher and the first sidelobe in elevation is 0.5 dB higher. For center frequency 10 kHz, the Christensen beamforming enables to suppress first and second sidelobes in azimuth from 0.6 to 2.9 dB but makes such sidelobes in elevation higher from 4.7 to 4.8 dB. For center frequency 18 kHz, the first and the second sidelobes are higher from 0.2 to 3.7 dB. The latest may be explained with indoor experimental conditions.

The difference of the AP beamwidth estimated with DAS and Christensen beamforming methods is given in Table 11.4. The beamwidth values depend on the wavelength (corresponding to center frequency) and the microphone array lengths L in corresponding angular directions as: $\beta^\circ \approx 57.3\lambda / L_\beta$ and as $\varepsilon^\circ \approx 57.3\lambda / L_\varepsilon$. Let us calculate the values β° and ε° for above-shown parameters: speed of sound $v_S \approx 344$ m/s; center frequency $f_C = 10$ kHz; the microphone array lengths $L_\beta \approx 0.32$ m and $L_\varepsilon \approx 0.33$ m. The calculated values are equal approximately to $6.1°$ and $5.9°$ in azimuth and elevation. The values comparison to those from Tables 11.2, 11.3 and 11.4 shows that in the indoor experiment, the DAS beamforming delivers beamwidth $1°$ wider than calculated one and Christensen beamforming narrows it less than $1°$.

11.4.3 Microphone Array Patterns Generated with the Modified Capon-Based Beamforming Method

The considered above recordings of signals were processed with modified Capon-based beamforming method (11.7). For center frequencies from 1, 3, 10 and 18 kHz, the estimated microphone array patterns parameters are given in Table 11.5.

For the lowest center frequency $\beta^\circ \approx 59°$ and $\varepsilon^\circ \approx 69°$. The time-bandwidth product of the narrowband noise signal with this center frequency is about 4 dB [18]. The equipment instability in the experiment rejects to obtain such result in long-time observations. Other methods did not enable to estimate beamwidth on the center frequency 100 Hz in the defined above field of view.

Table 11.5 The parameters of the microphone array pattern estimated indoor using modified Capon-based beamforming method

f_C, kHz	$\beta°$	$\varepsilon°$	Sidelobes levels, dB	
			From	To
1	48.3	49.3	−5.3	−3.8
3	16.2	18.8	−5.1	−4.1
10	2.8	2.8	−7.4	−6.8
18	1.25	1.3	−11.5	−10

For the center frequency 500 Hz, the estimated beamwidth equals to $\beta° \approx 50°$ and $\varepsilon° \approx 59°$. The latest values are 2.56 times narrower comparable to such estimated using Christensen method; the AP generated with DAS beamforming method does not cross the −3 dB level in the ±90° field of view, as it was noted above. The obtained sidelobes level and highest null level are equal to approximately −7.4 dB.

For center frequency 10 kHz, the peak sidelobe level is approximately −6.8 dB. Its angular position corresponds to the estimated with DAS and Christensen beamforming methods. The position in azimuth is approximately 30°.

Let us compare the unwanted penetration level estimated with the modified Capon-based and DAS beamforming methods. The first and the second sidelobes are under consideration. For center frequencies 1, 3, and 10 kHz, the Capon sidelobe level is higher. For center frequency 1 kHz it is higher from 2.5 to 4.4 dB; for center frequency 3 kHz it is higher from 3.3 to 7.4 dB; for center frequency 10 kHz it is higher from 1.1 to 9.2 dB, the latest can be affected by the existing multipath propagation of the acoustic signal inside the office room. The difference in sidelobe levels those obtained using Capon-based and DAS beamforming methods varies from 5.8 to −0.1 dB, for center frequency 18 kHz. These level lower values were obtained in the third sidelobe angular position for center frequencies 10 and 18 kHz. The improvement is from 1.5 to 3.4 dB.

Let us compare the unwanted penetration level estimated with the modified Capon-based and Christensen beamforming methods. The first and the second sidelobes are again under consideration. For center frequencies from 1 to 10 kHz, the Capon level is higher. For center frequency 1 kHz it is higher from 5.6 to 8.1 dB; for center frequency 3 kHz it is higher from 5.3 to 6.6 dB; for center frequency 10 kHz it is higher from 3.8 to 5.3 dB. For center frequency 18 kHz, the level can be 3.5 dB lower as well as 2.1 dB higher. The 3.5 dB improvement is obtained in the second sidelobe in azimuth.

The difference of the AP beamwidth estimated with the modified Capon-based beamforming method to such estimated with DAS and Christensen beamforming methods is given in Table 11.6. In the table, the ratio between beamwidth values those estimated with DAS and modified Capon-based beamforming methods is given in column k_1. The k_2 column shows the ratio between beamwidth values estimated with Christensen and modified Capon-based beamforming methods. The insignificant mainlobe width narrowing for center frequency 3 kHz (k_2 in Table 11.6) is connected to indoor conditions of the experiment.

Table 11.6 Beamwidth ratio of such estimated with delay-and-sum and Christensen beamforming methods to values estimated using modified Capon-based beamforming method

f_C, kHz	k_1		k_2	
	$\beta°$	$\varepsilon°$	$\beta°$	$\varepsilon°$
1	1.27	1.2	1.17	1.17
3	1.09	1.09	1.02	1
10	2.5	2.5	1.96	1.79
18	2.72	2.62	2.56	2.46

In the experimental study usage of the modified Capon-based beamforming method (11.7) instead the DAS beamforming method (11.5) enabled to narrow the mainlobe width from 1.17 to 1.27 times for center frequency 1 kHz; from 1.79 to 2.5 times for center frequency 10 kHz; from 2.46 to 2.72 times for center frequency 18 kHz (Table 11.6). The mainlobe width narrowing obtained by using modified Capon-based beamforming method instead of the Christensen beamforming method is from 1.17 to 2.56 times (Table 11.6). The results obtained in the center frequency range from 100 Hz up to 500 Hz show that acoustic maps generation in the field of view ±90° is suitable using modified Capon-based beamforming method only. The

Fig. 11.6 Acoustic map generated using the modified Capon beamforming method: **a** acoustic map; **b** slice of the map for elevation 1, deg.

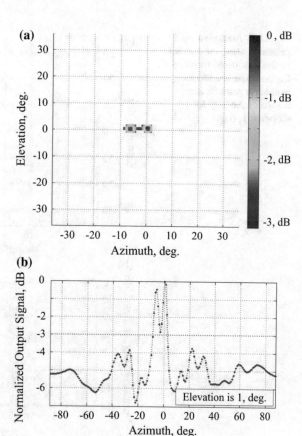

obtained AP beamwidth narrowing increases the microphone array angular resolution that improves the generated acoustic map quality.

11.4.4 Microphone Array Responses for Two Point-like Emitters

The acoustic maps of two point-like sources (Fig. 11.5) of acoustic noise signals were generated. The signal bandwidth, the signal source range, and other experiment parameters are similar to previous ones. The signals center frequency is 10 kHz. The sources are separated is azimuth plane approximately on 0.085 m. In the defined range, their azimuth angular separation is $\arctan(0.085/0.78) \approx 6.2°$. The latest equals approximately to the calculated AP beamwidth and is less than the beamwidth estimated using DAS beamforming method (Tables 11.2 and 11.3). For further experiments the acoustic maps field of view is equal to $\pm 35°$ in both azimuth and elevation directions; the acoustic map slices field of view is equal to $\pm 90°$ in

Fig. 11.7 Acoustic map generated using the heuristic method—squared modified Capon-based beamforming method: **a** acoustic map; **b** slice of the map for elevation 1, deg.

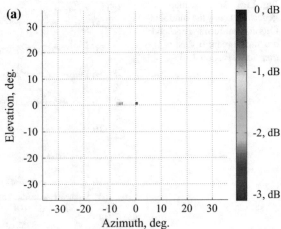

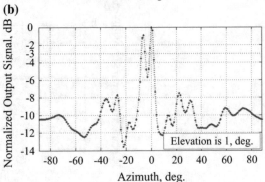

both angular directions. The power of the acoustic signal sources is approximately equal. Acoustic maps normalized are under consideration. These maps threshold is −3 dB. That means that the map nodes with power less than 3 dB comparable to the map peak are not shown. The normalized map slices have a dashed line for −3 dB level. The acoustic maps are generated with DAS (11.5), Christensen (11.6) and modified Capon-based beamforming methods (11.7).

The acoustic map generated by modified Capon-based beamforming method (11.7) is shown in Fig. 11.6. The level of the hollow between the two peaks approximately equals to −3 dB, so they will be resolved as two sources. The peak sidelobe level is approximately equal to −4 dB. For the center frequency, the grating lobes position in azimuth is approximately equal to ±30° as was noted above.

The heuristic, modified methods, like squaring of the method (11.7) enable to improve the obtained result. The noted peak sidelobe level can be suppressed to about −8 dB (Fig. 11.7).

Fig. 11.8 Acoustic map generated using the Christensen beamforming method: **a** acoustic map; **b** slice of the map for elevation 1, deg.

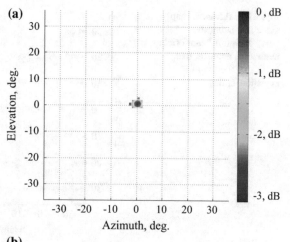

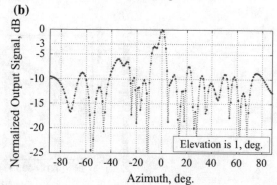

The acoustic map generated by Christensen beamforming method (11.6) is shown in Fig. 11.8. The same signal recordings were used to generate this map. The level of the hollow between peaks in Fig. 11.8 is lower than −3 dB. These peaks cannot be resolved. The unwanted penetration level is lower than such for modified Capon-based beamforming method.

The acoustic map generated by DAS beamforming method (11.5) is shown in Fig. 11.9. The sidelobes level rise to −3 dB level (dashed line). The map shows one peak only. The map responses at some angular coordinates from 20° to 40° correspond to acoustic signal multipath propagation inside the office room and to sidelobes level of the microphone array as well.

The level of the unwanted penetration of acoustic signals is about 3 dB higher in comparison with Christensen beamforming method (Fig. 11.8); the level is about 1 dB higher than such obtained using the modified Capon-based beamforming method (Fig. 11.6); the level exceeds the level obtained using the heuristic method (Fig. 11.7) by more than 3 dB.

Fig. 11.9 Acoustic map generated using the delay-and-sum beamforming method: **a** acoustic map; **b** slice of the map for elevation 1, deg.

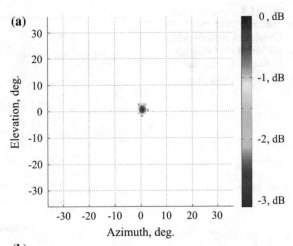

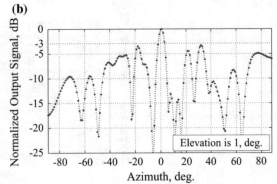

11.4.5 The Acoustic Camera Responses for Two Point-like Emitters

The acoustic camera was applied to check the results of the generation of the acoustic images of two point-like sources of acoustic noise signals. The experiment scenario (Fig. 11.5) was re-assembled, thus, the sources range, cross-range separation and angular coordinates are close to described above. The acoustic camera demo project enables to select the center frequency 10 kHz and the frequency range from 8.913 to 11.22 kHz. We assume that an acoustic noise source (Fig. 11.5) spectrum shape is uniform in the frequency range. The bandwidth 2.308 kHz is wider comparable to used above bandwidth 1 kHz. The "acquisition time" is set to 2 s. The B&K software "calculation setup" was defined as "default delay and sum". The software indicates that it includes calculation of cross spectra, principal component decomposition, and transducer electronic data sheet application. The generated acoustic image is given in the Fig. 11.10.

The experiment was repeated in order to record the Acoustic Camera signals for processing with the modified Capon beamforming method using the above approach. The frequency range for the modified Capon beamforming method is widened. The frequency range was from 8.912 to 11.22 kHz. The observation time 0.25 s is 8 times less than used by the demo project. The acoustic map threshold level is −5 dB (Fig. 11.11). The level of the hollow between the map peaks is

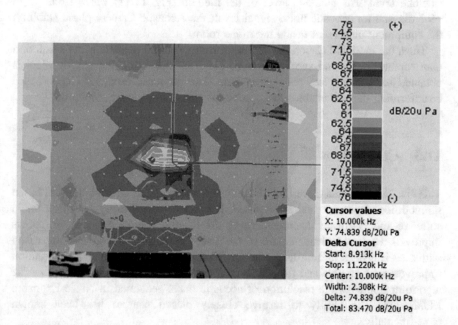

Fig. 11.10 The Acoustic Camera built-in software output: the acoustic map screenshot with corresponding color bar, center frequency, and frequency range

Fig. 11.11 Acoustic map
generated using the modified
Capon beamforming method
for comparison with the
Acoustic Camera built-in
software

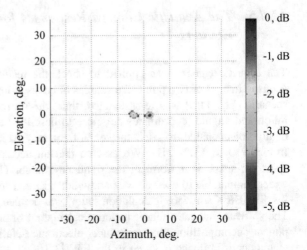

better than −3 dB (Fig. 11.11). This depth of hollow enables to resolve these two peaks. The shift of angular coordinates of these peaks in Fig. 11.11 corresponds to the sources shift affected by the mentioned above re-assembling of the experiment scenario.

The experiment with observation time 0.25 s was repeated several times that showed the resolution improvement repeatability as well as the variation of the sidelobe level. The sidelobe level of the method (Fig. 11.11) varies from −3 to −9.7 dB due to: acoustic noise signal level; the Acoustic Camera phase stability; the equipment placement inside the office room.

Thus, the modified Capon-based beamforming method delivers better resolution than "default delay and sum" regime of the B&K software. The improvement is obtained for 8 times less acquisition time. The proposed modified Capon-based beamforming method enables to improve the microphone array resolution.

11.5 Conclusions

This chapter discussed algorithms for acoustic map generation. An algorithm based on modified Capon method is proposed. It has been tested using acoustic camera software and hardware manufactured by B&K. The comparison analysis shows improved resolution characteristic of the newly proposed method in comparison with classical ones and built-in ones in B&K equipment. The angular resolution improvement was obtained for center frequency in the range of 0.1–18 kHz. The algorithm improves the resolution of acoustic maps generated in Noise Exposure Monitoring. The ability to resolve closely placed sources has been shown experimentally.

Acknowledgments The research work reported in the chapter was partly supported by the Project AComIn "Advanced Computing for Innovation", grant 316087, funded by the FP7 Capacity Programme (Research Potential of Convergence Regions).

References

1. Billingsley, J.: An acoustic telescope. Aeronautical Research Council 35/364 (1974)
2. Billingsley, J., Kinns, R.: The acoustic telescope. J. Sound Vib. **48**, 485–510 (1976)
3. Gerges, S., Fonseca, W.D., Dougherty, R.P.: State of the Art Beamforming Software and Hardware for Applications. Paper presented at the 16th International Congress on Sound and Vibration, Krakow, 5–9 July 2009
4. Greguss, P.: Ultrasonic Imaging: Seeing by Sound: the Principles and Widespread Applications of Image Formation by Sonic, Ultrasonic and Other Mechanical Waves. Focal Press Limited, London (1980)
5. Born, M., Wolf, E.: Principles of Optics, 7th edn. Cambridge University Press, Cambridge (1999)
6. Dougherty RP (2002) Beamforming in acoustic testing. In: Mueller T (ed) Aeroacoustic Measurements. Springer Berlin Heidelberg, New York: 62– 97
7. Sarradj, E., Schulze, C., Zeibig, A.: Einsatz eines Mikrofonarrays zur Trennung von Quellmechanismen, Vortrag auf der Tagung der Deutschen Arbeitsgemeinschaft für Akustik—DAGA, Technischen Universität München, München, 14–17 Marz 2005
8. Capon, J.: High-resolution frequency-wavenumber spectrum analysis. Proc. IEEE **57**(8), 1408–1418 (1969)
9. Marple, S.L.: Digital Spectral Analysis with Applications. Prentice-Hall Inc., Upper Saddle River (1987)
10. Munier, J., Delisle, G.Y.: Spatial analysis in passive listening using adaptive techniques. Proc. IEEE **75**(11), 1458–1471 (1987)
11. Ermolayev, V.T., Flaksman, A.G.: Metodi Otsenivaniya Parametrov Istochnikov Signalov i Pomeh, Prinimayemikh Antennoi Reschetkoi (Methods for Estimation of Parameters of Sources of Signals and Interferences Received by Antenna Array). Lobachevsky State University, Nizhny Novgorod (2007)
12. Stoica, P., Moses, R.: Introduction to Spectral Analysis. Prentice Hall, Upper Saddle River (1997)
13. Lukin, K., Kudriashov, V.V., Vyplavin, P., Palamarchuk, V.: Coherent imaging in the range-azimuth plane using a bistatic radiometer based on antennas with beam synthesizing. IEEE Aerosp. Electron. Syst. Mag. **29**(7), 16–22 (2014)
14. Lopez, E.S.: Sensor Array Optimization for Multiple Harmonic Sound Source Separation and DOA. MA Dissertation, Lund University, Lund (2013)
15. Christensen, J.J., Hald, J.: Beamforming. Brüel & Kjær Technical. Review **1**, 1–48 (2004)
16. Brüel & Kjær Sound and Vibration Measurement A/S: Product Data: PULSE Array-based Noise Source Identification Solutions: Beamforming—Type 8608, Acoustic Holography—Type 8607, Spherical Beamforming—Type 8606. Brüel & Kjær Sound and Vibration Measurement A/S, Nærum (2009)
17. IEEE Instrumentation and Measurement Society: 1588-2002—IEEE Standard for a Precision Clock Synchronization Protocol for Networked Measurement and Control Systems. IEEE, USA (2002)
18. Lukin, K., Vyplavin, P.: Integrated sidelobe ratio in noise radar receiver. Paper presented at the IEEE 13th International Radar Symposium (IRS), Warsaw, 23–25 May 2012

Printed in the United States
By Bookmasters